Functional Biology
of Plants

Functional Biology of Plants

Martin J. Hodson
Oxford Brookes University, UK

John A. Bryant
University of Exeter, UK

WILEY-BLACKWELL

A John Wiley & Sons, Ltd., Publication

This edition first published 2012 © 2012 by John Wiley & Sons, Ltd.

Wiley-Blackwell is an imprint of John Wiley & Sons, formed by the merger of Wiley's global Scientific, Technical and Medical business with Blackwell Publishing.

Registered office: John Wiley & Sons, Ltd, The Atrium, Southern Gate, Chichester, West Sussex, PO19 8SQ, UK

Editorial offices: 9600 Garsington Road, Oxford, OX4 2DQ, UK
                The Atrium, Southern Gate, Chichester, West Sussex, PO19 8SQ, UK
                111 River Street, Hoboken, NJ 07030-5774, USA

For details of our global editorial offices, for customer services and for information about how to apply for permission to reuse the copyright material in this book please see our website at www.wiley.com/wiley-blackwell

Library of Congress Cataloging-in-Publication Data

Hodson, Martin.
    Functional biology of plants / Martin Hodson, John Bryant.
        p. cm.
    Includes bibliographical references and index.
    ISBN 978-0-470-69940-9 (cloth) − ISBN 978-0-470-69939-3 (pbk.)
1. Plant physiology – Textbooks. 2. Botany – Textbooks. I. Bryant, J. A. II. Title.
    QK711.2.H63 2012
    571.2 – dc23
                                                            2011047547
A catalogue record for this book is available from the British Library.

Wiley also publishes its books in a variety of electronic formats. Some content that appears in print may not be available in electronic books.

Typeset in [9/12 Minion] by Laserwords Private Limited, Chennai, India
Printed and bound in Singapore by Markono Print Media Pte Ltd

First Impression   2012

MJH would like to dedicate this book to the three plant biologists who have been most influential in his career:

Dr. Helgi Öpik (Swansea University, Wales), who both taught me as an undergraduate and supervised my doctoral studies.

Dr. Dafydd Wynn Parry (Bangor University, Wales), who first introduced me to the delights of studying silicon in plants.

Prof. Allan Sangster (York University, Toronto, Canada), with whom I had my longest and most successful research collaboration.

Without their guidance and friendship, I would never have got as far as writing this book.

JAB dedicates this book to the memories of two inspirational teachers of Plant Biology:

Dr Cecil Prime (1909–1979) at Whitgift School, Croydon: a firm but caring school teacher whose love and knowledge of plants was infectious. This led me to study plants at university and I was grateful, as a 'first-generation' university student, for his continued interest and support during my undergraduate years.

Professor Tom ap Rees (1930–1996), University of Cambridge: a clear and enthusiastic teacher of undergraduates and a supportive, understanding PhD supervisor. His advice led me to pursue a career in plant science, a career that he followed with interest until his untimely death in a road accident.

Contents

Preface

As we complete the manuscript of *Functional Biology of Plants*, many thousands of refugees, driven by drought and famine from the Horn of Africa, have found their way to camps in Kenya. Nowhere is it more obvious that people need feeding, yet it is also true to say that, with appropriate land use, the continent of Africa could become self-sufficient in food production.

This is not the place to discuss the political and economic challenges that will need to be faced; rather, we state that plant growth has never been so important. It may be true in some developed countries that students seem relatively uninterested in botany or plant biology, but it is equally true that we need to know more about plants and how they work, at least partly in order to harness and, indeed, to increase their potential in human nutrition. Thus we hope that this book will engender interest in the functioning plant.

We have not set out here to write a book about plant biochemistry or cell biology or molecular biology or genetics. Instead, after an introduction to plant function at those levels, we have attempted to show how activities at molecular and cellular levels are integrated and coordinated in the functioning of whole organs and of whole organisms – the plants themselves. In the later parts of the book, we place plants into their natural environments as they deal with abiotic and biotic stresses before considering, in the final chapter, the importance of plants in relation to some of the pressing problems facing humankind in the 21st century.

Acknowledgements

It is our pleasure to thank our wives, Margot Hodson and Marje Bryant, for their roles in this enterprise. Their combination of patience, tolerance, support and encouragement has been truly saintly. The fact that this book has been finished owes a lot to them.

We thank all our colleagues who readily gave us help, advice and encouragement, especially Deborah Cannington, James Doyle and Tom Rost at the University of California (Davis), Jeff Duckett at Queen Mary, University of London, David Evans at Oxford Brookes University, Donald Grierson and Lizzie Rushton at the University of Nottingham, Steve Hughes and Nick Smirnoff at the University of Exeter, Yoav Waisel at Tel Aviv University, Michael Pocock at the University of Bristol, Peter Rudiak-Gould at McGill University, Sue Grahame at the University of Leeds and Darren Evans at the University of Hull.

We are also grateful to the institutions and farms that have graciously allowed us to use photographs taken on their grounds: Westonbirt Arboretum, UK; the University of Oxford Botanic Garden, UK; the University of Leeds, UK; Boyce Thompson Arboretum, Superior, Arizona, USA; the National Botanic Gardens, Glasnevin, Dublin, Ireland; Manor Farm, Haddenham, Buckinghamshire, UK (farmer Tom Bucknell); and Manor Farm, Warmington, nr. Banbury, UK (farmer John Neal). Many thanks also to those relatives, friends and colleagues who have supplied photographs or who have granted permission to use their previously published pictures.

Particular thanks are due to our editors at Wiley-Blackwell: in the early stages, Andy Slade, Celia Carden and Rachel Wade and then, for the final stages of manuscript preparation and for the publication process itself, Izzy Canning and Fiona Woods. Their patience has been phenomenal, their help and advice invaluable and their efficiency and effectiveness admirable.

CHAPTER 1

Origins

1.1 Plants – what are they?

We might simply define plants as photosynthetic eukaryotes – a description that would certainly include all the types of organisms that find their way into courses in botany or plant biology. However, as will become clear later in this chapter, such a definition brings together some very diverse groups whose common ancestor existed possibly as long ago as 1.6 billion years before the present time. These include glaucophytes (very simple unicellular aquatic organisms), all the different groups loosely known as algae and also the land plants, including the most advanced of these, the angiosperms (flowering plants), on which this book is mainly focused.

Charles Darwin, in a letter to Joseph Hooker, the Director of the Royal Botanic Gardens at Kew, described the origin of flowering plants as an 'abominable mystery'. They seemed at that time to appear in the fossil record without any obvious immediate precursors. Our understanding today, although somewhat more extensive than it was in Darwin's time, is still far from complete; the mystery is not yet completely solved. To appreciate this, it is necessary to go right back to the origin of cellular life and then of eukaryotes. It is a fascinating story.

1.2 Back to the beginning

For much of the 20th century, our knowledge of the history of life on Earth went no further back than the dawn of the Cambrian period – 'only' 550 million years ago. Fossils of quite sophisticated marine eukaryotes have been dated to that time and, during the Cambrian period itself, a very wide range of new lifeforms appeared. This flourishing of diversity in this period is known as the *Cambrian explosion*. However fascinating this is, it does not actually tell us of the earliest lifeforms.

Intense searches in pre-Cambrian rocks were conducted from the mid-1960s onward, but for many years failed to yield any fossils. However, one of those pivotal moments in science came when the American paleobiologist William Schopf identified fossil microorganisms dating back 3.5 billion (i.e. 3.5×10^9) years. Whether or not these represent the oldest living things on Earth is still not clear. Some paleogeochemists have suggested that there is chemical evidence of life processes in rocks dating back 3.8 billion years, while others are of the opinion that the chemicals that supposedly indicate some form of metabolism at that time could equally have arisen by non-biogenic processes. Nevertheless, Schopf's discovery unlocked the 'log-jam' and, since then, many more fossils have been found in pre-Cambrian rocks. Furthermore, paleogeochemical analyses have given us a good idea of what conditions on Earth were like during this period. To this we can add detailed knowledge of the molecular biology and genetics of organisms living today. All this has enabled scientists to build up a picture of the main features of the evolution of living organisms during the pre-Cambrian.

So, life originated around 3.5 billion years ago (and possibly slightly earlier). The predominant, indeed probably the only, organisms then were similar to modern prokaryotes. Earth's atmosphere contained no free oxygen at that time, so these early bacteria were inevitably all anaerobic. Indeed, study of the properties of amino acids in modern anaerobic and aerobic organisms indicates strongly that the genetic code evolved under anaerobic conditions.

A good case has been made that the earliest cells were similar to today's Gram-positive bacteria and gave rise to two further lineages – the Gram-negative bacteria and the Archaea (or archaebacteria). The origin of the Archaea has thus been dated as occurring very early in the history of

Functional Biology of Plants, First Edition. Martin J. Hodson and John A. Bryant.
© 2012 John Wiley & Sons, Ltd. Published 2012 by John Wiley & Sons, Ltd.

life. Fossil evidence indicates that photosynthetic bacteria (like modern cyanobacteria) first appeared about 2.8 billion years ago. The presence of photosynthetic organisms led to the 'great oxidation event' (between 2.2 and 2.45 billion years ago), which was bad news for anaerobic organisms because it generated free oxygen, which was (and still is to an extent) toxic to them. This selective pressure led to the evolution of aerobic organisms, capable of using oxygen in energy generation, probably at least two billion years ago.

1.3 Eukaryotes emerge

The idea that chloroplasts and mitochondria may have been derived from bacteria was first mooted in the 19th century, but it was not until the 1960s that the idea received wider attention. Based on her studies in cell biology, Lynn Margulis proposed specifically that mitochondria were derived in evolution from aerobic bacteria that had been engulfed by anaerobic bacteria, establishing the lineage that led to modern eukaryotes. According to this view, the inner membrane of the mitochondrion

represents the original plasma membrane of the engulfed bacterium and the outer mitochondrial membrane represents the plasma membrane of the original host cell (see Figure 1.1). A second engulfment, this time of a photosynthetic (cyano)bacterium, led to the lineage(s) of photosynthetic eukaryotes and eventually to plants.

It is fair to say that, although some scientists embraced it enthusiastically, the **endosymbiotic theory** was not widely accepted when Margulis originally proposed it. Nevertheless, there was interest in what was called the 'autonomy' of chloroplasts and mitochondria. DNA from these organelles was unequivocally identified, as was the whole range of protein synthesis 'machinery'. To all intents and purposes, these organelles appeared to be organisms within organisms – except that they had only a fraction of the number of genes needed to support independent life. If the endosymbiont hypothesis was correct, then transfer of genes from the endosymbiont to the host genome must have occurred during subsequent evolution.

Further analysis showed that a wide range of molecular biological features – including gene promoters, ribosome structure, sizes of particular types of RNA

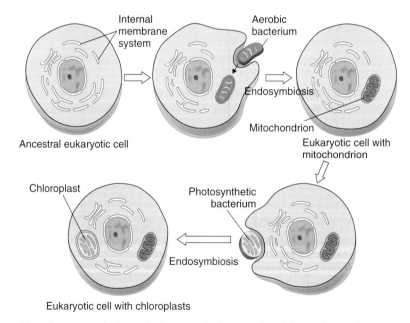

Figure 1.1 Diagram of 'engulfment' events leading to the formation of eukaryotic cells and then of photosynthetic eukaryotic cells. The original engulfing cell ('ancestral eukaryote') was almost certainly descended from an archaebacterium. It must have already possessed some features of eukaryotic cells, including a membrane system and possibly a nucleus (see text). Reproduced, with permission, from http://scienceisntfiction.blogspot.com/2011/04/endosymbiotic-origins.html

and the initiation of protein synthesis in plastids and mitochondria – resembled much more the equivalent features in bacteria than those of the major genetic system in the eukaryotic cells that contain the organelles. Further, the plastids of glaucophytes have a peptidoglycan wall, similar to the cell walls of cyanobacteria. All this is, of course, consistent with the endosymbiotic hypothesis and, by the time Margulis published her book *Symbiosis in Cell Evolution* in 1981, the hypothesis was accepted by the majority of biologists.

Further research during the past three decades has further confirmed the validity of the hypothesis, and it is now firmly stated that eukaryotes arose by the engulfment of an aerobic α-proteobacterium. Whether the 'host' cell was an archaean or a eubacterium is a matter for discussion. However, comparisons of biochemical mechanisms involved in DNA, RNA and protein synthesis, and of the sequences of genes and proteins, suggest a close relationship between the eukaryotic and archaebacterial clades. The authors of this book thus favour an archaebacterial origin for the eukaryotes, as shown in Figure 1.1, but there are some who believe that eukaryotes and archaebacteria are sister clades, having diverged from a common ancestor. Whichever of these two views one holds, there are still further problems to consider, of which we highlight three:

• First, there are some 60 clear differences between the organization, activity and structure of eukaryotic and prokaryotic cells. One of these differences is that prokaryotes are incapable of phagocytosis. However, the engulfment of a proteobacterial cell by an archaebacterial cell, a key part of the endosymbiont theory, would have been achieved by phagocytosis. So, either we envisage that a sub-group of ancient archaebacteria had already acquired some eukaryote-like features, such as phagocytosis, or that merger of two cells occurred by an unknown process.

• The second problem concerns another of these major differences, namely the sequestration of the main genome inside a complex organelle – the nucleus. With this came specific mechanisms for the division and segregation of the genome in the processes of mitosis and meiosis (the latter arising as part of the evolution of sexual reproduction). There has been much speculation on the evolution of the nucleus, but to date no really convincing hypothesis has emerged. The origin of this major feature of all eukaryotic cells remains totally mysterious.

• The third problem is that of the age of the eukaryotic lineage. The 'molecular clock' approach uses comparisons of sequences of genes and proteins in diverging lineages. Assumptions about rates of mutation, based on rates in living organisms, give an estimate of when lineages diverged from each other. This method places the origin of the eukaryotes at between 1.9 and 2.0 billion years ago, and there is some support for this dating from the fossil record. Most paleobiologists accept this dating, but there is a small group who contest it vigorously, suggesting that the eukaryotic lineage is much younger, dating back 'only' 800–900 million years. The authors of this book accept the majority view.

1.4 Photosynthetic eukaryotes – the first 'plants'

The emergence of photosynthetic organisms and the resulting 'great oxidation event' provided the selective pressure for the emergence of aerobic organisms and the establishment of the eukaryotic lineage. However, we can say with some justification that the arrival of photosynthetic eukaryotes was even more significant. This large and now diverse array of autotrophic organisms, ranging from simple single-celled organisms to huge forest trees, has had a greater effect on the world's ecosystems than any other, and thus the engulfment of a photosynthetic cyanobacterium by an early aerobic eukaryote was a key step in the development of life on Earth.

Eukaryotes had split relatively rapidly into two groups: the unikonts (with one flagellum[i]), which gave rise to animals and fungi; and the bikonts (with two flagella). It was among the latter that photosynthetic ability was acquired, approximately 1.6 billion years ago. The Australian cell biologists Geoffrey McFadden and Giel van Dooren leave us in no doubt about the significance of this event:

'This fusion of two cell lineages...brought the power of autotrophy to eukaryotes and descendants of this partnership have populated the oceans with algae and the land with plants, providing the world with most of its biomass'.

---

[i]The Greek word *kontos* actually means 'barge-pole' or 'punt-pole' and gave rise to the English word *quant.*

From this foundational step, there arose several of the groups that we included in our earlier loose definition of plants, including the green plants (see Box 1.1).

Box 1.1 Abundance of green plants

The role of plants in contributing to biomass is clearly seen by considering *cellulose* (Chapter 2, section 2.2.1). This polysaccharide component of the cell walls of nearly all photosynthetic eukaryotes is the most abundant organic compound on Earth.

Furthermore, the most abundant protein in the world and the most abundant naturally occurring polar lipid in the world are both associated with photosynthesis. The protein is the primary carboxylating enzyme, *ribulose bisphosphate carboxylase oxygenase* (also known as Rubisco; see Chapter 7, section 7.4.5), while the lipid, *monogalactosyl diglyceride* (MGDG), is an essential component of the chloroplast thylakoid membrane (see Chapter 2, section 2.5.2). It is ironic that many biologists are unfamiliar with these two important molecules.

However, the story does not end there. There are many photosynthetic eukaryotes, some of them loosely classified in the past as algae, in which the plastids do not have the 'classical' double membrane but instead have four (or in some groups, three) membranes round them. Where did these complex plastids come from? Detailed sequence analysis of their genes and the genes of 'conventional' plastids indicate strongly that *all* plastids arose from a single ancestral source – the originally engulfed cyanobacterial cell. Study of the extra membranes round these complex plastids shows that they originated when a non-photosynthetic eukaryote engulfed a photosynthetic eukaryote.

The extra membranes round these plastids thus represent the plasma membranes of the engulfed cell and of the host. The major event of this type was the engulfment of a red algal cell, which led to lineages that include cryptophytes (which still carry a relic of the nuclear genome of the engulfed cell, the nucleomorph, with approximately 500 genes in a much reduced genome), the dinoflagellates (which have lost the host-derived outer plastid membrane), the brown algae and the diatoms.

In some of the lineages arising from this secondary symbiosis, the plastid has been lost or is much reduced. The Apicomplexa, a phylum that includes the malaria parasites (*Plasmodium* species) provide examples of this. Until the evolutionary origin of this group was understood, the possession of plastids by these organisms seemed very bizarre. The organisms are, of course, non-photosynthetic; over the course of evolution, their plastids (known as apicoplasts) have lost all the components of the photosynthetic machinery. However, they still have an important role in fatty acid metabolism and are essential to the life of the organism.

Finally in this section, it is noted that there have certainly been more than one of these secondary symbioses. The current view is that three such events took place in total, the other two involving engulfment not of red but of green algal cells. One of these events gave rise to the euglenoids (e.g. *Euglena gracilis*), which, like the dinoflagellates, have lost the outermost of the four chloroplast membranes. The other event led to the emergence of the chlorarachniophytes, which, like the cryptophyte lineage arising from the 'main' secondary symbiosis, have retained the vestiges of the engulfed cell's genome in the form of a nucleomorph.

1.5 The greening of Earth – plants invade the land

The evolutionary 'journey' from the first living organisms to the emergence and initial diversification of photosynthetic eukaryotes, discussed here in the space of a few paragraphs, covered a period of well over two billion years (the secondary symbioses described above are dated by different authorities at some time between 1.2 and 0.55 billion years ago). All the events described took place in water and, even today, 40–70 per cent of the world's primary production (based on photosynthesis) occurs in marine environments (despite the fact that the total 'photosynthetic biomass' of marine photosynthetic organisms is only about 0.33 per cent of the total). Admittedly, photosynthetic prokaryotes – cyanobacteria – are responsible for a large proportion of the CO_2 of that fixed in marine environments, but marine algae of various lineages, and especially diatoms, are also very important.

As a habitat, water has one major disadvantage for photosynthetic organisms: the deeper the water, the less light there is. Light may be reflected off the water surface, it may be scattered by particles in the water and it is absorbed by the water. The speed at which the latter happens depends on the wavelength of the light; light at the red end of the spectrum is absorbed before light at the blue end of the spectrum. Thus, in clear water, red light penetrates only to about 15 metres, whereas blue light may reach 100 m. There is therefore a zone – the **euphotic zone** – in which light penetration is adequate to support photosynthesis. In general, shallow water occurs on the margins of land masses and, in this primal history of photosynthetic eukaryotes, the land represented a major niche (actually, of course, a wide array of niches), endowed with a much better light environment.

Although better access to light was an obvious advantage, there were also obvious disadvantages. The need for water in order to maintain life meant that the possibility of desiccation was a serious problem. Water is also the medium into which algae release their gametes. Sexual reproduction on land would be more difficult. Furthermore, immersion in water made for easy uptake of nutrients and also provided support for the larger organisms.

Successful conquest of the land needed solutions to these problems and, based on fossil evidence, this did not occur until between 450 and 490 million years ago. It was another defining event in the history of planet Earth, albeit an event that unfolded slowly. There are now at least 370,000 species of land plants. Their evolution and diversification led to dramatic changes in Earth's environment, including a reduction in the concentration of carbon dioxide in the atmosphere, which resulted in a lowering of the planet's surface temperature. Linda Graham refers to all this as a '*quiet but relentless transformation of terrestrial landscapes*' which initiated the development of new ecosystems and the provision of niches for the evolution of other organisms.

In the transition from water to land, we see a major change in the predominant lifestyle. The aquatic ancestors of the land plants, in common with the majority of modern aquatic photosynthetic eukaryotes, were protists. Most protists are single-celled; the relatively few multicellular forms have little in the way of cellular differentiation, even though some (such as kelps) are very large. Some more complex protists, including the kelps and other brown algae, possess a region of dividing cells, equivalent to the meristems of land plants. The organization of these protist meristem-like regions is simpler than it is in land plants, with fewer possible planes of division.

Simpler protists are capable of, and in many circumstances do undergo, asexual reproduction. In those forms that also reproduce sexually (i.e. by the fusion of gametes), a meiotic division is necessary somewhere in the life cycle.

In the simplest examples, this occurs in the zygote, straight after fertilization, but in many protists there is an alternation of generations in which a lifeform that produces gametes alternates with a lifeform that produces spores.

In contrast to the protist life style, we see in land plants the **embryophyte** lifestyle. Embryophytes are multicellular, with clear cellular and tissue specialization. Dividing cells are organized in regions known as **meristems;** meristematic cells possess more than two cutting planes and can thus generate three-dimensional structures. All embryophytes exhibit alternation of generations and possess antheridia (male gametophyte organs) and archegonia (female gametophyte organs) or the equivalent of these structures. Above all, their embryos are **matrotrophic**, meaning that for all or part of their period of existence they are closely associated with maternal tissues, from which they draw nutrients and signalling molecules.

The simplest, and probably the most primitive, embryophytes, the mosses and liverworts (Bryophyta) are still extensively reliant on water. They have no obvious means of restricting water loss and there are no specialized water-conducting cells. The plants also require water to enable the male gametes to swim to the female gametes within the archegonia in order to bring about fertilization. Modern bryophytes are desiccation-tolerant (i.e. they can recover from severe dehydration) and it is likely that this was also true of the earliest members of this group.

So how and when did these early land plants arise? Study of the cell biology and ultrastructure of modern green algae and bryophytes shows that the bryophytes resemble more the charophyte algae than the chlorophyte algae. For example, in both charophytes and bryophytes (and indeed in all embryophytes), the mitotic spindle is persistent and mitosis is open. The cell wall between daughter cells is laid down via a structure called the **phragmoplast** (see Chapter 2, section 2.12.2), involving a cleavage furrow with a microtubule array oriented at $90°$ to the plane of cell division.

There are also clear biochemical similarities between charophytes and embryophytes, while molecular phylogenetic analysis, based on gene sequences in nuclear, plastid and mitochondrial genomes, places the charophytes as a sister group to all embryophytes. Furthermore, extant charophytes have rudiments of the matrotrophic embryo, in that there are cellular

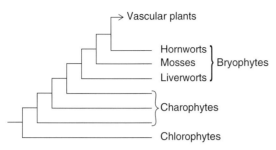

Figure 1.2 Diagram illustrating the positions of the chlorophytes and charophytes in the ancestry of embryophyte land plants.

Table 1.1 The geological periods.

Period	Years before present
Quaternary	1.8 million to present day
Tertiary	66.4 million to 1.8 million
Cretaceous	144 million to 66.4 million
Jurassic	208 million to 144 million
Triassic	245 million to 208 million
Permian	286 million to 245 million
Carboniferous	360 million to 286 million
Devonian	408 million to 360 million
Silurian	438 million to 408 million
Ordovician	505 million to 438 million
Cambrian	570 million to 505 million
Pre-Cambrian	4.5 billion to 570 million

interactions between haploid maternal cells and diploid zygotes that are thought to be involved in nutrient transfer. They also possess cell wall polymers that inhibit fungal degradation. In particular, a polymer laid down in charophyte zygotes resembles strongly the sporopollenin present in the cell walls of seed-plant pollen. All these data suggest that embryophytes and charophytes are descended from a common ancestor which itself had arisen by divergence from the chlorophytes (Figure 1.2).

Although the family tree for the earliest embryophytes appears clear enough from the data based on extant species, the fossil record is much less helpful. The main problem is that the earliest fossil evidence (consisting of tetrads of spores) for embryophyte land plants dates back about 450–490 million years, to the mid Ordovician period (see Table 1.1) whereas the earliest known fossil charophytes occur in rocks from upper Silurian strata, dating back about 414 million years. Thus we have no clear picture of the immediate ancestor of the embryophytes. We do not know whether the embryophyte lifestyle evolved in an aquatic environment, or whether charophytes invaded the land before the origin of embryophytes. The existence today of many species of both chlorophyte and charophyte algae that live in terrestrial habitats (albeit still needing water for sexual reproduction) certainly shows that the latter was possible. Nevertheless, from our point of view as we follow the journey from the earliest living organisms to flowering plants, the main point is clear: the land was invaded.

There is still some discussion about which of the three bryophyte lineages – hornworts, mosses or liverworts – represent the earliest land plants. Although there is some support (mainly from comparative anatomy and morphology) for the view that hornworts were the

earliest land plants, studies of genome structure, of gene sequences and of particular biochemical mechanisms in extant plants, point to the liverworts.

For example, in common with charophytes, the immediate progenitors of land plants, the mitochondrial DNA of liverworts lacks a particular type of intron[ii], the type II intron (see Chapter 3, section 3.2.1). All other bryophytes and all vascular plant groups possess three mitochondrial type II introns, although there have been subsequent losses in some lineages within these plant groups. Indeed, those who use molecular data in constructing phylogenies suggest that such data settle the question beyond doubt, so that is the position we take here: the earliest land plants were liverworts, from which mosses and hornworts diverged. The latter eventually gave rise to vascular plants (see next section).

1.6 Embracing the terrestrial lifestyle

While terrestrial habitats may indeed provide a good light environment, they also pose some strong challenges for living organisms. The lifestyle of modern bryophytes almost certainly typifies the way in which the earliest multicellular land plants dealt with those challenges. Such a lifestyle is successful in its own way, in its own ecological niches, but it can hardly be said to have conquered the land. Invasion is different from conquest.

[ii]An intron is a sequence of DNA that interrupts the coding sequence of a gene (see Chapter 3, section 3.2.1).

Nevertheless, the popular view of early land plants is one of conquest. We are very accustomed to reconstructions and artistic presentations showing a rich flora of vascular plants. The dominant forms differ according to which geological period is being portrayed, but the common feature is that it is *vascular plants* which make up these fossil forests. Conquest, rather than just invasion of the land, required a number of adaptations, including mechanisms or structures for prevention of water loss and for movement of water within the plant. Furthermore, the selective pressure to seek the light also led to the need for support as many plants evolved an upright stance.

In modern floras, symbiosis between green plants and soil-dwelling fungi features very strongly, as seen in different types of **mycorrhizae** (see Chapter 5, section 5.8). It now seems likely that mycorrhizae, and possibly other forms of symbiosis, were important in helping green plants to invade the land. Mycorrhizae identical in form to modern vesicular-arbuscular mycorrhizae have been discovered in association with *Aglaophyton major*, a very early Devonian land plant, suggesting that nutrient transfer mutualism (symbiosis) may have been in existence when plants invaded the land. This would have aided green plants in exploiting nutrient-poor substrates.

Evidence for the early evolution of vascular plants comes from fossils, from new, less destructive techniques for investigating fossil structure, from comparative anatomy and physiology of extant plants and from molecular phylogenetic studies. These studies provide strong evidence that the hornworts were the immediate ancestors of vascular plants. It is interesting that hornworts can exert some degree of control over water loss and gas uptake because they possess stomata, an important adaptation to life on land and a feature found in all vascular plants (see Chapter 9, section 9.4).

The evidence for a single origin ('monophyly') of the vascular plants comes both from comparative morphology and from an increasing array of DNA sequence data. What is not so clear is the position in the evolutionary tree of some fossil plants found in a remarkable assemblage in the Rhynie chert in Scotland. These fossils, which include *Aglaophyton, Horneophyton* and *Rhynia*, possess some features of vascular plants but also retain several bryophyte-like characteristics.

The earliest true vascular plants were the lycopsids or lycophytes. These first appeared in the late Silurian period. Modern members of the group include quillworts

Figure 1.3 *Lycopodium thyoides*.
Photograph by Dr Gordon Beakes © University of Newcastle upon Tyne. Image from Centre for Bioscience (Higher Education Academy) ImageBank.

(*Isoëtes*), *Selaginella* and club mosses (*Lycopodium*; see Figure 1.3). Today they are relatively scarce, but in the Carboniferous period they were a dominant group, with tree lycopods forming extensive forests. The ability to grow as trees reflects the dual function of vascular tissue, both as a means of conducting water and nutrients throughout the plant and as a means of support of large aerial structures (see Chapters 5 and 6). Tree lycopods eventually became extinct in the Permian period, but they left a legacy, providing the bulk of the material from which coal was formed.

Molecular phylogenetic evidence indicates strongly that lycopods gave rise to a lineage which then diversified into several groups, including the ferns and other fern-like plants, horsetails and eventually the various seed-plant groups. The horsetails, still represented in today's biosphere, are particularly interesting. Like lycopods, they produced dominant forests of tall plants. The ability to grow tall was related to the role of silica in supporting the stems, in contrast to today's tall plants, which are supported by lignin (see Chapter 2, section 2.2.4 and Chapter 6, section 6.5).

In summary then, the invasion of the land that started with bryophytes became a conquest as vascular plants appeared and then diversified. Indeed, the diversification of plant life on land (and its knock-on effects on the evolution of other organisms) known as *the Siluro-Devonian primary radiation*, is regarded as the terrestrial equivalent of the Cambrian explosion of marine life (as discussed in section 1.2).

Examination of fossil assemblages in strata of different ages reveals a succession of plant groups appearing, some of which became abundant for at least several million years. Many of these groups survive today, but there are some notable exceptions. We have already seen that tree lycopods, dominant in Carboniferous forests, became extinct in the Permian. The fossil record also contains a major phylum, the progymnosperms, that arose in the late Devonian and early Carboniferous and flourished for a time. The name is somewhat misleading, because they produced spores rather than seeds[iii] and did not give rise to modern gymnosperms. Nevertheless, the late Carboniferous/early Permian periods saw the emergence of gymnosperm groups which are still represented in extant floras. Indeed, gymnosperms were one of the dominant groups in late Triassic and early Jurassic forests – an indication of the selective advantages of the seed-based mode of reproduction (see Box 1.2).

Box 1.2 Advantages of seeds

Reproduction via seeds provides distinct advantages for life on land. Fertilization does not require water because the sperm does not have to swim to the egg. The one exception to this amongst seed plants is *Ginkgo biloba,** in which the sperm are motile. The seed that develops following fertilization is effectively an embryo held in a state of quiescence or dormancy, usually provided with a food store and surrounded by a. protective coat.

*Maidenhair tree: the sole extant member of a group of gymnosperms that arose in the Permian and were abundant through to the end of the Triassic. Ginkgo is illustrated in Figure 7.1, Chapter 7.

[iii]The name *gymnosperm* means 'naked seed', in contrast to *angiosperms*, in which seeds are enclosed in a structure called the carpel.

Today, the gymnosperms are represented by just four groups – the Gnetophyta or Gnetales (see below), the Coniferae, *Gingko* and the cycads (Cycadophyta; Figure 1.4). Except for the conifers, these groups are just relics in terms of their former abundance and dominance, For example, there are only a few species of cycads, while *Ginkgo biloba* is the sole living representative of a once more diverse group.

1.7 Arrival of the angiosperms

Modern angiosperms share with each other many features that are not represented at all in other groups (see Box 1.3) and on that basis they have been regarded as a single discrete group arising from one ancestral lineage – i.e. they are monophyletic. This view has been extensively confirmed by modern molecular phylogenetic analysis.

Box 1.3 Essential features of angiosperms

The term 'angiosperm' derives from two Greek words: *angeion*, meaning 'vessel' and *sperma*, meaning 'seed'. The angiosperms are those plants whose seeds develop within a surrounding layer of plant tissue, called the carpel, with seeds attached around the margins. This arrangement is easily seen by slicing into a tomato, for example.

Collectively, carpels, together with the style and stigma, are termed the ovary, and these plus associated structures develop into the mature fruit. The enclosed seeds and the presence of carpels distinguish angiosperms from their closest living relatives, the gymnosperms, in which the seed is not enclosed within a fruit but, rather, sits exposed to the environment.

Some defining characteristics of angiosperms include flowers, carpels and the presence of endosperm, a nutritive substance found in seeds, produced via a second fertilisation event. Angiosperms thus exhibit the phenomenon of **double fertilisation** (see Chapter 8, section 8.6.3).

But from where did the angiosperm lineage arise? What is the sister group to the angiosperms? If these questions could be answered, we would be making progress towards solving Darwin's abominable mystery. Prior to the availability of molecular techniques, morphological comparisons had led to the angiosperms being regarded as sister group to the Gnetales, a varied group of gymnosperms represented today by just three families. The most bizarre of these is the family Welwitschioideae, type genus *Welwitschia*, which produce flowers that rest on the ground (see Figure 10.9, Chapter 10).

Figure 1.4 The cycad *Encephalatos ferox*, native to coastal habitats in Mozambique. Photo: MJH. Used with the permission of Oxford Botanic Gardens.

Angiosperms and Gnetales were together known as the anthophytes, but it is now clear from molecular phylogenetic analyses that the anthophyte hypothesis is untenable. However, that is not to say such analyses have solved the problem. When molecular phylogenetic analysis first became available, it was widely thought that its careful application to seed plants would sooner or later lead to an understanding of angiosperm origins. However, this has not proved to be the case. Indeed, some plant scientists believe that the mystery is as deep now as it was in Darwin's day. The problem is that different analyses tell different stories, depending on which genes are used in the analysis and whether DNA or protein sequences form the basis for comparison. Thus, the distinguished evolutionary botanist, James Doyle, at Davis, California, wrote in 2008: '*Much of what we thought we knew 10 years ago about seed plant phylogeny . . . has been thrown into doubt by molecular analyses.*'

Doyle also wrote that: '*Resolution of these problems requires integration of molecular, morphological and fossil data in a phylogenetic framework.*'

The data from fossils include analyses of flowers and flower-like structures in presumed angiosperms such as *Archaefructus* from the early Cretaceous (currently the earliest known fossils of angiosperm-like flowers date from this period) and in seed ferns. The molecular data include molecular clock estimates that put the origins of angiosperms no earlier than the Jurassic period. This integration of approaches suggests that the divergence from the gymnosperms (and more specifically from the cycads) of the lineage that led to angiosperms happened probably as early as the Carboniferous. In other words, the last common ancestor between the two groups of extant seed plants – gymnosperms and angiosperms – was alive in the Carboniferous period. This divergence established the lineage known variously as the angiophytes or the pan-angiosperms. However, as pointed out by Doyle in a personal communication to JAB, the early members of this lineage: '*. . . need not have looked any more like modern angiosperms than pelycosaurs (in an early Permian branch from the mammalian stem lineage) look like mammals.*'

So, although the gymnosperms (and in particular the cycads) are the nearest living relatives to modern angiosperms, the actual sister-group to angiosperms is to be found among extinct groups within the pan-angiosperms, namely the seed ferns (Figure 1.5). Current

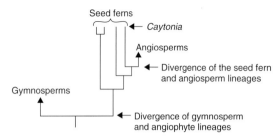

views are that the most likely seed-fern group to fulfil this role is that containing the genus *Caytonia*.

In summary then, angiosperms arose as the crown group of the pan-angiosperms in the late Jurassic (based on molecular clock data), or slightly later in the early Cretaceous (based on fossil evidence). Their evolution as a distinct group, after divergence from seed-ferns, involved the adaptation and development of pre-existing structures to form, among other things, the characteristic angiosperm flower. The double fertilization involved in endosperm formation (see Chapter 4, section 4.3 and Chapter 8, section 8.6.3) also evolved at this time (double fertilization also occurs in the Gnetales, mentioned above, but in that group it leads to the formation of two embryos).

Until relatively recently, the Nymphaeales (including present-day water lilies and probably also the fossil *Archaefructus*) were regarded as the most primitive angiosperms. However, based on extensive phylogenetic analysis, *Amborella trichopoda* (Figure 1.6), a semi-climbing shrub only found in the rain forests of New Caledonia, is now regarded as sister to all extant angiosperms and is therefore at the base of the angiosperm phylogenetic tree (Figure 1.7)[iv].

Thus *Amborella* is at the base of the very diverse taxon, extant angiosperms. Early divergence brought into

existence two other primitive groups, the Nymphaeales (water lilies) and Austrobaileyales; these groups, together with *Amborella* are often known as the **ANITA grade**, based on the genera *Amborella*, *Nymphaea*, *Illicium*, *Trimenia* and *Austrobaileya*. All other angiosperm groups are often termed the mesangiospermae. The more primitive mesangiosperms include the magnolids (see Figure 1.7), but the most obvious indications of the extensive radiation of the angiosperms are the **monocots** (monocotyledones) and **eudicots** (eudicotyledones). The latter term, meaning effectively 'good dicots' or 'true dicots' distinguishes these from the 'paleodicots' represented by the ANITA grade and by the magnolids and other more primitive groups.

The ecology of *Amborella* and of other primitive angiosperms suggests that the group first arose in shady, damp or wet and possibly disturbed habitats. The subsequent radiation of the angiosperms occurred mainly between 100 and 65 million years ago, and a large proportion of currently living groups had appeared by the end of the Cretaceous period.

This very rapid radiation is certainly worthy of comment. It was one of the features that caught Darwin's attention; flowering plants seemed to him to appear from nowhere (although we are now beginning to understand something of their origins). The rapidity and extent of the angiosperm radiation is indeed astonishing, such that we are justified in speaking of a 'Cretaceous explosion'. This radiation has seen angiosperms progress from being a relatively minor component of the biosphere to becoming the major vascular plant group, totalling between 250,000 and 300,000 species[v], occupying the widest possible range of ecological niches and dominating the vegetation in many terrestrial and some aquatic ecosystems. In attaining such dominance they ousted the gymnosperms from their previously dominant position, and although angiosperm distribution over the Earth has changed with the changing form and climate of the planet, they have remained the dominant plant group for the past 65 million years.

In morphology and growth form, the angiosperms vary between the tiny *Wolffia*, a genus in the duckweed family (Figure 1.8a) to very large and long-lived trees (Figure 1.8b) (although, admittedly, the tallest, the largest

---

[iv] Interestingly *Amborella* (or any plant similar to it) has not been found in fossil form – an indication of the incompleteness of the fossil record.

[v] The current estimate for the total number of species of all land plants is 370,000 (see Chapter 12, section 12.2.1).

Figure 1.6 Flowers of *Amborella trichopoda*, the most primitive living angiosperm. Photo: Scott Zona, Florida International University.

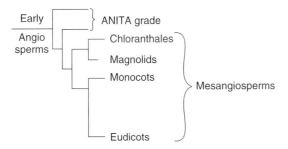

Figure 1.7 Simplified diagram illustrating the divergence of angiosperm groups.

and the longest-lived trees are all gymnosperms: see Chapter 6, section 6.6). As a group, angiosperms exhibit a wide range of interactions with other members of the biosphere, of which arguably the most famous is that many species rely on insects for pollination. Indeed, several authorities regard co-existence and co-evolution with insect pollinators as being one of the factors that contributed to the angiosperm rise to dominance. This relationship is discussed further in Chapter 8, section

8.6.2, but at this point we discuss a more basic aspect of reproduction.

1.8 Sex and the alternation of generations

The evolutionary history of the angiosperms, traced back to the origins of life, incorporates another story, namely the evolution of reproductive mechanisms. Indeed, the angiosperms are named for one aspect of their reproduction, namely the enclosure of the developing seeds in the carpel. It is therefore appropriate at this point to consider another facet of sexual reproduction – namely the need to alternate between the haploid and the diploid states.

In section 1.1 we noted the emergence of eukaryotic cells as cells with defined sub-cellular organelles, including a nucleus that harbours the genetic material, DNA. Among the 60 or so differences between prokaryotes and eukaryotes is the existence in the latter of complex cell division mechanisms to ensure the segregation of the genetic material between daughter nuclei and, hence, between daughter cells. One of these cell division

(a)

(b)

Figure 1.8 (a) Individual plants of *Wolffia arrhiza* (Watermeal) are 1–2 mm wide, and it is the smallest vascular plant. Photo: Aaron Woods. (b) Oak trees (*Quercus robur*) coming into leaf in spring. Fallow deer grazing under the trees. Photograph taken by JAB at Ripley, Yorkshire, UK.

mechanisms is meiosis, the division that produces haploid cells from diploid cells (i.e. halves the number of genome copies in a cell). This is an absolute requirement for sexual reproduction, without which the number of copies of the genome per cell would double with each generation.

Although it is not entirely clear, it is likely that both mitosis and meiosis evolved before the endosymbiont engulfment that produced the first true eukaryote. The

acquisition of these activities was part of a process known as *eukaryogenesis*. What is clear is that sex is a eukaryotic activity. Prokaryotes cannot undertake a reduction division and therefore cannot indulge in sexual reproduction.[vi] The evolutionary significance of sex is enormous. Not only can genetic variation be generated by mutation and horizontal gene transfer, but also by the mixing of the genetic variation of the two sexual parents.

Sexual reproduction has been incorporated into eukaryotic lifestyles in a number of different ways but all inevitably involve an alternation between a haploid and a diploid phase. Eukaryotic organisms exhibit several basic types of sexual life cycles, differing in the ploidy of adult organisms and in the site of meiosis. The simplest sexual lifestyle we can envisage would involve the meiotic reduction division occurring immediately after the sexual fusion of two haploid cells. This is seen in the single-celled green alga *Chlamydomonas*, in which the haploid vegetative cells produce haploid gametes. Gametes fuse to form diploid zygotes, which undergo meiosis to form new haploid vegetative cells. The zygote is therefore the only diploid cell in the lifecycle.

Animals provide a complete contrast. They exist as diploid organisms and the only haploid cells are the gametes; meiosis occurs during gametogenesis. The fusion of the two gametes restores the diploid state, thus initiating the next generation. Admittedly there are many variants of this basic pattern. For example, some animals have larval stages and attainment of the adult form may involve quite a dramatic metamorphosis. At the other end of the scale, in reptiles, birds and mammals, the young that hatch from the egg, or that are born, grow and develop 'seamlessly' into the adult. Nevertheless, among all this variety of animal life progressions, it remains true that the only haploid cells are the gametes.

Land plants (and some non-vascular aquatic plants), however, have adopted a completely different pattern, in which there are two different multicellular generations, one haploid and one diploid. In other words, there is an *alternation of generations* (also known as a diplobiontic life cycle; see Figure 1.9). The diploid phase is the **sporophyte** or spore-producing generation. Spores are produced by meiosis; the haploid spores germinate, undergo cell

---

[vi]Prokaryotes are able to exchange genetic material in processes such as conjugation, but these processes are not equivalent to sexual reproduction.

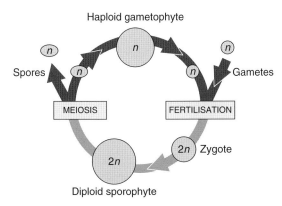

Figure 1.9 Diagram illustrating the basic features of alternation of generations. Note that some organisms are *heterosporous* (the spores germinate to form single-sex gametophytes); male gametophytes release sperm, while female gametophytes carry the egg cells. In *homosporous* organisms, the spores germinate to give only one type of gametophyte, which produces both sperm and egg cells.

division, differentiation and morphogenesis to produce the mature **gametophytes**. As the name implies, the gametophyte phase produces gametes which fuse to produce a diploid zygote which is the start of the new sporophyte generation. In land plants, the two generations differ in appearance; the two generations are *heteromorphic*. However, in the green alga *Ulva* which exhibits alternation of generations, the two life cycle phases look the same; they are thus *isomorphic*.

In the simpler land plants, bryophytes, the dominant generation (the one that we see and recognize as a moss or liverwort) is the gametophyte generation. The sporophyte generation is short-lived and generally dependent on the gametophyte. In vascular plants, by contrast, the sporophyte is the dominant generation. Gametophytes of vascular plants are much smaller than their sporophytes and are either free-living or retained within sporophytic tissues. For example, in ferns, spores develop within clusters of sporangia on the sporophyte plant; spores germinate to produce gametophytes that are, like the sporophyte, photosynthetic and free-living but small and inconspicuous. The gametophytes produce sperm cells and egg cells within specialized structures called antheridia and archegonia; the gametes fuse to produce a diploid zygote which develops into the new sporophyte.

In gymnosperms and angiosperms, it is again the sporophyte generation that we recognize as the plant. The gametophytes are very small, non-free-living and non-photosynthetic. The female gametophyte is retained on the parent plant within an ovule. Pollen grains are immature male gametophytes which produce sperm cells. The details of sperm cell generation and of fertilization mechanisms differ between gymnosperms and angiosperms but, essentially, fertilization of the egg cell occurs within the female gametophyte, as do the early stages of growth of the new sporophyte generation, namely embryogenesis and seed development. These processes in angiosperms are discussed more fully in Chapters 4 and 8.

Selected references and suggestions for further reading

Doyle, J.A. (2006) Seed ferns and the origin of angiosperms. *Journal of the Torrey Botanical Society* **133**, 169–209.

Doyle, J.A. (2008) Integrating molecular phylogenetic and paleobotanical evidence on origin of the flower. *International Journal of Plant Sciences* **169**, 816–843.

Field, T.S., Brodribb, T., Jaffre, T. & Holbrook, N.M. (2001) Acclimation of leaf anatomy, photosynthetic light use and xylem hydraulics to light in *Amborella trichopoda* (Amborellaceae). *International Journal of Plant Sciences* **162**, 999–1008.

Graham, L.E. (1996) Green algae to land plants: an evolutionary transition. *Journal of Plant Research* **109**, 241–251.

Lenton, T. & Watson, A. (2011) *Revolutions that Made the Earth*. OUP, Oxford.

Margulis, L. (1981)

McFadden, G.I. & van Dooren, G.G. (2004) Evolution: Red algal genome affirms a common origin of all plastids. *Current Biology* **14**, R514–R516.

Qui, Y-L., Cho, Y., Cox, J.C. & Palmer, J.D. (1998) The gain of three mitochondrial introns identifies liverworts as the earliest land plants. *Nature* **394**, 671–673.

Sanderson, M.J., Thorne, J.L., Wikström, N. & Bremer, K. (2004) Molecular evidence on plant divergence times. *American Journal of Botany* **91**, 1656–1665.

CHAPTER 2
Introduction to Plant Cells

The previous chapter described the emergence in evolution of the angiosperms, the flowering plants. Much of the rest of this book deals with angiosperm function at the levels of organ and whole organism; we discuss the integration of growth and development, the angiosperm life cycle and the inter-organism interactions involved in various angiosperm lifestyles.

However, in order to understand the plant as a functioning organism, it is necessary to have some knowledge of plant biology at the cellular and sub-cellular levels. Therefore, in this chapter and the next, we provide introductions to plant cells and to the major molecular activities in which the cells participate.

2.1 Plant cells

There is a sense in which there is no such thing as a 'typical' plant cell. Cell structure varies extensively according to the function of the cell in question. Nevertheless, it is helpful at this point to consider the main features of plant cells before looking at those features in more detail in subsequent sections. The features are illustrated diagrammatically in Figure 2.1.

First, plant cells are characterized by being contained within a *cell wall* (section 2.2), composed mostly of polysaccharides and whose structure varies according to cell age and function. Inside the cell wall is the cell's outer membrane, the *plasma membrane* (section 2.3). In older cells, the next most obvious feature is the cell *vacuole* (section 2.8), a large aqueous space bounded by another membrane, the *tonoplast*. The vacuole's main functions are storage of particular solutes and the sequestration of hydrolytic enzymes. In vacuolated cells, the *cytosol* or *cytoplasm* is confined to a narrow zone between the vacuole and the plasma membrane (Figure 2.1) but, in

non-vacuolated cells, the cytosol occupies much of the space bounded by the plasma membrane.

Within the cytosol, three membrane-bound organelles are very apparent. The first is the *nucleus* (section 2.7), a feature of all eukaryotic cells (although some cells, such as red blood cells in mammals and phloem sieve tubes in plants, lose their nuclei during cell differentiation). Most of the genetic material, DNA, is located in the form of chromosomes within the nucleus, and all of the biochemical activities associated with gene expression and DNA replication occur there (sections 2.7 and 2.13.3 and Chapter 3, section 3.1.2).

The other two obvious organelles are the *chloroplasts/plastids* (section 2.5) and the *mitochondria* (section 2.6). The former are primarily associated with photosynthesis and starch storage and the latter with energy conservation (as ATP and NADH) during respiration, although chloroplasts actually carry out a much wider range of biochemical reactions. *Microbodies* (Section 2.10) are often located in the vicinity of chloroplasts and mitochondria. These organelles, which are sometimes known as *peroxisomes*, participate in photorespiration (Chapter 7, section 7.5). A particular class of microbody, the *glyoxysome*, is involved in the mobilization of the lipid reserves during germination of fat-storing seeds (Chapter 4, section 4.11.2).

The cytosol is permeated by an extensive endomembrane system, the *endoplasmic reticulum* or ER (section 2.9), which is involved in transport within and out of cells, in the sequestration of calcium ions and in the synthesis of (among other things) proteins that are destined for export. The ER is continuous with the outer envelope of the nucleus and also interacts with the *Golgi apparatus*, also known as *Golgi bodies* or *dictyosomes* (section 2.9). The Golgi bodies appear as stacks of flattened sacs with vesicles located around them. These flattened sacs are

Functional Biology of Plants, First Edition. Martin J. Hodson and John A. Bryant.
© 2012 John Wiley & Sons, Ltd. Published 2012 by John Wiley & Sons, Ltd.

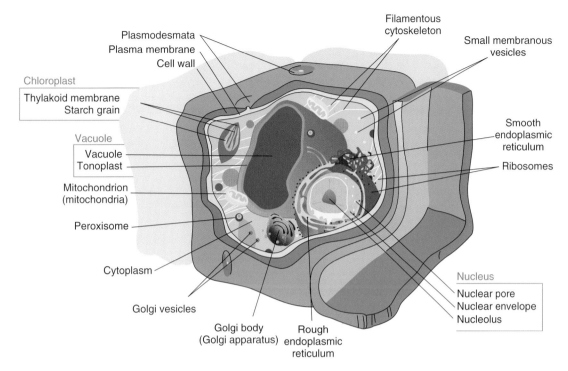

Filamentous cytoskeleton

Small membranous vesicles

Plasmodesmata
Plasma membrane
Cell wall

Chloroplast

Thylakoid membrane
Starch grain

Smooth endoplasmic reticulum

Vacuole

Vacuole
Tonoplast

Ribosomes

Mitochondrion (mitochondria)

Peroxisome

Cytoplasm

Nucleus

Nuclear pore
Nuclear envelope
Nucleolus

Golgi vesicles

Golgi body (Golgi apparatus)

Rough endoplasmic reticulum

Figure 2.1 Diagram of a plant cell.
Author: Mariana Ruiz.

the sites of synthesis of cell wall polysaccharides; 'shuttle vesicles' are budded off from the Golgi bodies for transport of the polysaccharides to the cell wall. Vesicles budded off the ER and carrying proteins for export merge with the Golgi bodies, which then transfer the proteins to shuttle vesicles for further movement.

The cytosol also contains millions of *ribosomes* (section 2.11), some of them located on the surface of the ER (regions of ER with associated ribosomes are known as 'rough ER'). These particles, consisting of RNA and protein, are the sites of protein synthesis. Finally, there is the *cytoskeleton* (section 2.12), a network of microtubules (made of the protein tubulin) and actin filaments (consisting, as the name implies of the protein actin). Among other things, the cytoskeleton is involved in the organization of the plane of cell division, the orientation of cellulose microfibrils (section 2.2.2), the channelling of Golgi vesicles to the plasma membrane (section 2.2.2) and the organization and orientation of chromosomes during cell division (section 2.13).

2.2 Cell walls

2.2.1 General structural features

As was noted in Box 1.1 in Chapter 1, *cellulose* (Figure 2.2) is the most abundant organic compound in the world, because it is a major component of the cell walls of nearly all photosynthetic eukaryotes. In primary cell walls (see section 2.2.3), it makes up between 15 and 30 per cent of the dry mass of the wall. In secondary but unlignified walls (see below), the proportion is even greater.

However, cellulose is only one of several different types of molecule that make up the cell wall: the cellulose, organized as microfibrils (see below) is embedded in a matrix of other polysaccharides. In order to understand this, it is necessary to go back to the earliest phase in the deposition of the plant cell, namely synthesis of the new wall immediately after cell division.

The first cell wall that separates the two daughter cells after cell division (see section 2.13) is known as the *cell plate*. It can be seen in more mature cells as the *middle*

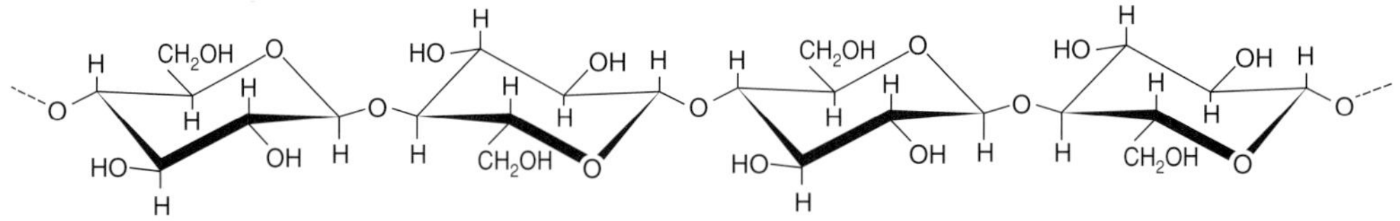

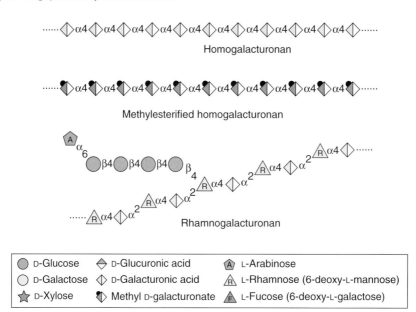

Figure 2.2 Structure of part of a cellulose molecule. Cellulose is a polymer of glucose units joined by β1→4 linkages. In the cell wall, the polymers are aligned in large parallel arrays called microfibrils.

Homogalacturonan

Methylesterified homogalacturonan

Rhamnogalacturonan

⬤ D-Glucose	◆ D-Glucuronic acid	⬠A L-Arabinose
○ D-Galactose	◇ D-Galacturonic acid	△R L-Rhamnose (6-deoxy-L-mannose)
☆ D-Xylose	◖ Methyl D-galacturonate	△F L-Fucose (6-deoxy-L-galactose)

Figure 2.3 Diagram showing structures of the main pectic polysaccharides. Based on Figure 1 of Fry, SC *et al.* (2011)The Biochemist 33, 14–19.

lamella between adjacent cells. The cell plate/middle lamella consists almost entirely of a group of polysaccharides known collectively as *pectins* (Figure 2.3). These are gel-forming polysaccharides (as is well-known by anyone who makes jam) made up of acidic sugars (especially galacturonic acid) and neutral sugars such as arabinose, galactose and rhamnose. Some of the polysaccharides are relatively simple, such as polygalacturonic acid, which is a polymer of α-D-galacturonic acid joined by 1→4 glycosidic linkages. It is often methyl-esterified (i.e. in the form of methyl-galacturonic acid), as shown in Figure 2.3.

Some molecules incorporate an occasional rhamnose residue which kinks the chain at that point. However, the bulk of pectic polysaccharides are more complex, as is typified by the rhamnogalacturonans – large polymers whose 'backbone' consists of alternating galacturonic acid and rhamnose residues. Particular regions of

these molecules carry complex oligo/polysaccharide side-chains, which are joined to the backbone via the rhamnose units. The most abundant side chains are branched *arabinans* (oligosaccharides consisting of arabinose units), *galactans* and *arabinogalactans*. In the latter, the backbone of the side chain consists of galactose units, some of which themselves carry a short side chain of a single arabinose unit. The presence of side chains limits the extent to which individual polysaccharide chains can align with each other, and thus extensive branching makes for a very open structure. Conversely, adjacent pectin molecules may be cross-linked by Ca^{2+} ions bridging between two carboxyl groups. This bridging is inhibited if the carboxyl groups are esterified with a methyl group (see above).

Onto this middle lamella the primary wall is deposited, with cellulose now embedded into the background matrix of firstly pectins (as described above) and then hemicelluloses (see below and Figure 2.4). Cellulose is a β-glucan, a polymer of β-D-glucose units joined via 1→4 glycosidic linkages (Figure 2.2). The linear cellulose molecules, each consisting of several thousand individual glucose units, can hydrogen-bond with each other to make microfibrils comprising many molecules lying parallel to each other. These are the main strengthening components of the unlignified wall.

There are on average, 36 cellulose molecules at any one place in a microfibril, all lying in the same 'chemical orientation'. However, there are many more than 36 cellulose chains in a typical microfibril. Chains do not all start and finish in the same place, but instead overlap with each other. Thus an individual microfibril may contain several thousand individual cellulose chains and may be several hundred μm in length.

The presence of cellulose and its organization into microfibrils is a very important feature of plant cell walls, contributing very significantly to plant cell form and function. Indeed, Canadian plant scientists Luc Duchesne and Doug Larson suggest that: '*The presence of cellulose microfibrils in cell walls may be one of the most critical factors in the evolution of modern plant life.*'

As well as hydrogen-bonding with each other, cellulose molecules can form hydrogen bonds and other non-covalent linkages with the polymers that form the *hemicellulose* component of the matrix in which the microfibrils are embedded. The major polysaccharides of the hemicellulose fraction, some of which are shown in Figure 2.4, are *xylans* (polymers of xylose), *xyloglucans* (with a backbone of β-D-glucose units joined via 1→4 glycosidic linkages and carrying individual xylose molecules as side groups), *arabinoxylans* (polymers of xylose with arabinose side chains) *glucomannans* (mixed polymers of glucose and mannose units), *galactomannans* (β-D-mannose unit joined via 1→4 glycosidic linkages and carrying individual galactose units as side-groups) and mixed-linkage glucans (β-D glucose units joined by either 1→3 or 1→4 linkages). These polymers coat the cellulose microfibrils (Figure 2.5) and hydrogen-bond both with themselves and with the cellulose.

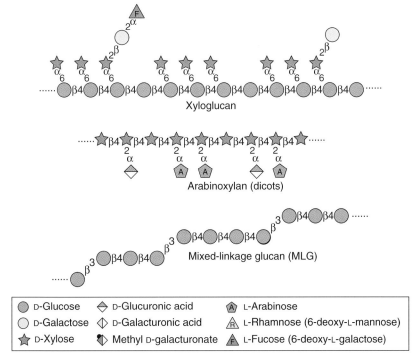

Figure 2.4 Diagram showing structures of some of the commoner polysaccharides in the hemicellulose fraction. Based on Figure 1 of Fry, SC *et al*. (2011) The Biochemist 33, 14-19.

In addition to polysaccharides, the cell wall also contains proteins. The protein component changes as the wall is built up from the middle lamella, but in general these proteins are rich in proline or hydroxyproline or glycine. Some of the hydroxyproline-rich proteins are glycosylated, i.e. they carry carbohydrate side-chains, consisting in this instance of arabinose and galactose. They are thus known as *arabinogalactan proteins* or AGPs.

The specific roles of all these proteins at different stages of plant cell development have not been established. However, it is likely that the hydoxyproline-rich **extensin** (which was the first to be discovered) is involved in cross-linking cellulose microfibrils after cell wall extension and that **expansins** actually participate in the expansion process (see section 2.2.3).

As well as these structural proteins, the cell wall also contains *enzymes*. They are mostly hydrolases of various types, possibly involved in defence and in the recycling or scavenging of nutrients. The enzymes involved in polymerization of lignin precursors are also located in the wall.

It must be noted that the proportion of these different polymers varies during the development of an individual cell (as already noted) and between species. Thus, the grasses, including the economically important cereals, have low proportions of pectic polysaccharides and of xyloglucans. The latter are replaced by *mixed-linkage glucans* (polymers of β-D-glucose units joined via either 1→4 or 1→3 glycosidic linkages).

The cell wall is often referred to as 'rigid'. However, this is not true of the primary cell wall. Cell walls are hydrated dynamic structures (the matrix component contains up to 75 per cent water).

As Stephen Fry and his colleagues at Edinburgh University put it so clearly:

'Although it is true that they are often strong (resisting breakage) and may be inextensible (resisting stretching and thus limiting cell expansion), most primary walls are highly flexible ... The phrase "rigid cell wall" should be expunged except in discussions of secondary walls.'

2.2.2 Cell wall synthesis

Discussion of cell wall expansion is deferred until the next chapter. Here we briefly consider the synthesis of the cell wall polysaccharides. Cell wall polysaccharides are no exception to the general rule that the donors for building up polymers of monosaccharides are nucleotide-sugars, as shown in these general equations:

1. $M\text{-}1\text{-}P + NTP \rightarrow NDP\text{-}M + PP_i$
2. $M_n + NDP\text{-}M \rightarrow M_{n+1} + NDP$

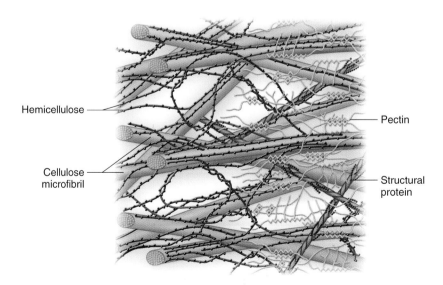

Figure 2.5 Cellulose in the cell wall. Cellulose microfibrils are embedded in a matrix of pectic polysaccharides and cross-linked (mainly via H-bonds) in a network of hemicelluloses polymers and of the arabino-galactan proteins, the extensins (the latter are shown in purple). From Buchnan, B. *et al.* (2002) Biochemistry and Molecular Biology of Plants. ASPB, Rockville, MD, p 81.

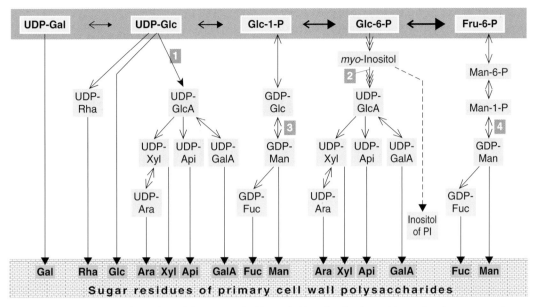

Figure 2.6 Biochemical pathways for synthesis of the monosaccharides needed for cell wall polymers. The numbered enzymes are: 1. UDP-glucose dehydrogenase; 2. myo-inositol oxygenase; 3. GDP-glucose-2-epimerase; 4. GDP-mannose pyrophosphorylase. Based on a diagram in Sharples SC, Fry SC (2002) Plant Journal 52, 252–262.

Key: M = monosaccharide unit; NTP = nucleoside tri-phosphate; NDP-M = nucleoside di-phosphate linked to monosaccharide; PP_i = inorganic pyrophosphate; M_n = growing polymer

The range of sugar interconversions and sugar-nucleotide syntheses required for pectin and hemicellulose synthesis is shown in Figure 2.6. It should be noted that some sugar interconversions occur when the monosaccharide unit has already been complexed with a nucleoside diphosphate. Nearly all of these reactions, including the formation of the polymers themselves, take place in the Golgi apparatus. The polymers are then transferred to the cell wall via vesicles budded off from the Golgi (sometimes called the trans-Golgi network–TGN) which discharge their contents by fusion with the plasma membrane in a process called **exocytosis**. Prior to formation of these vesicles, the sites from which they will be budded off become coated temporarily with a protein known as coat protein (COP I[i]).

Cellulose, however, is not synthesized in the Golgi. The cellulose synthesis complex, consisting of many individual cellulose synthase enzymes, is located on the plasma membrane and is visible as a rosette-like structure in electron micrographs. It is currently thought that polymerization of the glucose units involves first the synthesis of sterol glucosides (three glucose units joined to a sterol molecule), which act as the initial acceptors in order to start chain growth. As each chain nears completion, the sterol is removed and the chain is extruded through the plasma membrane, where it associates with other newly synthesized chains to form the microfibrils. The orientation of the microfibrils is guided by microtubules that are aligned adjacent to the plasma membrane.

The glucose donor for cellulose synthesis is UDP-glucose and there is some evidence that this is derived from sucrose via the enzyme sucrose synthase[ii] associated with the cytosolic side of the cellulose synthase. The reaction pathway is thus:

1. G-F + UDP → UDP-G + F
2. G_n + UDP-G → G_{n+1} + UDP

Key: G-F = sucrose; G = glucose; F = fructose

---

[i]Confusingly, COP is also the abbreviation by which one of the proteins in the phytochrome signalling pathway is known (Chapter 3, sections 3.7.3 and 3.7.4).

[ii]Note that, despite its name, this enzyme generally participates in sucrose breakdown in reactions that transfer the glucose unit direct to UDP to form UDPG.

Note that much of the energy from the hydrolysis of sucrose is conserved in the direct synthesis of UDP-G (compare this with the general reaction shown earlier).

2.2.3 Cell extension

Although the cell wall is not rigid, it does resist elongation because of the orientation of the cellulose microfibrils and the cross-linking between cellulose and other cell wall components. Nevertheless, cells elongate dramatically as part of normal plant growth. How can this happen?

First, the vacuole increases in size, generating increased turgor pressure (section 2.3.4), which pushes against the cell wall, similar to what would happen if a balloon were blown up inside a cardboard box. The resistance of the cell wall to the turgor pressure is reduced markedly, mainly because of the action of the plant hormone auxin (Chapter 3, section 3.6.2). Auxin stimulates a proton-pumping ATPase, which is located in the plasma membrane; hydrolysis of ATP is linked to the pumping of protons from the cytosol to the cell wall, which thus becomes acidic (see Chapter 5, section 5.7.3).

In the more acid environment, expansin proteins are able to loosen the hydrogen bonding between cellulose and the hemicellulose polysaccharides. Further, enzymes that can break and rejoin hemicellulose chains, such as *xyloglucan endotransglycosylase* (XET), are active at acid pH. Both of these features mean that cross-linking between cellulose and other cell wall components is very much reduced. Turgor pressure is thus able to push the microfibrils apart (Figure 2.7) and, eventually, new microfibrils are inserted between the older ones.

There is also a very interesting link with anti-oxidant metabolism here in that ascorbate oxidase, which converts ascorbate (see section 2.14.7) to dehydro-ascorbate, increases cell extension by up-regulating a gene or genes encoding one or more expansins.

2.2.4 Thickening, lignification and silicification

As is discussed above and in other chapters, cell expansion leads to a very large increase in cell volume. However, growth eventually stops and, at this time, the cell wall does become rigid, firstly because cellulose microfibrils become more firmly cross-linked by extensins (see above) and secondly because the proportion of cellulose deposited into the wall is increased. The wall is now a secondary

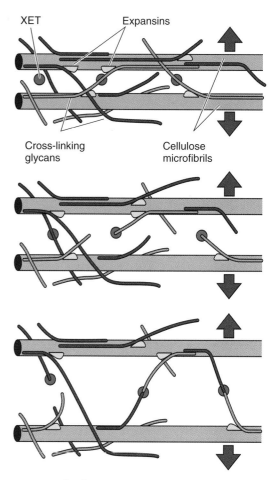

Figure 2.7 Cell wall extension. Starting at the top, the acidification of the cell wall (see text) leads to weakening of H-bonds (this process is aided by expansin proteins) and activation of XET. The cross-linked network of hemicelluloses and extensin proteins (the latter not shown in this diagram) is loosened, and cell turgor pressure causes the cellulose microfibrils to move apart (bottom panel). Later, more cellulose is synthesized to fill the gaps. Diagram from Buchanan, B *et al.* (2002) Biochemistry and Molecular Biology of Plants. ASPB, Rockville, MD, p 96.

wall, much thicker than the cell wall that characterizes the growing cell.

It may be helpful to visualize the cell as now having a three-layered wall. The first layer is the pectin-rich middle lamella; the second layer is the primary wall, containing pectins, increasing amounts of hemicelluloses and up to 30 per cent (by dry weight) cellulose; and the third layer is the secondary wall with very little pectin and increasing amounts of cellulose.

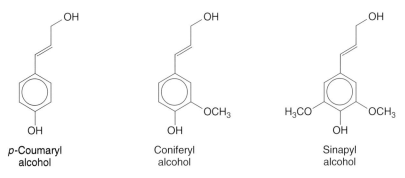

Figure 2.8 Lignol alcohols – lignin precursors.

Many cells remain in this state for the rest of their lives. However, others – in particular the water-conducting and strengthening cells of the xylem – undergo further modification in the form of lignification. Lignin is a very hydrophobic mixed polymer of monolignol alcohols (Figure 2.8) and, once formed, it is highly insoluble.

This obviously presents problems in relation to its synthesis. Currently it thought that, after their synthesis from the amino acid phenylalanine, the soluble monolignol alcohols are deposited into the cell wall, where they are oxidized by peroxidases located in the wall, thus generating reactive free radicals that can couple with each other. The polymerization process usually starts in the region of the wall that is furthest from the plasma membrane, i.e. in what was the primary wall[iii], and then works back towards the plasma membrane. This ensures that the whole wall becomes lignified, with the polysaccharides and other components completely locked up in the thickened wall. There is no set pattern for the polymerization of the different monolignols; each can join to any other. Indeed, it is quite conceivable that all the monomers involved in lignification of an individual cell are linked – the cell is thus effectively enclosed in one large insoluble molecule!

In some plants, including the grasses and cereals, silicon in the form of amorphous silica (SiO_2) is incorporated into the cell wall, often in large amounts. The deposited silica takes the shape of the cell wall and the deposits are known as phytoliths. These phytoliths have uses in plant taxonomy and, increasingly, as markers in archaeological and palaeoecological research, where they are used to determine past agricultural practices, human

diet and environment. Deposited silica also increases the strength of cell walls, giving plants mechanical support (Chapter 1, section 1.6), and in Chapter 11 we discuss its role in defence against herbivory (section 11.2.1) and plant pathogens (section 11.3.7).

2.3 The plasma membrane

2.3.1 Introduction

Immediately internal to the cell wall is the cell's bounding membrane, the plasma membrane. Its basic structure is a lipid bi-layer with the hydrophilic 'heads' facing outwards and the hydrophobic 'tails' facing inwards (Figure 2.9). Individual lipid molecules are not linked with each other, so lateral movement is possible, giving the membrane a degree of fluidity.

Proteins and protein complexes of various types are embedded in the bi-layer, and the overall lipid-protein array is often referred to as a *fluid mosaic*. Many of the proteins form interfaces with the cytosol on one side and the cell wall on the other. The cellulose synthase complex, discussed above, is one example of this, but there are many others, including pumps, carriers and channel proteins of various types, and proteins that bind signalling molecules such as hormones. Others, however, may be embedded in or attached to one side of the lipid bi-layer.

The plasma membrane also contains carbohydrates, mostly as oligosaccharides attached to proteins. Thus, several plasma membrane proteins are glycoproteins, including several AGPs (see section 2.2.1) which appear to have at least a minor role in the regulation of plant development. In addition, there are proteins attached by glycosidic linkages to sterols (section 2.3.2).

[iii] Some primary walls, especially in wood, may become lignified without the deposition of a secondary wall.

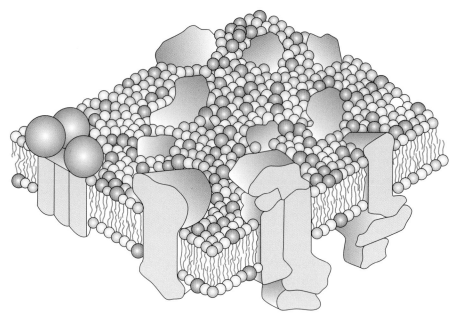

Figure 2.9 The fluid mosaic model for membrane structure.
From Lea, P and Leegood, R (1999) Plant Biochemistry and Molecular Biology, Wiley-Blackwell, Chichester, UK, p 133.

This overall structure (Figure 2.9) enables the plasma membrane to fulfil six important roles. First, it is the outer boundary of the cytosol, and that role defines the other five, namely:
• initial detection and transduction of some hormonal and environmental signals (see Chapter 3, section 3.6);
• regulation of transport in and out of the cytosol (section 2.3.3 and Chapter 5, section 5.7.3);
• synthesis of cellulose (and in cooperation with micro-tubules, assembly of the microfibrils – section 2.2.2);
• formation of inter-cellular cytoplasmic connections known as **plasmodesmata**;
• interaction with and attachment to the cell wall via **Hechtian strands**[iv].

2.3.2 Plasma membrane lipids

The overall composition of plant membranes is 40 per cent lipids, 40 per cent proteins and 20 per cent carbohydrates. The lipid fraction is comprised of phospholipids, glycolipids and sterols so in general resembles the lipid fraction of animal cell membranes. However, the actual composition of each fraction is different from that in animals. Unlike animals, it also differs between species and, within a particular species, between the different organs (Table 2.1). Thus, plant plasma membrane proteins are able to function effectively in a range of lipid environments.

In most plant cells, the majority of the plasma membrane lipids are **phospholipids**, although there are exceptions (Table 2.1). Free **sterols**, mainly campesterol, sitosterol and stigmasterol but with very little cholesterol, are also present in significant amounts. Other components may include, in different organs or in different species, *steryl glycosides* (a sterol linked to a glucose residue), *acylated steryl glycosides* (steryl glycosides linked to a fatty acid molecule) and **glucocerebrosides** (glucose linked to sphingosine, to which is attached a fatty acid molecule). The fatty acid component of plant plasma membrane lipids (Table 2.1) mainly consists of palmitic acid (16:0), linoleic acid (18:2) and linolenic acid (18:3)[v]. There is thus

---

[iv]The attachment appears to be mediated by a specific arabino-galactan protein (AGP – see section 2.2.2) that is located on the outer surface of the plasma membrane.

[v]We are using standard notation for fatty acids: the number before the colon is the number of carbon atoms; the number after the colon is the number of double bonds (i.e. 'unsaturated' linkages).

Table 2.1 Fatty acid composition of plasma membrane phospholipids.

Species	Percentage fatty acid composition								Percentage unsaturation
	14:0	16:0	16:1	18:0	18:1	18:2	18:3	Others	
Solanum tuberosum, potato, leaf	0	34.0	1.3	2.9	2.3	36.5	18.2	4.8	57.6
Oryza sativa, rice, cell culture	0.3	31.0	0	2.2	1.7	49.2	15.5	0.4	59.5
Ramonda serbica*, leaf	1.3	33.5	3.2	12.7	10.5	35.0	3.8	0	52.6

*Ramonda serbica is a 'resurrection plant', able to recover after extreme dehydration. The fatty acid composition of membrane lipids, already with a lower than average percentage unsaturation, changes during dehydration to give an even lower level of unsaturation.

a high level of unsaturation, imparting extra fluidity to the membrane, in contrast to the membranes of animal cells.

2.3.3 Transport across the plasma membrane

This topic is dealt with in detail in Chapter 5, section 5.7.2. Here we simply note firstly that, in addition to any chemical gradients that may exist at any time across the plasma membrane, there is also a charge difference or electrical potential of about -80 to -100 mV. This affects the entry of negatively-charged ions into the cell.

Secondly, according to the substance in question, there are various ways in which a solute may enter or leave a cell, some of which involve specific carrier proteins or specific channel proteins.

Thirdly, some large molecules, such as pectic and hemi-cellulosic polysaccharides, are delivered to the cell wall by exocytosis or *vesicle-mediated export*. In this process, the membrane of the Golgi-derived vesicle merges with the plasma membrane, thus extending it. The reverse – **endocytosis** – can also occur, with uptake from outside of the plasma membrane mediated by invagination of the membrane, leading to vesicle formation. Indeed, it appears that there is traffic in both directions between the Golgi and the plasma membrane.

Fourthly, for delivery of proteins into and through the plasma membrane, there is a specific pathway of synthesis, transport and delivery. This is discussed in section 2.8.2 and in Chapter 3.

2.3.4 Cellular water relations

Water movement into cells occurs via diffusion across the plasma membrane and via water-specific gated channels, the **aquaporins**. Water enters the cell down an osmotic gradient because solute concentrations inside the cell are greater than outside the cell. In order to understand this, some terminology is necessary.

The **water potential** (φ) of pure water is 0 MPa (megaPascals – pressure units). The presence of solutes decreases the osmotic potential. In formal terms, the water potential becomes *negative* and is equal to the **osmotic potential** (φ_s). The presence of solutes inside the cell therefore creates a gradient, allowing water to move from a zone of less negative water potential (outside) to more negative water potential (inside). However, there comes a point at which the pressure exerted by the cell wall, **turgor pressure (P)**, cancels out the decrease in water potential caused by solutes. The water potentials inside and outside the cell are now the same, and there is no longer any net water uptake.

As mentioned in section 2.2.3, turgor pressure is the driving force for cell expansion and, if expansion does indeed occur, water uptake will continue. Further discussion of plant water relations will be found in Chapters 5, 6, 9 and 10.

2.4 Cell compartmentation

In common with all eukaryotic organisms, plant cells are compartmented. The evolutionary origin of several of the sub-cellular compartments is still a matter for discussion. However, as we have seen in Chapter 1 (section 1.3), the evolutionary origins of two compartments are very clear: mitochondria and chloroplasts were derived from cells engulfed by other cells, as delineated in the endosymbiont hypothesis. The first engulfment of an aerobic proteobacterium by an archaebacterium led to the evolution of the mitochondrion and the establishment of the various eukaryotic lineages. The second engulfment, of a cyanobacterium, led to the establishment, within the eukaryotes, of the photosynthetic lineages, to the formation of chloroplasts and, thus, to plants as we know them today. So it is the chloroplast that is the special organelle of plants, and it is with the chloroplast that we start our discussion of the internal structure of plant cells.

2.5 Chloroplasts

2.5.1 Introduction

The greening of the land that we noted in Chapter 1 is entirely down to chloroplasts. They effectively provide the background pigment, chlorophyll, against which many (although not all) land animals lead their lives. Indeed, without those chloroplasts, biodiversity as we now know it would not have evolved. The existence of a mechanism for converting light energy from the sun into chemical energy for CO_2 fixation made possible the evolution of many lifeforms that are directly or indirectly dependent on that mechanism.

However, it is equally clear that chloroplasts themselves have evolved since that primary engulfment event. Modern chloroplasts differ significantly from

their cyanobacterial ancestors. One of the most obvious changes is the transfer of genes from the original symbiont to the nucleus of the host. Thus, while a chloroplast retains many of the metabolic activities of the cyanobacterial cell, many of the enzymes that mediate those activities are encoded in the nucleus. An angiosperm chloroplast has about 120 genes but contains at least 2,200 proteins (some recent proteomic analyses have suggested that the true number may be as high as 5,000). This has implications for the genetic control of chloroplast development and activity, as discussed in the next chapter.

2.5.2 Chloroplast membranes

Another change that has occurred in chloroplast evolution is the elaboration of the photosynthetic membrane system in which the light reactions occur. Extensive folding of the membranes has produced structures called **thylakoids**, simpler versions of which may be seen in the photosynthetic lamellae (sometimes also called thylakoids) of some modern cyanobacteria. Thylakoids are in places stacked like piles of pennies (Figure 2.10) to form *grana* (singular, *granum*) – although, unlike a pile of pennies, the thylakoids are inter-connected.

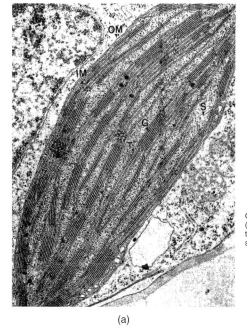

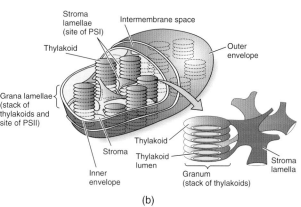

(a)   (b)

Figure 2.10 **(a)** Electron micrograph of chloroplast. From Lea, P and Leegood, R (1999), Plant Biochemistry and Molecular Biology, Wiley-Blackwell, Chichester, p 249. **(b)** Diagram of chloroplast structure.

The space inside the thylakoid membranes is known as the lumen, and many of the thylakoid-located proteins span the membrane so as to protrude into the lumen. External to the thylakoids is the **stroma**, the chloroplast equivalent of the cytosol.

The lipids of thylakoid membranes are very different from those of non-photosynthetic organelles but have several striking similarities to the composition of cyanobacterial thylakoids. About 55 per cent of the lipids are *monogalactosyl diglycerides*[vi] (MGDG: diacylglycerols with a galactose residue in the third position on the glycerol backbone); about 28 per cent are *digalactosyl diglycerides* (*DGDG*); and nearly 10 per cent are *sulfolipids* (especially *sulfoquinovosyl acyl glycerol*).

The fatty acids themselves are also very unusual. The highly unsaturated $18:3^{\Delta 9,12,15}$ and $16:3^{\Delta 7,10,13}$ fatty acids[vii] make up about 70 per cent of the fatty acid component of thylakoid lipids in general and 90 per cent of the MGDG fraction. As noted in Chapter 1, MGDG is the most abundant class of lipid in the world (and is often known as 'the plant lipid') – another indication of the central role of chloroplasts in the biosphere. The high level of unsaturation implies that the membranes are fluid, a feature that is regarded as aiding membrane function at low temperatures (see also Chapter 10, section 10.2.1). Indeed, studies of *Arabidopsis* mutants show that the presence of fatty acids with fewer double bonds (i.e. less unsaturation) impairs thylakoid function at low temperature. However, it is puzzling that the same mutants show impaired function at temperatures above the normal range experienced by the plant.

In angiosperms, the chloroplast envelope has two layers, each consisting of a lipid bi-layer. The inner envelope is thought to have been derived from the cell membrane of the engulfed cyanobacterial cell (see Chapter 1, section 1.3) and the outer envelope from the plasma membrane of the engulfing cell. Certainly, the lipids of the *inner envelope* support this view. The overall lipid content and the fatty acid components of those lipids are similar to those of the thylakoids, with one exception – namely that the 16:3 fatty acid is present in the envelope at only a very low level. This is at least

consistent with the idea that the thylakoid membranes have originated in evolution from the cyanobacterially-derived inner envelope. Indeed, as with the thylakoids, there is a strong similarity to cyanobacterial membranes.

However, the *outer envelope* presents more of a puzzle in that it seems to be a 'hybrid' between a cyanobacterial type membrane and a 'normal' plant membrane. Although the outer envelope contains both MGDG and DGDG, there is very much less of the former (as a percentage of total lipids) than in the inner membrane or in the thylakoids. The total sulfo-lipid content is also very much lower, whereas the phospholipid content is much higher. The amount of 18:3 fatty acid (linolenic acid) is lower than in thylakoids or than in the inner envelope, but it is still much higher than, for example, in the plasma membrane.

2.5.3 Organization for biochemical function

The light reactions of photosynthesis are described in detail in Chapter 7, section 7.4. Here we note that the thylakoid proteins involved in the light reactions are organized as several different groups (Figure 2.11).

Firstly, there are the *light-harvesting* or *antennae* proteins. They are complexed with carotenoids and with chlorophylls *a* and *b*. Different light-harvesting proteins are associated with photosystem I and photosystem II. Light energy is transferred from carotenoids to chlorophyll *b* to chlorophyll *a* and thence to the second group of proteins, those of the *reaction centres* (including the oxygen-evolving complex of photosystem II).

Then there are the proteins of the two *electron transport chains* associated respectively with photosystem II and photosystem I (and including the cytochrome $b_6 f$ complex of photosystem II).

Finally, there is the *ATP synthase complex* (the chloroplast $F_0 F_1$ ATP synthase: see section 2.6). The proton gradient between the thylakoid lumen and the stroma, generated by the electron transport chains (see Chapter 7, section 7.4) is dissipated via the ATP synthase, leading to the synthesis of ATP which discharged on the stromal side of the membrane.

It is in the stroma that the carbon reduction phases ('dark reactions') of photosynthesis take place (Chapter 7, section 7.4), leading eventually to the export of triose phosphate from the chloroplast and the temporary deposition of starch within the chloroplast. The starch is

---

[vi]See Chapter 1, Box 1.1.
[vii]The superscript numbers show the positions of the first of the two carbons involved in each unsaturated linkage. The positions of the double bonds in the chloroplast 16:3 fatty acid, namely 7, 10 and 13, are very unusual.

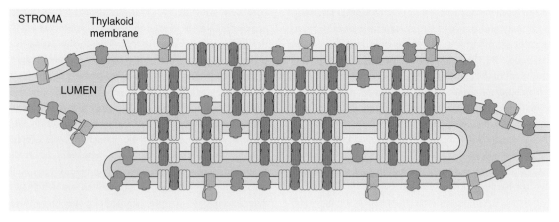

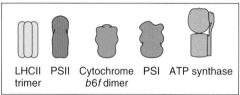

Figure 2.11 Organisation of proteins in the thylakoid membranes.
Key: LHC – light-harvesting complex; PSI – photosystem 1; PSII – photosystem II. From Taiz, L and Zeiger, E (2002), Plant Physiology, Sinauer, Sunderland, MA, p122.

degraded during the night and the carbon is transported as sucrose (synthesized in the cytosol) to other parts of the plant. Starch biosynthesis and the flow of carbon from starch to sucrose are shown diagrammatically in Figure 2.12.

However, the reduction of carbon dioxide with the associated sugar interconversions and the biosynthesis of starch are far from being the only metabolic pathways located in the stroma. Other metabolic activities include part of the photorespiration pathway (Chapter 7, section 7.5); the synthesis of fatty acids (described in more detail in section 2.14.5), the synthesis of chlorophyll (and its breakdown during leaf senescence), synthesis of carotenoids and terpenoids (including abscisic acid precursors of the gibberellins), synthesis of purines and pyrimidines, reduction of nitrite ions (section 2.14.8), synthesis of several amino acids and all of the activities associated with genes and gene expression – DNA replication, RNA synthesis and protein synthesis (Chapter 3).

2.5.4 Other plastids

Chloroplasts are the most obvious examples of a class of organelle called *plastids*. The different members of the

class are, to varying extents, interconvertible, and in an individual plant all are derived from the plastids present in the egg cell prior to fertilization. Those egg-borne plastids are, of course, not photosynthetic, and neither are those present in the embryo; they are actually *proplastids*. After germination, proplastids located in the aerial organs of the plant develop into chloroplasts. If seeds are germinated in the dark, these proplastids only develop part of the way to chloroplasts and become *etioplasts*.

Proplastids are located throughout the plant and, in some types of cell, may remain in this state throughout life. However, in many cells, non-photosynthetic plastids become specialized for starch synthesis and deposition; these are called *amyloplasts*. They function as storage organelles in storage organs such as potatoes and in the nutrient reserves of seeds.

The amyloplasts of the root cap are sedimentable under gravity. They are called **statoliths** and are involved in gravity perception (Chapter 5, section 5.11.1). In many ripening fruit, exemplified by tomatoes, chloroplasts that were previously photosynthetic lose their chlorophyll and synthesize the pigments associated with ripening. These plastids are called *chromoplasts*. This transition also occurs

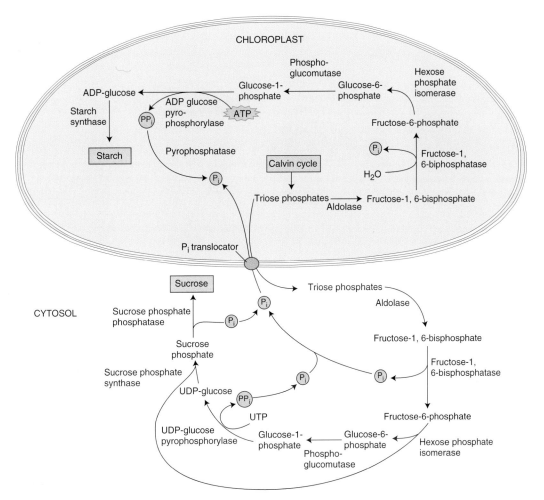

Figure 2.12 Carbon flow between chloroplast and cytosol. The export of triose phosphate from the chloroplast is partly regulated by the availability of inorganic phosphate (Pi) in the cytosol. This is because triose phosphate is exported from the chloroplast in a one-for-one swap with import of Pi.
From Taiz, L. and Zeiger, E (2002) Plant Physiology, Sinauer, Sunderland, MA, p163.

during the senescence of many types of leaves (Chapter 7, section 7.11). Finally, there are non-pigmented plastids known as *leucoplasts*, which seem to be specialized for synthesis of particular terpenoids. The various forms of plastid and their interconversions are shown diagrammatically in Figure 2.13.

2.6 Mitochondria

The organelle derived from the initial endosymbiotic process that led to the emergence of eukaryotes is the mitochondrion, the site of the major energy conservation

pathways in respiration. As with chloroplasts, mitochondria are bounded by two membranes: the inner membrane (derived from the outer membrane of the engulfed cell) and the outer membrane (derived from the outer membrane of the original host cell). However, unlike chloroplasts, mitochondria do not have an extensive internal membrane system equivalent to the thylakoids. Instead, the inner membrane is extensively folded so as to produce *cristae* which protrude into the mitochondrial matrix (Figure 2.14).

The mitochondrial matrix contains the enzymes of the *tri-carboxylic acid cycle* (TCA cycle) or *Krebs cycle*,

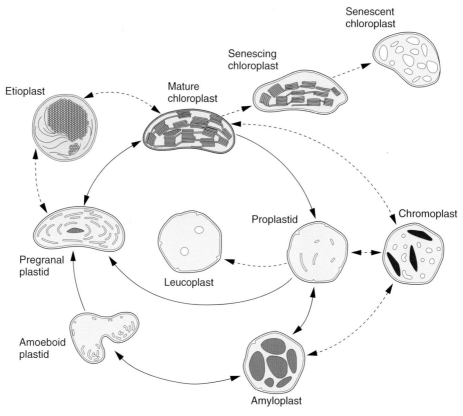

Figure 2.13 Plastid interconversions.
From Buchanan, B. *et al.* (2002) Biochemistry and Molecular Biology of Plants, ASPB, Rockville, MD, p38.

while the proteins of the electron transport chain are embedded in or actually span the inner membrane. As described in more detail in section 2.14, the passage of electrons along the electron transport chain leads to the build-up of a proton gradient across the inner membrane, i.e. from the inter-membrane space to the matrix. The gradient is dissipated by passage of protons through the membrane-spanning *ATP synthase complex* (the mitochondrial F_0F_1 ATP synthase) leading to the formation of ATP. This *chemi-osmotic* mechanism for ATP synthesis is thus directly equivalent to that in the chloroplast, depending on a charge separation process that delivers protons to one side of a membrane (section 2.5.3 and Chapter 7, section 7.4).

The extensive folding and, hence, much increased surface area of the cristae, mean that the inner mitochondrial membrane contains about 90 per cent of the mitochondrion's membrane lipids and membrane-embedded

proteins. Among the lipids is the rather unusual **cardiolipin** (given this name because it was first isolated from animal hearts) or diphosphatidyl glycerol (see Glossary). The configuration of cardiolipin changes according to the ionic and pH environment, and it is suggested that this helps it to maintain within the membrane the right conditions for function of the membrane proteins. It is also suggested that it acts as a proton trap at its inner face, thus helping to ensure that protons remain in the inter-membrane space until they return to the matrix through the ATP synthase complex. Interestingly, cardiolipin is confined to the mitochondrial *inner* membrane; it is also found in the cell membranes of bacteria, again supporting the endosymbiont hypothesis.

Mitochondrial membranes, in contrast to chloroplast membranes, are rich in phospholipids, with phosphatidylcholine being the most abundant. There is a

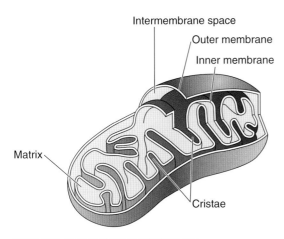

Figure 2.14 Diagram of mitochondrial structure.

Table 2.2 Percentage fatty acid composition of the lipids in plant mitochondrial membranes.

	16:0	18:0	18:1	18:2	18:3
Inner membrane	10	1	7	13	69
Outer Membrane	50	4	20	8	18

marked difference in the level of fatty acid unsaturation between the inner and outer membranes. The inner membrane is rich in 18:3 while the outer membrane is rich in 16:0 (see Table 2.2). The outer mitochondrial membrane lipids are thus less unsaturated than the lipids of most other plant membranes. It is probable that this is related to maintenance of the proton gradient generated by passage of electrons down the respiratory chain, and recent work with *Arabidopsis* mutants suggests that inserting more 18:3 into the outer membrane promotes the leakage of protons and thus 'uncouples' respiration (see section 2.14.2).

Mitochondria are primarily associated with respiration and, in plants, do not house any other major metabolic pathway[viii]. Thus, fatty acid oxidation, which in animals takes place in the mitochondria, takes place in plants within the peroxisomes/glyoxysomes (section

---

[viii]One small puzzle is that some of the enzymes involved in fatty acid synthesis – which occurs in chloroplasts – have been found in plant mitochondria. Their specific function is unknown.

2.10). Mitochondria participate with chloroplasts and peroxisomes in photorespiration (Chapter 7, section 7.5).

Intriguingly, the final step in ascorbate biosynthesis takes place in the mitochondrial inter-membrane space, mediated by an enzyme located in the inner membrane (see section 2.14.7). Mitochondria also export metabolites from the TCA (Krebs) cycle for use in other metabolic pathways. In addition, they possess all the biochemical machinery associated with DNA replication, gene expression and protein synthesis.

As with chloroplasts, mitochondrial genomes are much smaller than those of their endosymbiotic ancestors; extensive transfer of genes from the mitochondrion to the nucleus has occurred during evolution. Current estimates suggests that angiosperm mitochondrial DNA contains genes for about 35 proteins, plus ribosomal and transfer RNAs. A fully functional mitochondrion needs several hundred proteins, implying that, as with chloroplasts, products of nuclear-located genes must be transferred into the organelle (Chapter 3, section 3.3.3).

2.7 The nucleus

Although the evolutionary origin of the nucleus remains a mystery, as was noted in Chapter 1 (section 1.3), its presence is one of the hallmarks of eukaryotic cells. Indeed, it may have evolved, along with other essentially eukaryotic features, before the first engulfment event. It is undoubtedly a complex organelle and its envelope alone has been the subject of scientific conferences and books. As shown diagrammatically in Figure 2.15, the *nuclear envelope* (NE) may be thought of as consisting of three membranous components. The first of these is the outer envelope, which is connected to the ER and, especially, to the rough ER (section 2.9.1). Indeed, like the rough ER, there may be ribosomes on the surface of the outer NE. As expected from this arrangement, the proteins and lipids of the outer envelope are similar to those of the rough ER. The outer NE is thus linked to or part of the cell's endomembrane system.

The inner NE is separated from the outer NE by the lumen, which is about 30 nm across. On its inner surface, the inner NE is closely associated with the filamentous proteins that make up the *nuclear lamina*. The latter is the main component of the *nuclear cage or matrix* that

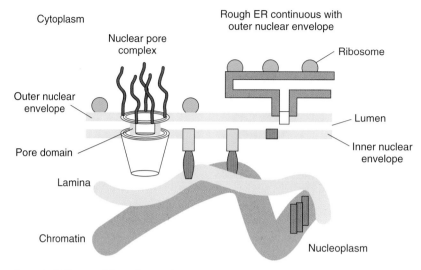

Figure 2.15 Diagram of the nuclear envelope.
From Evans, DE *et al*. (2004) The Nuclear Envelope, BIOS, Oxford, p2.

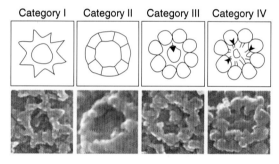

Figure 2.16 Nuclear pore complexes in tobacco cell nuclei.
From Figure 2 in Fiserova, J *et al*. (2009) Plant Journal 59, 243–255.

is obtained when the envelope membranes are gently removed with detergents (see below).

The third membrane component, the pore membrane, links the inner and outer NEs. It is part of the *nuclear pore complex* (NPC), through which any macromolecule that enters or leaves the nucleus must pass.

The nuclear pore complex is itself a complicated structure, but many biologists also find it very beautiful (Figure 2.16).

Electron microscopy reveals a clear eight-fold symmetry in a structure that has one side protruding into the nucleoplasm and the other into the cytosol. On the cytosolic side, the eight-fold symmetry takes different overall shapes, including stars and various types of

rings (Figure 2.16), according to the age and function of the cell. The cytosolic side is characterized by outward-extending rod-like filaments, while the nucleoplasmic side is a basket-like structure. About 30 different *nucleoporin* proteins, many of them glycosylated by attachment of up to five *N*-acetylglucosamine residues, make up the pore complex, through which all entry to and exit from the nucleus must occur. This is discussed more fully in Chapter 3; here we emphasize the dynamic nature of the process. It has been estimated that an average of ten import and ten export events occur at each pore complex per second; for example, in pea root meristem cells, about 15,000 ribosomes leave the nucleus every minute (see section 2.11).

The nature of the lamina that underlies the inner NE and to which the inner NE is attached is still a matter of debate. In animals, it consists of a specific class of filamentous proteins – the lamins – but lamins have not been detected in plants at either protein or gene level. Nevertheless, plants do have a filamentous network of proteins underlying the inner NE, and this network serves the same functions as the animal cell nuclear lamina. Martin Goldberg at Durham University has coined the term **plamina** to distinguish this plant-specific structure from its functional homologue in animals.

As recently as 2009, we had no information on the identity of plamina proteins. However, two candidates have now been tentatively identified. These are the Nuclear Matrix Constituent Proteins (NMCP), also known as LINC proteins (based on *Little Nuclei* mutants in *Arabidopsis*).

Inside the nucleus, the most obvious component is chromatin, consisting mainly of DNA and histones, which is attached via *scaffold-associated regions* (SARs) to the nuclear matrix or scaffold. Within chromatin, denser regions – the *nucleoli* – are apparent. These are the sites of transcription of the ribosomal RNA genes, which actually takes place in the fibrillar centre of each nucleolus (with the genes themselves looped out from the backbone of chromosome). Around the fibrillar centre is a denser region in which the newly synthesized ribosomal RNA is processed (see Chapter 3, section 3.2.2). In the outer granular region of the nucleolus, the RNA is packaged with proteins to form ribosomal subunits for export via the nuclear pore complexes.

2.8 The vacuole

Another very obvious feature of most plant cells is the vacuole. It is not apparent in meristematic cells, nor in cells that have not started to expand. However, meristematic cells do contain several small vacuoles that merge as the cell starts to expand. The resulting single vacuole, bound by a membrane called the *tonoplast*, continues to expand as the cell grows and, in mature cells, it may occupy as much as 90 per cent of the cell volume. Indeed, enlarging the vacuole has been described as a 'cheap way' of achieving cell enlargement. There is some debate as to the origin of the pro-vacuoles; they arise either from the ER or from the Golgi, although, in view of the relationship between these two membrane systems (section 2.9), in practice it makes little difference as to which is the primary source. Indeed, it is possible that both routes are operative, according to cell type and specific vacuole function.

Vacuoles serve several functions, again according to the type of cell under consideration. Almost universally, they store water and water-soluble metabolites, including sugars, various ions and organic acids. An example of the latter is the transport, via a specific carrier, of malate into the vacuole during the night in CAM plants and its movement back into the cytosol during the next morning (Chapter 7, section 7.8).

Some vacuoles store defence chemicals which are released if the cell is damaged, and the same is true of hydrolytic enzymes (usually with an acid pH optimum) that are pushed into the vacuole from ER/Golgi-derived vesicles in a process similar to exocytosis. The hydrolytic enzymes may also function in the turnover of cell components, even as large as chloroplasts and mitochondria, that are deposited in the vacuole. In this way, the vacuole acts as a lytic compartment, similar to animal cell lysosomes. Vacuoles may also store pigments, especially anthocyanins; also, as described in detail in Chapter 4 (sections 4.5.1 and 4.5.2), protein reserves in seeds are stored in modified vacuoles. Finally, vacuoles have a role in salt tolerance, as potentially cytotoxic sodium and chloride ions are stored there in halophytes (Chapter 10, section 10.5.2).

The vacuolar membrane, the tonoplast, is similar in lipid composition to other major cellular membranes and also carries a range of proteins, including channel proteins and specific transporters. Movement of water in and out of the vacuole is involved in controlling turgor pressure, while transport of some solutes is integrated into particular metabolic pathways. (see above, re CAM).

2.9 Endomembrane systems

2.9.1 Endoplasmic reticulum

We have already encountered the endoplasmic reticulum (ER) in discussing cell wall biogenesis and in consideration of the nucleus. It is effectively a membranous reticulum that permeates the cytosol. Structurally, it consists of two membranes separated by a lumen, and it is continuous with the outer envelope of the nucleus. Parts of it are coated with ribosomes (rough ER), and proteins synthesized on those ribosomes are destined for export outside the cytosol or to other membrane-bound compartments, including the Golgi and the vacuole (but *not* chloroplasts, mitochondria or microbodies/peroxisomes). The proteins accumulate in the lumen of the ER (as described in more detail in Chapter 3, section 3.2.4) and the ER buds off vesicles which merge with Golgi vesicles; these are channelled by the cytoskeleton to the appropriate membrane (see section 2.8.2).

The ER is also used for the storage and release of calcium ions. Some aspects of cell signalling involve rapid changes in cytosolic calcium concentrations, and the ability to move calcium ions quickly in and out of the ER is important for this.

In addition to these functions, the ER is involved in lipid biosynthesis (sections 2.14.5). Fatty acids synthesized in the chloroplast are transferred to the ER for further metabolism, including combination with glycerol to form triacylglycerols (also known as triglycerides: one glycerol molecule carrying three fatty acid chains). These are deposited in the cytosol as oil bodies or oleosomes[ix], bounded not by a lipid bi-layer but by a single layer of lipid with its hydrophilic heads on the outside. Proteins called oleosins are embedded in the monolayer. Small oil bodies occur in nearly all plant cells, but they are very abundant in the storage tissues of lipid-storing seeds (as described in Chapter 4, section 4.5.1).

Finally, very recent research has shown that ER can traverse plasmodesmata, the cytoplasmic connections between cells. This means that the ER of adjacent cells may be continuous and this provides another means of transport of, for example, proteins between cells.

2.9.2 The Golgi Apparatus

The relationship between Golgi and ER becomes very obvious as we look at the pathways for protein sorting and onward transport. Most of the proteins which are imported into the lumen of the ER are actually destined for another cell compartment or membrane component. In the Golgi, the proteins are sorted and glycosylated (i.e. are modified by the addition of carbohydrates) and are then either retained in the Golgi or transferred to another compartment or membrane via shuttle vesicles (trans-Golgi network), as shown diagrammatically in Figure 2.17.

There is evidence that vesicles destined for particular locations are marked by specific proteins. Thus, clathrin-coated vesicles are targeted to the tonoplast for discharge into the vacuole. This sorting and glycosylation also involves proteins that end up in the lumen of the ER, including, for example, the enzymes involved in lipid synthesis. In the course of their synthesis and processing, these proteins thus shuttle from the ER to the Golgi and then back again.

In addition to glycosylation of proteins, the Golgi also transfers carbohydrate residues to particular lipids, thus making glycolipids (see section 2.8). Further, as already discussed, most of the cell wall polysaccharides are made in the Golgi. The sugar nucleotide donors for all these carbohydrate syntheses and transfers are delivered to the Golgi via specific carrier proteins that span the Golgi membrane.

The cell's endomembrane system is thus not a static set of separate entities. ER and Golgi are dynamically interrelated. The ER is also continuous with the outer nuclear envelope; the Golgi bodies' export system leads to merging between Golgi membranes and plasma membrane and also between Golgi membranes and tonoplast. The picture that emerges is one of interconnection and dynamic flux.

2.10 Microbodies/peroxisomes

Small, more or less spherical organelles were first observed in electron micrographs of plant cells in the 1960s. Similar structures had previously been seen in animal cells, and for both plants and animals the term 'microbody' was coined. However, it is now known that they contain a very active catalase and participate in the detoxification of H_2O_2 (for example, during photorespiration – see Chapter 7, section 7.5). For this reason, the term 'peroxisome' is now more widely used.

The organelle itself is bounded by a single membrane (a conventional lipid bi-layer) and ranges in diameter from $0.2\mu m$ to about $1.7\mu m$. Although they do not possess DNA, nor any protein-synthesizing machinery, new peroxisomes arise by division of pre-existing peroxisomes. Elongated peroxisomes divide in the G2 phase of the cell cycle (section 2.13), and subsequent expansion requires the accumulation of lipids, proteins and other molecules synthesized elsewhere in the cell (see Chapter 3, section 3.2.6).

Although the enzyme *catalase* is a particular marker for peroxisomes, they may, in specific types of cell, exhibit other metabolic activities. Thus, during the mobilization of lipids in germination of fat-storing seeds, the glyoxylate cycle takes place in peroxisomes which, at this stage, actually lack catalase. The organelles are thus known as glyoxysomes (Chapter 4, section 4.11.2). After lipid mobilization has been completed, the enzymes of the

---

[ix]These have been also been known as spherosomes and this term still occurs in some texts.

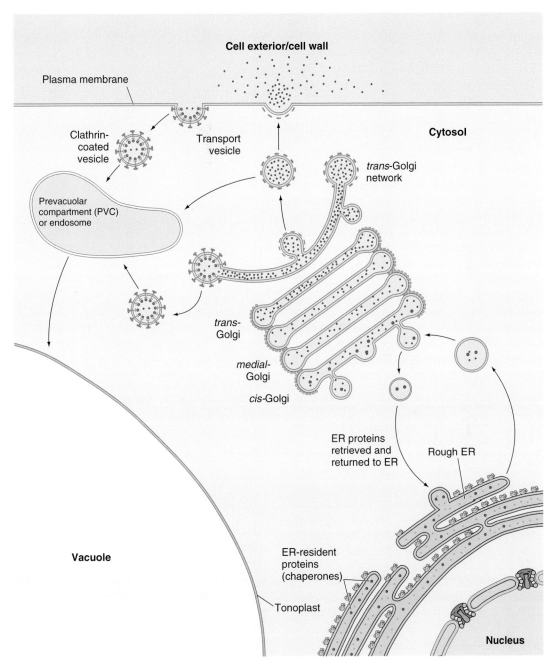

Figure 2.17 Protein sorting and delivery in the cell's endomembrane system.
From Buchanan, B. *et al.* (2002) Biochemistry and Molecular Biology of Plants, ASPB, Rockville, MD, p 176.

glyoxylate cycle are lost and the organelle regains its typical peroxisome enzyme complement. Furthermore, in the root nodules of some leguminous plants (section 2.14.8 and Chapter 5, section 5.3), peroxisomes participate in the 'down-stream' metabolism of fixed nitrogen, leading to export of N-containing compounds from the nodules.

2.11 Ribosomes

Electron micrographs of plant cells reveal that they contain several million ribosomes, the ribonucleoprotein particles on which proteins are synthesized. Estimates of the actual numbers vary according to the type of cell being studied. In general, cells with high metabolic activity and cells that make abundant protein (e.g. during the development of seeds – see Chapter 4, sections 4.5.1 and 4.5.2) have more ribosomes than less metabolically active cells.

Each ribosome consists of two sub-units, often defined by their sedimentation rate, in *Svedberg units*, when subjected to ultra-centrifugation. Thus, the 80S ribosomes of plant cells (with the exception of those in chloroplasts and mitochondria – see Chapter 3, sections 3.3.2 and 3.3.3) are composed of 60S and 40S sub-units (Figure 2.18). These values are typical for the ribosomes that occur in the cytosol of eukaryotic cells. Each contains a specific subset of ribosomal RNA molecules (Chapter 3, sections 3.2.2 and 3.2.3) and ribosomal proteins.

There are about 20 proteins in the small (40S) sub-unit and about 30 in the large sub-unit. The two sub-units are assembled separately in the nucleus and they also exit the nucleus separately. Indeed, they do not associate together until after the initiation complex for protein synthesis has been assembled on the small sub-unit (see Chapter 3, section 3.2.1). The rates of assembly of each sub-unit in rapidly dividing cells are remarkable: 15,000 per minute is not unusual.[x]

In the cytosol, there are two populations of ribosomes, namely those that are 'free' and those that are on the surface of the ER. The two populations of ribosomes synthesize different sets of proteins (see section 2.9). The mechanism by which ribosomes making particular proteins become associated with the ER is discussed in Chapter 3, section 3.2.4.

[x]Readers should beware of those plant science/botany websites that give the total number of ribosomes per plant cell as 15,000.

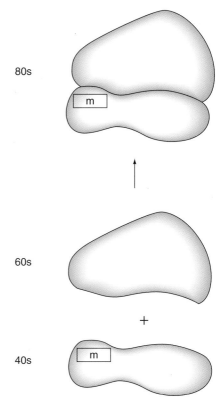

Figure 2.18 Diagram of ribosome subunits. The 80S ribosomes of plant cells are made up of two subunits – the 40S subunit and the 60S subunit. The mRNA binding site (indicated by m) is on the small subunit; the two subunits do not associate together until after the mRNA is bound.

2.12 The cytoskeleton

2.12.1 Introduction

Plant cells, like the cells of all eukaryotic organisms, possess an internal network of proteinaceous filaments that extends throughout the cytosol. This network is involved in an a range of important cellular processes, including channelling of vesicles (see sections 2.2.2, 2.8.1 and 2.8.2, controlling the plane of cell division, separation of chromosomes, intra-cellular movement of organelles and directional organization of cellulose microfibrils. There are two main components of the cytoskeleton: *microtubules* and *actin filaments*.

Animal cells contain a third type of cytoskeletal element – intermediate filaments. Different types are made up of different proteins (e.g. keratin and, in the nucleus, lamins). There have been claims over the last ten years or so that plants also possess intermediate filaments – claims usually based on the presence of proteins recognized by antibodies raised against animal intermediate filament proteins. Intermediate filaments themselves have not been unequivocally identified in plant cells and, in the one structure that is known, the plant nuclear lamina ('plamina' – see section 2.7), the proteins are different from those in the equivalent structure in animals.

2.12.2 Microtubules

Microtubules are hollow tubes made from *tubulin*. Each tubulin unit is actually a heterodimer of α- and β-tubulin, very similar proteins each with a molecular weight of about 55 kDa. The tube is built as a shallow helix with 13 tubulin units per turn (Figure 2.19), giving an external tube diameter of 24 nm. A 'vertical' row of tubulin units (looking along the axis of a vertically arranged microtubule) is called a **protofilament**.

An individual microtubule consists of hundreds of thousands of tubulin units, all oriented in the same direction. During formation, which takes place at specific cellular locations called *microtubule organizing centres*, one end of a microtubule grows faster than the other, giving *plus* and *minus* ends.

Microtubules exhibit dynamic instability in that, once formed, they can be disassembled and re-assembled. The rate of assembly and disassembly is at least partially regulated by the ratio of polymerized to unpolymerized tubulin. Assembly requires the binding of GTP (energetically equivalent to ATP) to the tubulin dimer; the polymerization of the dimers is driven by GTP hydrolysis. The resulting GDP remains bound to the tubulin and is released during disassembly.

Associated with microtubules are two types of *motor protein* which use the energy from the hydrolysis of ATP to 'walk' along the microtubules while carrying 'cargo' such as membrane-bound vesicles and even organelles. **Dyneins** transport cargo towards the minus-ends and **kinesins** towards the plus-ends of microtubules.

Microtubules have several roles in cell division. Prior to the start of mitosis, the main population of microtubules depolymerizes and then re-assembles as the **pre-prophase band**. This encircles the nucleus in the position at which the new cell plate (section 2.2) will eventually form, and it is thus thought to determine the plane and position of division. As prophase gets under way, two further populations of microtubules are assembled on opposite sides of the nucleus, thus forming the prophase spindle.

During the transition from prophase to metaphase, the pre-prophase band is disassembled, implying that the position at which the new cell plate will form is now 'remembered'. The nuclear envelope breaks down (plants have an 'open mitosis') and the *mitotic spindle* is formed by assembly of further populations of microtubules radiating out from zones at opposite ends of the cell. Some of these spindle microtubules are involved in pulling the replicated chromosomes apart during anaphase. In order to do this, they become attached to the chromosome centromeres (see section 2.13) at positions known as *kinetochores*. Finally, the process of cell division is completed by the laying down of the cell plate by the **phragmoplast**. This is formed from the ER and from microtubules that are re-assembled at that position after disassembly of the spindle (Figure 2.20).

2.12.3 Actin filaments

Actin filaments (sometimes known as microfilaments), each consist of two chains of polymerized *actin* (a protein of molecular weight 42 kDa) wrapped round each other in a tight helix of about 7 nm diameter (Figure 2.20). Polymerization requires hydrolysis of ATP and, as with GDP in microtubules, the ADP remains bound in the polymer. Again, as in microtubules, disassembly readily occurs and, in actin filaments, this can give rise to a

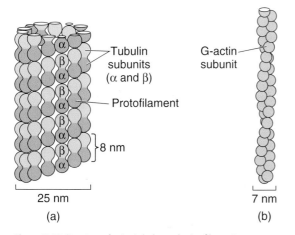

Figure 2.19 Structure of microtubules and microfilaments.

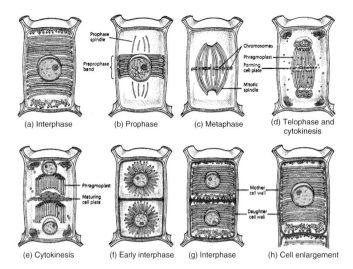

(a) Interphase   (b) Prophase   (c) Metaphase   (d) Telophase and cytokinesis

(e) Cytokinesis   (f) Early interphase   (g) Interphase   (h) Cell enlargement

Figure 2.20 Behaviour of microtubules during mitosis of wheat root meristem cells.
From Raven PH *et al.* (2005) Biology of Plants, 7th edition, Freeman, NY, p 66

process called 'treadmilling', in which the rate of assembly at the plus end of the filament is matched by the rate of disassembly at the minus end.

Actin filaments are involved in cytoplasmic streaming and in the movement of organelles during streaming. They have a role in anchoring the nucleus within the cytosol and they participate with microtubules in the positioning of other organelles and in delivering vesicles to the new cell plate and, more generally, to the plasma membrane. They also act specifically in delivering vesicles for exocytosis in cells that exhibit tip growth, i.e. pollen tubes and root hairs. The motor protein associated with actin filaments is *myosin*. When bound to an actin filament, this is able to hydrolyse ATP to provide energy for propulsion along the filament.

2.13 The mitotic cell cycle

2.13.1 Introduction

Mention of the roles of the cytoskeleton in cell division leads on to consideration of the mitotic cell cycle. Throughout their lives, plants maintain populations of embryonic cells called meristems, as discussed in more detail in Chapters 4, 5, 6 and 7. Furthermore, many non-dividing cells retain the capacity to re-enter the cell division cycle in response to wounding or to pathogen attack. As this pattern of regulation of cell division is a key aspect of plant morphogenesis, it is surprising that, for several decades of the 20th century,

the plant cell division cycle received relatively little research attention. For example, in the late 1970s there were only about five research groups worldwide working on the biochemistry of plant DNA replication. However, the situation changed dramatically in the late 1980s, following very significant discoveries on cell cycle control in other organisms, coupled with increased interest in plant molecular biology in general and increased sophistication of available techniques. Our knowledge of the mitotic cell cycle is now very extensive, and thus we present here an overview that covers the key features.

2.13.2 Phases of the cell cycle

Early microscopic studies of cell division revealed the intricate movements of chromosomes during mitosis. They became visible as they condensed during prophase, then disappeared again after mitosis was complete. The process of cell division was thus divided into *mitosis*[xi] and *interphase*, the latter being the gap between successive cell divisions in which chromosomes were not visible.

However, the advent of radioactive labelling as an investigative tool in biology in the 1950s led to the discovery that DNA replication (as detected by the incorporation into DNA of radioactive thymidine) occurs during a specific period within interphase. This is now

[xi]Strictly speaking, mitosis refers to the events of nuclear division which are followed by *cytokinesis*, division of the cell itself. However, in general usage, mitosis is usually taken to include cytokinesis.

known as the *S-phase* (*S* standing for *synthesis of DNA*). Based on this finding, the cell cycle was divided into four phases: G1, S, G2, M (*G* stands for *gap* and *M* for *mitosis*). Obviously, the term *gap* does not do justice to the array of cellular activities that go on throughout the cell cycle but, nevertheless, this terminology does cover the key events: the replication of the genetic material and then its distribution into two daughter cells (Figure 2.21). Further, it has turned out that this essentially intuitive division of the cell division cycle is actually reflected in the major control mechanisms.

2.13.3 Regulation of the cell cycle

An overall picture of cell cycle regulation requires attention at three levels. First, there is the level of regulation involving organ and cell identity – in particular, meristem identity and patterning. In other words, this is the specification of the cell populations in which cell division occurs and which ensures that the genes necessary for the cell cycle are active. This is discussed in Chapters 4, 5, 6 and 7.

Second, in a dividing cell, there is regulation of the entry into, and subsequent progress through, the cell cycle. We might call this the internal regulation of the cell cycle, but this level of regulation clearly links both upwards to higher level regulation and downwards to the third level, which is the regulation of the enzymes and other proteins that mediate each step in the cycle. In fact, the second and third levels are so intertwined that we deal with them together.

It was Jack Van't Hof, working at the Brookhaven Laboratory on Long Island, New York, who first showed that the plant cell cycle has two major control points, leading him to propose the *principal control point* hypothesis. Thus, meristematic cells that are deprived of sucrose arrest either at the G1-S transition or at the G2-M transition. Similarly, cells that cease to divide prior to differentiation may arrest in G1 or G2.

Subsequent research has shown that the major cell cycle events are controlled by reversible protein phosphorylations involving **protein kinases** and phosphatases. At the heart of these mechanisms are pairs of proteins: *cyclin-dependent kinases* (CDKs) and *cyclins*. In simple organisms such as fission yeast (*Schizosaccharomyces pombe*), one CDK, the product of the *CDC2* gene (the homologous gene in budding yeast, *Saccharomyces cerevisiae*, is *CDC28*) is involved in both control points (Figure 2.22). Indeed, the G1-S control point, sometimes known as *START*, is the major control point in these

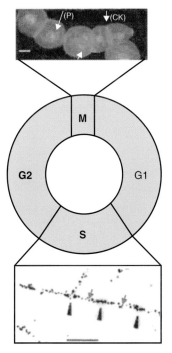

Figure 2.21 The plant cell cycle. The middle panel shows the phases of the cell cycle; the upper panel shows tobacco cells going through mitosis; the bottom panel shows replicating DNA visualised by fibre autoradiography.
From Francis, D (2007) *New Phytologist* **174**, 261–268.

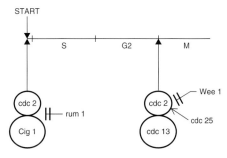

Figure 2.22 Diagram showing the basic control mechanisms in the fission yeast (*Schizosaccharomyces pombe*) cell division cycle. The same cyclin-dependent kinase, CDC2, is involved both in entry into S-phase ('Start') and in the G2-M transition. At Start, CDC2 associates with cyclin cig1. The complex is inhibited by rum1 but the inhibition is lifted when rum1 is inactivated by phosphorylation. At G2-M, the cyclin partner of CDC2 is CDC13. The complex is regulated by the balance between the inhibitory protein kinase, WEE1, and the activating protein phosphatase, CDC25, which compete for the same two phosphorylatable sites on CDC2.

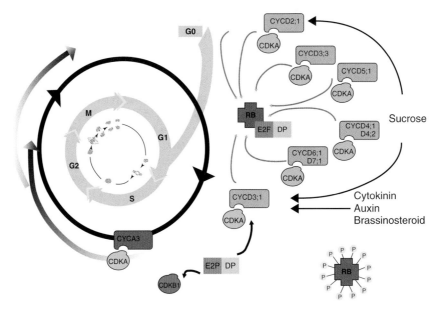

Figure 2.23 Control of the G1-S transition in plants. The key regulatory factors in the G1-S transition are the E2F-DP transcription factor complexes. These are inactive in G1 because they are bound by the Rb (retinoblastoma) protein. The CYCD-CDKA cyclin-kinase pair phosphorylates Rb. This releases E2F-DP, enabling the transcription of genes required to move from G1 to S. The process is controlled by several higher level factors including hormones and sucrose, as indicated in the diagram.
From Bryant, J. and Francis, D (2007) The Eukaryotic Cell Cycle, Taylor and Francis, Abingdon, p15.

organisms: arrest at G2-M is extremely rare. However, association of the CDK with different cyclins confers some specificity on the activity of the CDK.

How does this work in plants? The first point to note is that plants have a range of CDKs, grouped into seven families, and an even wider range of cyclins, more than 60 in all, grouped into six families. This gives rise to a bewildering array of possible CDK-cyclin interactions. Nevertheless, it remains true that the cyclin-dependent kinase activity of the CDKs plays a key role in the regulation of the plant cell division cycle.

We focus first on cells not in the cell cycle but, nevertheless, capable of proliferation. These cells are referred to as being in G0 and, if they enter the cell cycle, they will progress from G0 to G1 to S-phase (Figure 2.23).

In G0, genes whose activity is required for the G1-S transition are down-regulated because of the inhibition by Rb (*retinoblastoma*) protein of the transcription factors in the E2F family. Movement into G1 is initiated when CDKD, in combination with a member of the cyclin-D family, phosphorylates CDKA, enabling it to bind to its activating cyclin, cyclin D3;1. CDKA can then phosphorylate the Rb protein causing it to release

the E2F transcription factor; this in turn, in a complex with DP ('dimerization protein'), up-regulates the genes involved in S-phase, thus facilitating the G1-S transition (Figure 2.23)

The G1-S transition itself is complex[xii]. First we need to note that plant chromatin, in common with that of all eukaryotes, is composed of multiple replication units called **replicons**. Replicons are organized as 'time-groups', or families, and different families are active at different times during the S-phase (see Figure 2.24). It is not yet clear as to whether all replicons are 'prepared' for replication together at the G1-S transition, or whether the replicons that 'fire' later in S-phase are only partially prepared at the beginning of S-phase.

Replication in any one replicon starts in a zone called the *origin of replication*, and origins are marked by the binding of a group of six proteins called the *origin-recognition complex* (ORC). During G1, the *pre-replicative complex* (pre-RC) is assembled at each ORC-marked origin. The pre-RC consists of several proteins and its

---

[xii]One of us (JAB) has written in detail about this – see Further Reading list at the end of the chapter.

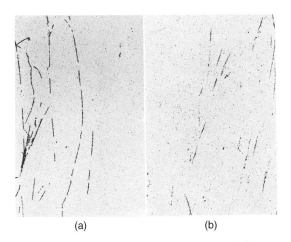

Figure 2.24 Replicating DNA in root meristem cells revealed by fibre autoradiography after labelling with radioactive thymidine. In the left-hand panel (a), it is clear that some tracts of DNA have been replicated during the labelling period, while others have not. This is also clear in the right-hand panel (b), where, halfway through the labelling period, the specific activity of the radioactive thymidine was changed from high to low. This 'step-down' technique reveals the outward movement of the replication forks from the replication origin within each replicon.
Autoradiographs supplied by Dr Jack Van't Hof, formerly of the Brookhaven National Laboratory, USA.

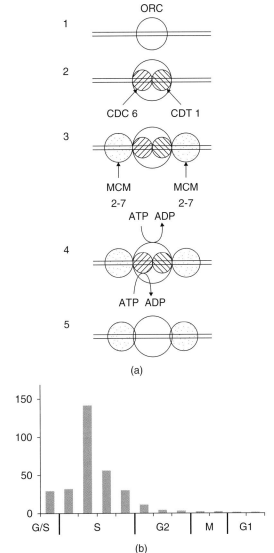

Figure 2.25 (a) Licensing of the replication origins. 1, The origin-recognition complex, ORC, is bound to a replication origin. 2, CDC6 and CDT1 are recruited to the origin followed by (in 3) two copies of the MCM2-7 helicase complex. 4, CDC6 and the ORC are phosphorylated by the CDC7-DBF4 protein kinase. 5, CDC6 and CDT1 are displaced from the origin (and are degraded). The origin is now licensed for replication. Re-licensing cannot occur within the same S-phase because the key regulator CDC6 has been broken down and is not renewed until transcription of its gene is initiated late in the next G1-phase. Only the main steps are shown. For further details see Bryant J (2010) Progress in Botany 71: 25–60. (b) Transcription pattern (assayed by the amount of mRNA, arbitrary units) of CDC6 in the cell cycle of tobacco BY2 cells.
From Dambrauskas G et al. (2003) Journal of Experimental Botany 54: 699–706.

assembly is dependent on at least two phosphorylation events. One of these is mediated by a kinase called CDC7, in concert with its partner DBF4, and the second by an S-phase-specific CDK in partnership with cyclinA3.

At the end of this process, strand separation has been started by the **helicase** consisting of the six proteins of the *MCM2-7* complex, and the replication origin is said to be 'licensed' (Figure 2.25). Re-licensing within one cell cycle is made impossible because some of the proteins, including CDC6 and CDT1, involved in setting up the pre-RC are destroyed at this stage via the ubiquitin-proteasome system (see Chapter 3, section 3.4.5).

Following S-phase, the cell goes through G2. Some events that are part of DNA replication, such as the final joining of long tracts of newly replicated DNA, may be delayed until this phase of the cycle. As soon as this is complete, the cell will be able to pass the DNA replication checkpoint and, provided it has attained an appropriate size for division, it will be able to move through the G2-M transition.

It is at this transition that we again encounter the range of CDKs and cyclins present in plants. In the simplest model, as developed for fission yeast, one CDK, working with an M-specific cyclin, is involved in this transition. In plants, however, at least two CDKs, one from the CDKA family and one from the CDKB family, together with three (or possibly four) cyclins – CYCB1;3, CYCA1, CYCA2 (and possibly CYCD4) – appear to participate in controlling the G2-M transition. However, it has proved difficult to assign specific roles to all of these.

Returning to the simplest model, as shown in Figure 2.22, the Cyclin-CDC2 complex (remember that CDC2 is the CDK in fission yeast) is regulated by phosphorylation/dephosphorylation of the CDC2 kinase. WEE1 protein kinase phosphorylates CDC2 at adjacent threonine and tyrosine residues. This inactivates CDC2 as part of the overall size controller that prevents cells entering mitosis at too small a size. By contrast, CDC25 phosphatase removes the phosphate groups from the threonine and tyrosine residues, thus activating CDC2 and enabling it to phosphorylate, among other things, histone H1, which is involved in chromosome condensation.

However, it is not at all clear that the transition works in the same way in plants. Plants certainly possess a WEE1 homologue that acts as a size controller when inserted into fission yeast cells. They also possess a truncated form of CDC25, but it is unable to rescue CDC25 mutants in fission yeast. Furthermore, mutagenesis of both the WEE1 and CDC25-like genes of *Arabidopsis* does not have any effect on the growth and development of the plants, suggesting that these genes have no role in cell division, at least under normal conditions.

This has led to the proposal by Dennis Francis at Cardiff University of a different model (see Figure 2.26) based on one of the proteins called *Interactors with CDKs* (ICKs)[xiii]. It is suggested that ICK1 binds to the CDKA-cyclin complex, thus inhibiting it (the CDKA is already dephosphorylated – see above). In order to initiate mitosis, CDKB (with its cyclin partner) phosphorylates ICK1, causing it to release CDKA-Cyclin and hence become available to phosphorylate its target proteins. This is certainly consistent with our current knowledge, although further evidence is needed to confirm this model.

At the beginning of mitosis itself, the cell possesses a full set of replicated chromosomes, each now consisting

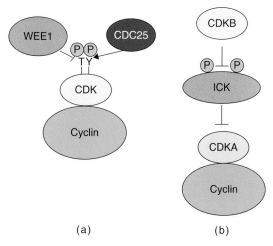

(a)   (b)

Figure 2.26 (a) The conventional model for regulation of the G2-M transition. (b) An alternative model for regulation of the G2-M transition in plants.
From Francis, D (2011). Annals of Botany, 107, 1065–1070.

of two *chromatids*. As the chromosomes condense during prophase, and thus become visible in light microscopy, it appears that the two chromatids in each chromosome are held closely together for much of their length; indeed, they are held together by **cohesin**, which is a complex of several proteins.

During metaphase, spindle microtubules become attached to the centromeres of each chromatid at the kinetochores, as mentioned in section 2.12.2. This is dependent on another protein kinase, the quaintly named AURORA kinase. At the same time, **separase** is released from its chaperone protein **securin**. The latter is degraded and separase is able to separate the chromatids, thus leading to anaphase when the two chromatids of each replicated chromosome are pulled to opposite poles of the cell (Figure 2.27).

However, progress into anaphase is dependent on the *anaphase-promoting complex* (APC, sometimes known as the cyclosome). This is a ubiquitin ligase and thus marks proteins for degradation by the 26S proteasome (see Chapter 3, section 3.4.5). The APC itself is tightly regulated so that it only comes into play at this point in mitosis. Its specific roles here are to initiate the destruction of securin (see above) and of the cyclin partners of mitotic CDKs, thus removing CDK activity. This loss of CDK activity is necessary for the completion of the cell cycle; in mutants in which the mitotic cyclins are not

---

[xiii]These proteins have several names. Sometimes they are known as *Inhibitors of CDKs*; there are also several abbreviations: *ICKs, KRPs* (and for animals, *CKIs*).

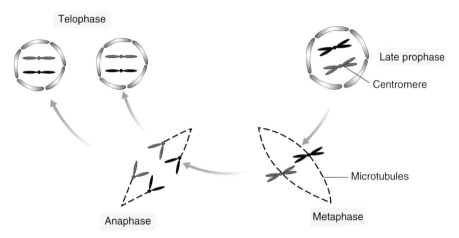

Figure 2.27 Diagram illustrating the behaviour of chromatids during mitosis. DNA is replicated during S-phase. The replicated molecules condense as chromatids during the prophase of mitosis. They are still held together at the centromeres. At metaphase, the chromatid pairs line up along the 'equator', and at anaphase the chromatid pairs separate and are pulled apart. At telophase, the nuclear envelope is reformed, completing the segregation of the replicated chromosomes into two daughter nuclei.
Based on a diagram in Brown, TA (2006) Genomes 2, BIOS, Oxford, p137, by permission of the publisher.

destroyed, anaphase is not completed and this, in turn, also affects telophase (the phase at which the nuclear envelope reforms and the chromosomes decondense) and cytokinesis.

Cytokinesis itself, the division of the cell to make two daughter cells, each with its own nucleus, starts with formation of the phragmoplast, as already described in section 2.12.2. The new cell plate is built up by the deposition of pectic polysaccharides from Golgi vesicles and the plasma membrane grows across the division plane as the membranes of the Golgi vesicles merge together. Cell division is complete.

2.13.4 Hormonal control of cell division

Plant hormones play key roles in controlling major developmental processes, in integrating those processes into the plant's developmental programme and in coordinating the activities of plant's organs. It is thus not surprising to know that hormones are implicated in the regulation of the cell division cycle. In relation to S-phase, cytokinins can increase and abscisic acid can decrease the number of replicon origins that are activated. At the G1-S transition, cytokinins and brassinosteroids (and possibly auxin) up-regulate the expression of the CYCD3;1 gene, while sucrose up-regulates both the CYCD2;1 and CYCD3;1 genes. Expression of these cyclin genes is essential for the

release of the E2F transcription factor from Rb protein, as described in the previous section.

Auxin and cytokinin are also involved in regulation of the G2-M transition and of mitosis itself. The details have yet to be fully worked out, but there are some pointers. Auxin seems to be necessary for the synthesis of M-phase CDKs, while cytokinins are required for the activity of these kinases. Auxin is also involved in the regulation of the anaphase promoting complex, possibly via control of gene expression. Finally, abscisic acid up-regulates the expression of ICK1, leading to an over-abundance of this protein and hence an inhibition of mitosis (see previous section). Since ABA is involved in several stress responses (Chapters 9 and 10), it has been suggested that this is mechanism for preventing entry into mitosis under stress conditions.

2.13.5 Incomplete cell cycles

In many of the types of organism that have been studied, passing 'Start' and initiating DNA replication normally leads to completion of the entire mitotic cycle. However, in plants, incomplete cell cycles are not unusual. Indeed, they occur as part of normal development and differentiation as discussed for example in Chapter 4.

Although there are several types of truncated cell cycle, by far the commonest are those in which mitosis is by-passed and S-phase is re-initiated, leading to extra rounds of DNA replication. This process is known as DNA **endoreduplication**. In cell suspension cultures, it can often be induced by changing the ratio of auxin to cytokinin in the growth medium (e.g. in tobacco BY2 cells by omitting auxin[xiv]). It is thus probable that in whole plants, endoreduplication is at least partly controlled by the concentration ratio of these two hormones.

The function of endoreduplication within the plant remains a matter for conjecture. It certainly leads to the formation of large cells, although in the tobacco BY2 cells mentioned above, the cell size increases in auxin-depleted cells before endoreduplication is initiated. There is also recent evidence that some plants initiate endoreduplication after being damaged by herbivores (Chapter 11, section 11.2.1); it is thought that the increased cell size may be an aid to re-establishment of biomass. However, it is not known how widespread this response is among plants. Therefore, we focus here not on its function (see Chapter 4, sections 4.2, 4.5 and 4.9 for further discussion), but instead need to draw attention to particular features of its regulation.

In order for endoreduplication to occur, first, mitosis must not occur. Failure to enter mitosis occurs because of failure to synthesize mitotic cyclins, thus preventing the activity of the relevant CDKs. But the lack of mitosis is not enough. Normally this would lead to cell arrest in G2, as is seen in a range of differentiated cells (see earlier). What must happen is that the cell bypasses the checkpoint which normally ensures that DNA cannot be 're-licensed' for replication until mitosis has occurred.

Two of the proteins involved in setting up the pre-replicative complex (section 2.13.3) – CDC6 and CDT1 – are implicated here. Over-expression of either, but especially of CDC6, in genetically modified cells leads to endoreduplication. Up-regulation of CDC6 is one of the first indications that cultured cells of tobacco will shortly initiate endoreduplication. This implicates, in turn, the regulation of the E2F-DP transcription factors (the involvement of cytokinin in this has already been noted). Furthermore, in cells undergoing successive rounds of endoreduplication, CDC6 protein is more stable than in cells undergoing 'normal' DNA replication: the protein is not degraded by the ubiquitin-proteasome system (see Chapter 3, section 3.4.5). This ensures that

this protein, and probably others involved in setting up the pre-replicative complex, remain available whether or not mitosis has occurred.

2.14 Metabolism

2.14.1 Introduction

This is not a book that is primarily about plant biochemistry. Indeed, detailed consideration of plant biochemistry could not be achieved in one book[xv]. However, in order to understand the plant as a functional unit, some knowledge of metabolism is necessary. Some aspects of this are dealt with in particular places. In this chapter for example, we have already discussed polysaccharide synthesis in the context of the cell wall; in Chapter 5, section 5.9, we discuss the mechanisms of nitrogen fixation in the *Rhizobium*-legume symbiosis, and in Chapter 7, sections 7.4 and 7.5, we describe photosynthesis and photorespiration. In this chapter, we deal with the important features of respiration, lipid metabolism, ascorbate synthesis and nitrogen metabolism.

2.14.2 Respiratory metabolism

Respiratory pathways have two main functions: first to transfer the chemical energy 'locked up' in respiratory substrates (hexose sugars) into forms in which it may be used in cellular syntheses; and second to produce initial substrates for synthesis of more complex molecules. The main energy-carrying molecule produced in respiration is ATP, and small amounts of NADPH are also produced (in the pentose phosphate pathway). In some plants, there is a third function of respiration, namely to produce heat (for example in *Arum* flowers).

Three main groups of reactions constitute the conventional respiratory pathway. First, there is the *EMP (Embden-Meyerhof-Parnas) pathway*, often known as *glycolysis*, in which the initial hexose substrates are converted to pyruvate. This takes place in the cytosol. Further metabolism of pyruvate involves a decarboxylation and then entry into the *TCA (tri-carboxylic acid) cycle* (also known as the Krebs cycle or the citric acid cycle), which takes place in the mitochondria. Also in the mitochondria, the NADH produced by the stepwise oxidation of

---

[xiv]These cells are unusual in that they do not require added cytokinin.

[xv] *The Biochemistry of Plants*, published by Academic Press in the late 1980s, ran to 16 volumes.

respiratory substrate is itself re-oxidized in a series of reactions – the *electron transport chain* – that take place in the inner mitochondrial membrane, leading eventually to the reduction of oxygen to form water, linked to the synthesis of ATP. In the absence of oxygen, the electron transport chain cannot function, and the pyruvate produced in the EMP pathway does not enter the mitochondrion but is instead further metabolized to form ethyl alcohol in a process called *fermentation*.

Finally, some of the hexose respiratory substrate does not initially enter the EMP pathway. Instead, it is metabolized via the *pentose phosphate pathway* (PPP), in which it undergoes two reductions and one decarboxylation before entering a series of sugar interconversions (similar to those in the reductive pentose phosphate pathway in photosynthesis – see Chapter 7, section 7.4), after which it may re-enter the EMP pathway.

One of the functions of the PPP is the generation of different monosaccharide sugars for use in various biosynthetic pathways – e.g. the five-carbon sugar ribose is used in synthesis of nucleotides, while the four-carbon sugar erythrose is involved in the synthesis of phenolic compounds via the *shikimic acid* pathway.

Another function is the generation of *NADPH*. NADP (nicotinamide adenine dinucleotide phosphate), rather than NAD (nicotinamide adenine dinucleotide), is the cofactor for the oxidations in the PPP. In general, NADPH is used to provide reduction potential in biosynthetic reactions rather than being used to generate ATP (depending on the cell's metabolic needs, a small proportion may be re-oxidized by the electron transport chain, thus generating some ATP). Between 5 and 15 per cent of the hexose substrate is metabolized by the PPP and, as shown by the British biochemist Tom ap Rees, the actual amount depends on the cell's biosynthetic needs, with differentiating cells showing greatest throughput.

The substrate for entry into the cytosolic phase of respiration is, as already stated, a hexose monosaccharide, namely glucose. The actual mode of entry varies according to whether the glucose is derived from sucrose or from breakdown of stored starch (Figure 2.28a).

Breakdown of sucrose by invertase produces free glucose and free fructose, both of which may be phosphorylated by hexokinase to form G6P and F6P respectively. Breakdown of sucrose by sucrose synthase again produces free fructose and also UDP-glucose (see section 2.2.2), which is readily converted to G1P and then to G6P.

Starch breakdown by amylase and glucosidase produces free glucose, which is phosphorylated to form G6P; starch breakdown by starch phosphorylase produces G1P, which is again converted to G6P. Thus, at the start of the pathway there are both G6P and F6P and, depending on the source of these hexose phosphates, some ATP may have already been expended in producing them. The G6P may enter the pentose phosphate pathway, or it may be converted to F6P to enter the EMP pathway (Figure 2.28a).

The next step is important in the regulation of the pathway. It is a second phosphorylation of F6P by *phosphofructokinase* (PFK); in the 'conventional' (and by far the better known) reaction, the source of the phosphate group is ATP. However, plants also contain a form of PFK that uses pyrophosphate as the phosphate donor (Figure 2.28a). Whatever the source of the phosphate group, this step effectively gives a hexose with a phosphate at both ends, and in this form it is split into two triose molecules, each phosphorylated. It is at this stage that, when necessary, photosynthetic product may directly enter the respiratory pathways, in the form of triose phosphate exported from the chloroplast.

The remaining steps of the EMP pathway convert triose phosphate to pyruvate (via phosphoenolpyruvate). These steps are the energy-conserving steps of the pathway; two oxidations occur, giving two NADH molecules per triose, and two ATP molecules are also generated for each triose molecule (Figure 2.28a). However, the bulk of the energy available in the original hexose molecule is still 'locked up' in pyruvate. Under aerobic conditions (which are normal conditions for most plant tissues), the pyruvate is transferred to the mitochondrion for entry into the TCA cycle (Figure 2.28b).

However, as indicated in Chapter 9 (section 9.3), plant cells may be exposed to low oxygen tension. Under such circumstances, the oxidative steps of the TCA cycle cannot occur; the NADH generated in the oxidative reactions cannot be re-oxidized in the electron transport chain for which the final step is the reduction of oxygen by cytochrome oxidase. Thus, under these conditions, pyruvate and NADH would accumulate. In the absence of NAD, the oxidation further 'upstream' in the EMP pathway would not occur and the pathway would come to a halt.

However, this situation is remedied by operation of the fermentation pathway. Pyruvate is decarboxylated to form acetaldehyde, which is then reduced by NADH

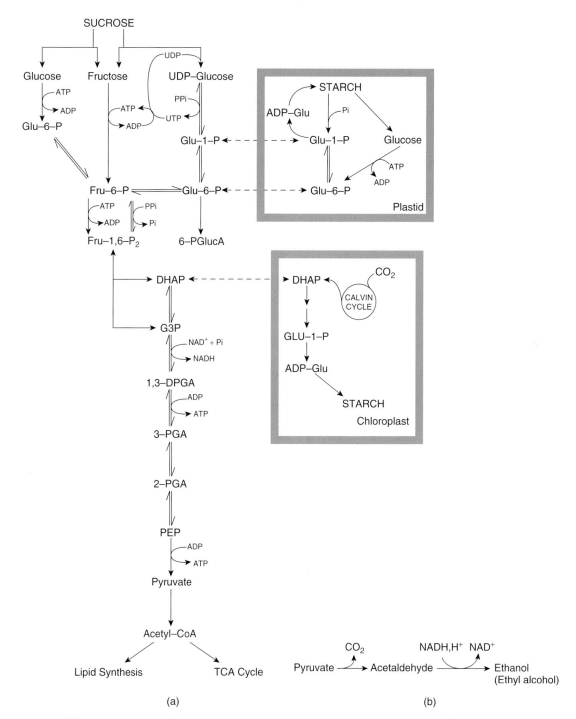

(a)   (b)

Figure 2.28 **(a)** The EMP pathway or glycolysis. Note that G-6-P can be shunted from glycolysis into the pentose phosphate pathway (PPP; not shown), as described in the text. F-6-P or triose-P may re-enter glycolysis from the PPP. **(b)** Recycling of NADH by fermentation under anaerobic conditions.

Taken from Lea, P and Leegood, R (1999) Plant Biochemistry and Molecular Biology, Wiley, Chichester, page 84.

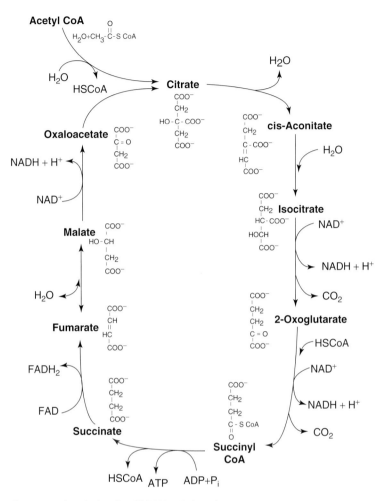

Figure 2.29 The Tricarboxylic acid (TCA) or Krebs cycle.

to form ethanol (ethyl alcohol). The NAD can then be recycled for use in the glyceraldehyde-P dehydrogenase reaction. The EMP pathway and fermentation conserve only a small fraction of the energy available from hexose oxidation and, in general, we should regard this is as a temporary survival mechanism. Indeed, both acetaldehyde and ethanol are toxic and the plant cannot tolerate a build-up of either. However, some organisms, such as the yeast (*Saccharomyces cerevisiae*), are able to thrive under anaerobic conditions. They do so by metabolizing very large quantities of hexose and excreting the resulting ethanol (to the delight of all those who enjoy alcoholic beverages!).

Returning now to normality, i.e. to aerobic conditions, several oxidative steps take place in the TCA (Krebs) cycle (Figure 2.29), and these play a central role in energy conservation.

The first oxidation is a complex one, catalysed by pyruvate dehydrogenase, which is actually formed of several proteins. The oxidation is a three-step reaction. First, the pyruvate is decarboxylated and then it is oxidized (with linked reduction of NAD to NADH). The resultant two-carbon acid acetate is then joined to coenzyme-A (CoA) via a thiol-ester linkage to form acetyl-CoA. The energy conserved in the thiol ester linkage is used to drive the combination of the acetate with the four-carbon acid

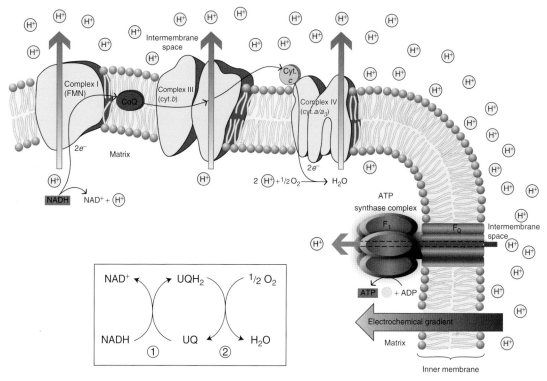

Figure 2.30 The mitochondrial electron transport chain, illustrating the creation of a proton gradient (chemiosmosis) to provide the energy to synthesise ATP. Inset: The cyanide-resistant pathway. 1: an alternative NADH oxidase, situated on the inner side of the inner mitochondrial membrane directly reduces ubiquinone (CoQ). 2: the alternative terminal oxidase (a 33 kDa protein situated in the inner mitochondrial membrane) then uses the reduced ubiquinone to reduce oxygen, thereby forming water. No ATP is synthesized.
Figure from Raven PH *et al* (2005) Biology of Plants, 7th edition, Freeman, NY, p 111.

oxaloacetate to form six-carbon citrate and thus enter the cycle. Two more combined decarboxylation-oxidation steps lead to the restoration of the four-carbon state in the form of succinate.

The enzyme responsible for the second decarboxylation-oxiation – 2-oxogluarate dehydrogenase – is similar to pyruvate dehydrogenase. Thus it catalyses the formation of a thiol ester linkage between succinate and CoA to form succinyl-CoA. In this instance, the energy released by breaking the thiol ester linkage is conserved in the formation of ATP[xvi]. Two more oxidations lead to the formation of oxaloacetate, which is available to start the cycle again. The first of the oxidations, catalysed by succinate dehydrogenase, is linked to the reduction of FAD (flavin adenine dinucleotide) rather than NAD.

[xvi]Note that in animals, GTP rather than ATP is synthesized at this step.

Up to this point, the respiratory pathways have generated a small excess of ATP molecules, plus large amounts of reduced coenzymes and, in particular, NADH. This leads to our brief consideration of the electron transport chain. The proteins of the chain are embedded in the inner mitochondrial membrane in a series (Figure 2.30).

The proteins are arranged so that the passage of electrons from NADH dehydrogenase to the final electron acceptor, cytochrome oxidase, is accompanied by passage of protons across the membrane into the inter-membrane space, thus building a proton gradient across the membrane (just as happens in photosynthesis – see Chapter 7, section 7.4). The membrane, unless it is damaged, does not permit re-entry of the protons, which can only return to the mitochondrial matrix by passage through the F_0F_1 ATP synthase complex. As shown by the elegant work of John Walker and his research team at Cambridge, the

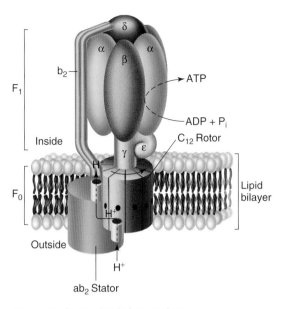

Figure 2.31 The mitochondrial ATP synthase.

Table 2.3 Examples of the use of respiratory metabolites in other areas of metabolism.

	Metabolite	Examples of use
EMP/PPP	G-6-P/G-1-P	Cellulose synthesis
		Synthesis of other cell wall polysaccharides
		Ascorbate synthesis
PPP	NADPH	Reductive biosynthesis
	Ribose-5-P	Nucleotide synthesis
	Erythrose-4-P	Shikimic acid pathway for synthesis of phenolics
EMP	Phosphoenolpyruvate	Shikimic acid pathway for synthesis of phenolics
	Pyruvate	Alanine synthesis
	Acetyl-CoA	Lipid synthesis
		Terpenoid synthesis (including GA and ABA)
TCA cycle	2-Oxoglutarate	NH_3 uptake into glutamate (and then to chlorophyll, cytochromes and to other amino acids)
	Oxaloacetate	Aspartate

subunits in the centre of the ATP synthase form a tiny turbine (Figure 2.31) that is driven by the passage of the protons to generate the energy to form ATP from ADP and Pi. The energy initially conserved in the form of reduced coenzymes is now conserved as ATP.

Two further brief points need to be made. First, NADH (and NADPH) generated in the cytosol and FADH arising from succinate dehydrogenase donate their electrons to ubiquinone, thus bypassing one of the proton-transfer steps. In the cell's energy budget, less ATP is generated from these re-oxidations than from NADH that enters the chain via the major NADH dehydrogenase.

Second, as in all aerobic organisms, plant cytochrome oxidase is inhibited by cyanide. However, plant cells exhibit significant oxygen uptake in the presence of cyanide, although conservation of energy in the form of ATP is very inefficient. This is because plants possess an *alternative oxidase*, through which electrons may be diverted, thus bypassing the cytochromes and cytochrome oxidase (Figure 2.30). Much of the energy initially conserved in reduced coenzymes is dissipated as heat.

It is this alternative oxidase pathway that is responsible for elevating the temperature of flowers in the *Arum* family (see Chapter 8, section 8.6.2 and Chapter 9, section 9.2), but it may also function under certain stress conditions. Plants also possess another method

for uncoupling electron transport from ATP synthesis in the form of an uncoupling protein that makes the inner mitochondrial membrane leaky to protons. Whether this works alongside the alternative oxidase, or is involved in separate stress responses, is not clear.

2.14.3 Linkage of respiration to other metabolic pathways

As has already been mentioned, another important function of respiratory metabolism is the synthesis of metabolites that can be 'tapped off' for use in a range of biosynthetic reactions. Some examples are given in Table 2.3. It is clear that, through the various phases of respiratory metabolism, there are branch points, with the 'choice' of which route to follow being regulated by metabolic needs.

One of the problems caused by this inter-linkage of metabolic pathways is at that at times of high demand, particular respiratory intermediates may become limiting and thus slow down the rate of respiration. This problem is countered by the operation of subsidiary pathways which top up the main pathways and thus 're-balance' respiratory metabolism. These are termed **anaplerotic reactions**. Examples include:

• Synthesis of malate from PEP by PEP carboxylase. The malate can be used to top up the TCA cycle or can be converted to pyruvate by NAD-linked malic enzyme.

• Uptake of citrate from the cytosol, with subsequent metabolism of the resulting 'extra' malate by malic enzyme to give pyruvate.

2.14.4 Regulation of respiration

In plants, as in animals, the enzyme phosphofructokinase (PFK) is a major player in the control of respiration. However, unlike animal PFK, the plant PFK is not strongly up-regulated by AMP nor strongly down-regulated by ATP; these compounds have only minor effects on the enzyme. The main negative regulator of plant PFK is phosphoenolpyruvate (PEP). Further, the inhibition of PFK by PEP is strongly ameliorated by inorganic phosphate (P_i) which means that the PEP:P_i ratio is important. PEP itself is converted in the EMP pathway to pyruvate by pyruvate kinase, or is diverted out of the EMP pathway in an anaplerotic reaction catalysed by PEP carboxylase (as mentioned above). Both of these enzymes are subject to feedback inhibition by metabolites from the TCA cycle, especially citrate, malate and 2-oxoglutarate. Thus the rate at which the TCA cycle operates has an effect on the concentration of PEP, which in turn affects throughput via PFK.

The first step in the TCA cycle is subject to regulation via protein modification. Pyruvate dehydrogenase is inactivated when it is phosphorylated by pyruvate dehydrogenase kinase and reactivated when the phosphate group is removed by pyruvate dehydrogenase phosphatase. The availability of ATP for the kinase reaction is thus an important controlling factor for slowing down entry of metabolite into the TCA cycle. Pyruvate dehydrogenase is also inhibited by NADH, as are several of the other dehydrogenases in the cycle (succinate dehydrogenase is an exception). The rate at which the electron transport chain is working (which is also partly dependent on the availability of ADP) thus has an effect on the operation of the TCA cycle.

2.14.5 Lipid synthesis

The term *lipid* covers a range of hydrophobic, water-insoluble compounds that dissolve in organic solvents. Earlier in this chapter (sections 2.3.2, 2.5.2 and 2.6), we encountered the various types of lipids that make up plant membranes. These are *glycerolipids* and *phospholipids*, glycerol molecules which carry two fatty acid chains and have a range of possible moieties at the third position, including galactose, phosphatidylcholine and phosphatidylethanolamine. Triacylglycerols (triglycerides) have also been mentioned (section 2.9.1) and are also discussed in Chapter 4, section 4.5.1. These consist of glycerol molecules carrying three fatty acid chains and are primarily storage compounds, found in oleosomes (oil bodies). Fatty acids are synthesized in chloroplasts/plastids but, as already mentioned, modification and further metabolism takes place in the ER.

The starting point for fatty acid synthesis is acetyl-CoA. In respiratory metabolism, this is produced as the respiratory substrate from the EMP pathway and enters the mitochondrion. We now know that the EMP pathway also operates in chloroplasts (often referred to as *chloroplast glycolysis*). Although there was initially some controversy over this, it is currently thought that this pathway, including the acetyl-CoA synthase component of the pyruvate dehydrogenase complex, is the source of acetyl-CoA for fatty acid synthesis. In photosynthetic cells, the substrate for the chloroplast EMP pathway is generated in photosynthesis; in non-photosynthetic cells, it is derived from sucrose arriving via the phloem and is imported into the plastids as hexose phosphate or as pyruvate.

Fatty acids are built up two carbons at time, a process that involves two enzymes – acetyl-CoA carboxylase[xvii] and fatty acid (FA) synthetase. The carboxylase is a tightly regulated enzyme whose activity is a major influence on the rate of FA synthesis. The synthetase is a large complex of six polypeptides, each with a separate role in the process. The reaction sequence is shown in Figure 2.32.

Using the energy derived from hydrolysis of ATP, acetyl-CoA carboxylates biotin[xviii]. From there, the CO_2 is transferred to acetyl-CoA to form malonyl-CoA. The malonyl moiety is then transferred to *acyl carrier protein* (ACP). Malonyl-ACP donates its original acetate unit to the 'primer', acetyl-CoA, releasing both CO_2 and CoA and leaving the four-carbon acetoacetate joined to ACP. While still attached to ACP, this four-carbon acid is successively

---

[xvii]In eudicots, acetyl-CoA carboxylase is a complex of four polypeptides each with its own catalytic site; in monocots it a large (240 kDa) polypeptide with four catalytic sites.
[xviii]Vitamin B7, a co-factor that participates in several CO_2 transfer reactions.

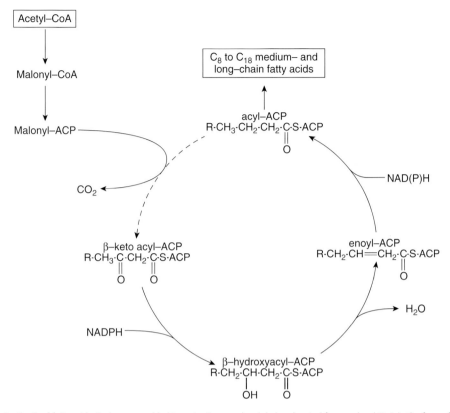

Figure 2.32 Synthesis of fatty acids. Carbons are added two at a time, each pair being donated from malonyl-CoA in the form of acetyl-CoA. From Lea, P and Leegood, R (1999) Plant Biochemistry and Molecular Biology, WIley, Chichester, page 125.

reduced (with NADPH as the hydrogen donor: see section 2.14.2), loses a water molecule and is again reduced by NADPH. The cycle then starts again with the donation of two more carbons from malonyl-ACP.

The main end-product of the synthesis cycle is 18:0-ACP, with some 16:0-ACP. One double bond may be inserted into 16:0-ACP and 18:0-ACP by a chloroplast desaturase enzyme to give 16:1-ACP and 18:1-ACP. Although the four FAs generated so far may contribute to the assembly of chloroplast membrane lipids (see section 2.5.2), the majority of the FAs need further modification before incorporation into any of the cell's membranes or into triacylglycerols. These further modifications can only occur in the ER. Thus, 16:0, 16:1, 18:0 and 18:1 FAs are exported from the chloroplast/plastid after transfer of the FAs from ACP to CoA (Figure 2.33).

In the ER, a combination of desaturase and elongase enzymes produces a range of unsaturated FAs, including 16:3, 18:2, 18:3, 20:1 and 22:1. Some of these are re-exported (as acyl-CoAs) to the chloroplast. The ER also assembles the glycero- and phospholipids for various cell membranes and triacylglycerols for lipid storage (see Figure 2.32). Triacylglycerols are formed in the *Kennedy pathway*, in which acyl transferase transfers successively two FA molecules to glycerol-3-phosphate. The phosphate is then removed and the third FA is then joined to the glycerol backbone. Acyl transferases exhibit different levels of activity and specificity with different FAs, and thus have a role in determining the FA composition of membrane and storage lipids.

2.14.6 Lipid oxidation

There are several circumstances in which lipids may be oxidized: membranes are turned over, cells undergo programmed cell death, storage lipids are mobilized and so on. However, the general principles are the same for all of them, so we focus here on the oxidation of triacylglycerols, the lipids used for storage.

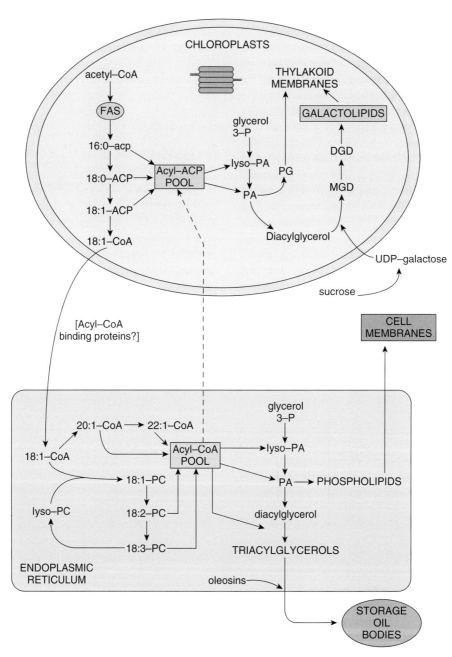

Figure 2.33 Further metabolism of newly synthesised fatty acids.
From Lea, P and Leegood, R (1999) Plant Biochemistry and Molecular Biology, Wiley, Chichester, page 128.

The first step is the removal, one by one, of the FAs from the glycerol with which they are combined; this is mediated by lipase, an enzyme that is particularly active during the germination of fat-storing seeds (note that in breakdown of membrane lipids, lipases with a higher degree of specificity are also required, including phospholipases and galactolipases). The glycerol released by the action of lipase may enter the EMP pathway as triose phosphate, or may be re-used in further lipid synthesis (for example, during turnover of membrane components). Free fatty acids, by contrast, are potentially harmful in that they may damage membranes, so as soon as they are released by lipase, they are transferred to coenzyme-A to form acyl-CoAs. In this form, they enter the *peroxisomes* to be broken down by β-*oxidation*.

As with FA synthesis, oxidation deals with two carbons at a time, starting at the thiol ester link with CoA (Figure 2.34).

The first step is an oxidation mediated by acyl-CoA dehydrogenase, the co-factor of which is FAD. The reduced FAD is re-oxidized in plants by molecular oxygen (O_2), giving rise to hydrogen peroxide (H_2O_2); this is immediately degraded by catalase. The following two steps are an addition of water (compare with FA synthesis, above) and a second oxidation, this one with NAD as the co-factor. Finally, the oxidized portion is removed as acetyl-CoA, while the remainder of the molecule – two carbons shorter – is transferred to another CoA molecule to start the process again.

The acetyl-CoA may be transferred to the mitochondrion for entry into the TCA cycle but, when lipid reserves are being mobilized (e.g. during seed germination), it enters the glyoxylate cycle (see Chapter 4, section 4.11.2 for details). For every two acetate residues that enter this cycle, three out of the four carbons are preserved for conversion into sugars (e.g. sucrose for transport to the growing regions of the seedling).

2.14.7 Biosynthesis of ascorbate

Ascorbic acid ('vitamin C'), which is the most abundant anti-oxidant in plants, occupies a very special place in the relationship between humans and plants. *Homo sapiens* is one of only relatively few species of mammal that is unable to synthesize ascorbic acid[xix], but we still need it

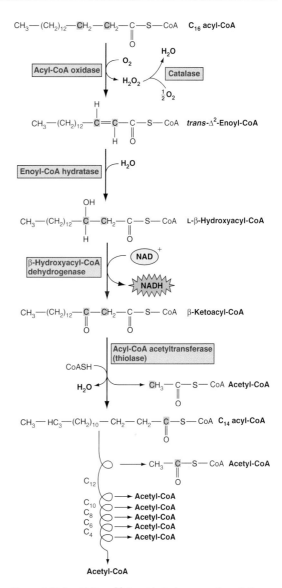

Figure 2.34 β-oxidation of fatty acids. Each passage through the pathway releases two carbons in form of acetyl-CoA.

as an essential co-factor in several biosynthetic reactions and as an important anti-oxidant.

Based on comparative biochemistry, and more recently on proteomic and genomic analysis, it appears that biosynthesis of ascorbic acid is a very ancient metabolic

[xix] The others are bats, guinea pigs, capybaras, and three primate groups: tarsiers, monkeys and apes. The latter three groups of

primates diverged from a common ancestor about 60 million years ago and it is presumed that it was the common ancestor that lost the ability to make vitamin C.

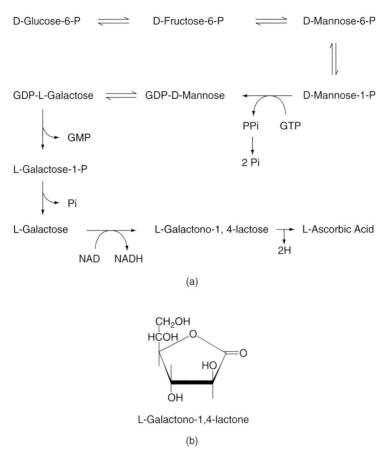

Figure 2.35 (a) Diagram illustrating the Smirnoff-Wheeler pathway for synthesis of ascorbic acid. (b) L-galactono-1,4-lactone, the immediate precursor of ascorbic acid.

character. Certainly, all living cells require it, which means that organisms that are unable to make it need to obtain it in their diet. Furthermore, because ascorbate cannot be stored or accumulated, dietary intake must be regular.

For humans, one of the best sources of ascorbate is fresh plant material. All plant tissues appear to have the ability to make ascorbate and it is present throughout the plant, although not in dry seeds. However, the highest concentrations are found in photosynthetic tissues: leaves contain between 1 and 5 mM ascorbate, and in the chloroplasts the concentration may be as high as 25 mM.

Despite it being essential in our diets and despite its importance to plants themselves (see Chapters 9, 10 and 11), the metabolic route by which plants[xx] make ascorbate

---

xxThe biosynthetic route by which animals make ascorbate has been known for many years.

remained unknown for many years. Indeed, it was the last of the major plant products to yield the secrets of its biosynthetic pathway. It was not until the late 1990s that very careful biochemical and enzymological analysis by Nicholas Smirnoff and his research team at the University of Exeter, UK, led to elucidation of the main biosynthetic route (Figure 2.35a). Prior to their work, it had been proposed, based on evidence from feeding experiments, that the most efficient precursor for ascorbate synthesis in plants would be L-galactono-1,4-lactone (Figure 2.35b). However, this had not been detected in plants and, in any case, no route for its synthesis had been established.

Smirnoff's team showed that supplying leaves with either L-galactose or L-galactono-1,4-lactone led to a significant increase in the concentration of L-ascorbic acid. They went on to show that GDP-L-galactose can be made from GDP-D-mannose; from GDP-L-galactose it is just four steps to L-ascorbic acid via L-galactono-1,4-lactone. The penultimate step in the pathway, the conversion of L-galactose to L-galactono-1,4-lactone, is catalyzed by L-galactose dehydrogenase, an enzyme that had not before been detected in plants.

Overall, therefore, starting with the common metabolite G6P (see section 2.14.2), it takes nine enzymic steps to make ascorbate. All of the enzymes involved in this pathway, known as the Smirnoff-Wheeler pathway, have now been isolated. The genes that encode them have been cloned and sequenced, giving rise to the possibility of using genetic modification techniques to increase ascorbate content.

The location of the enzymes is interesting. Most of them are located in the cytosol, but the final enzyme in the pathway, L-galactono-1,4-lactone dehydrogenase, is located in the inner mitochondrial membrane, facing into the inter-membrane space. L-galactono-1,4-lactone must therefore be imported through the outer mitochondrial membrane, and ascorbate itself must be exported through the same membrane. Finally, GDP-L-galactose phosphorylase occurs in the nucleus as well as the cytosol. Its role there is not known, but it may have a secondary or 'moonlighting' function, as has been shown for several other primarily cytosolic enzymes (Chapter 3, section 3.2.5).

2.14.8 Aspects of nitrogen metabolism

Space does not allow us to discuss the whole range of nitrogen metabolism in plants. Indeed, it would take another book to complete such a task. The focus here is on the mechanisms involved in the uptake of nitrogen in various forms and its entry into general metabolism. For most plants, nitrogen is available externally as nitrate or as ammonia (in the form of ammonium salts), while some exploit the capability of particular microorganisms to fix atmospheric nitrogen.

Nitrate (NO_3^-) is mainly taken up by roots, and in many species it is also reduced in the roots. However, when nitrate supply is abundant, it is also translocated to the shoots before it is reduced, with different plant species varying in the extent to which nitrate is translocated before reduction. The enzyme that catalyses the reduction is *nitrate reductase*. The main form of the enzyme, found in both green and non-green tissues, uses NADH as the reductant (Figure 2.36), whereas the minor form of nitrate reductase, located in non-green tissues, can use NADH or

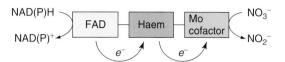

Figure 2.36 Nitrate reductase. The enzyme's three prosthetic groups/co-factors (FAD, haem and Mo) form a short electron transport chain between NAD(P)H and nitrate.

NADPH. The enzyme has three tightly bound co-factors: FAD, haem and molybdenum (Mo) – the latter in the form of a complex with a pterin.[xxi] These co-factors act in effect as a small electron transport chain; electrons are transferred from NADH to nitrate via FAD, haem and Mo (Figure 2.36), and nitrate is thus reduced to nitrite (NO_2^-).

Nitrate reductase is another tightly regulated enzyme. First, it is regulated at the level of gene transcription by nitrate itself. For example, supplying nitrate to plants previously grown without any will lead rapidly to transcription of the nitrate reductase genes and then to the accumulation of the enzyme throughout the plant. In leaves, this induction is dependent on functional chloroplasts and sucrose (this product of photosynthesis may also be involved in transcriptional up-regulation).

The enzyme itself is subject to control by phosphorylation/dephosphorylation. In the dark, and/or when carbohydrate supply is limiting, nitrate reductase is phosphorylated by a calcium-dependent protein kinase. This makes the protein available for a magnesium-dependent binding of an inhibitory protein. The binding event makes the nitrate reductase more liable to degradation but is reversible in the light, when the concentration of free Mg^{2+} ions is lowered. Light also inhibits the protein kinase, while soluble carbohydrates activate a phosphatase that removes the phosphate groups. All this makes sense when we consider that the final assimilation of nitrogen from nitrate into amino acids requires a good supply of 'carbon skeletons'.

The product of nitrate reductase, nitrite, is toxic and highly reactive. It is transferred to chloroplasts or to non-photosynthetic plastids where it is reduced to ammonium (NH_4^+) by *nitrite reductase*. The reduction is a complex one, as shown in Figure 2.37. The electron donor is reduced ferredoxin and, again, there is a short electron

---

[xxi] A heterocyclic N-containing organic compound.

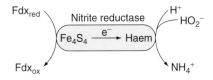

$$NO_2 + 6\ Fdx_{red} + 8H^+ + 6e^- \longrightarrow NH_4^+ + 6\ Fdx_{ox} + 2H_2O$$

Figure 2.37 Nitrite reductase. The top half of the diagram shows the short electron transport chain built into the structure of the enzyme by virtue of its prosthetic groups (an iron-sulphur centre and haem). The bottom half shows the stoichiometry of the reaction. Note that the oxidized ferredoxin is re-reduced by NADPH. Key: Fdx = ferredoxin; red = reduced; ox = oxidized.

chain through the enzyme's prosthetic groups (an iron-sulphur cluster and a haem group) between reduced ferredoxin and the substrate. In chloroplasts, the initial reduction of ferredoxin depends on the light reactions of photosynthesis. In non-green plastids, the ferredoxin is reduced by NADPH from the pentose phosphate pathway (see section 2.14.2).

As with nitrate reductase, nitrite reductase is controlled at the level of gene transcription. Because nitrite is toxic, it is important that, when nitrate reductase is active, there should be enough nitrite reductase to remove the nitrite quickly. Nitrite reductase gene expression is thus controlled by nitrate supply and by light. When nitrate reductase is present in a cell, there is always an excess of nitrite reductase.

The uptake of nitrate thus leads to the formation of ammonium ions and, as stated earlier, these ions can also be taken up from the soil. However, ammonium ions are also toxic and have the potential to damage membranes. They are rapidly incorporated into the amino acid pool by the action of two enzymes working in concert (Figure 2.38). The first of these is *glutamine synthetase* (GS), which transfers the ammonium ion to glutamate to form glutamine. This is driven by the hydrolysis of ATP.

The second enzyme is known either as *glutamate synthase* or as *glutamine:2-oxoglutarate aminotransferase* (GOGAT, discovered independently by two UK research groups in the mid-1970s). The longer name describes exactly what it does: 2-oxoglutarate (from the TCA cycle) is aminated by the transfer of the amide (i.e. secondary amino) group from glutamine. This results in the formation of two glutamate molecules, one of which remains in the GS-GOGAT cycle and the other of which enters

into the amino acid pool and hence into the large range of processes involving organic nitrogen. There are two forms of GOGAT: in non-green plastids, the main form uses NADH as an electron donor, while the enzyme in chloroplasts (which is also present as minor form in non-green plastids) uses reduced ferredoxin. As with nitrite reductase, this requirement links nitrate assimilation with photosynthesis, and we note that the same GS-GOGAT cycle is also involved in photorespiratory ammonium assimilation (Chapter 7, section 7.5.1).

Overall, there is a strong linkage between nitrogen assimilation and carbohydrate supply because of the need for a respiratory intermediate (2-oxoglutarate) as an amino group acceptor. Thus, the GS-GOGAT cycle is activated under conditions of high light and ample soluble carbohydrates.

Atmospheric nitrogen can be fixed by a range of micro-organisms, some which live in mutualistic associations with plants. The best known of these associations is symbiosis between plants in the family Leguminosae and bacteria in the genus *Rhizobium*. This symbiosis, involving the formation of root nodules, and the biochemistry of nitrogen fixation are discussed in Chapter 5, section 5.9. Here we note that the product of N-fixation is ammonia (NH₃). This is toxic and must be rapidly incorporated into organic compounds before export from the root.

In many legumes, mostly those living in temperate or cool-temperate regions, the ammonia is exported in the form of the secondary amino or amide group of asparagine or glutamine. The enzymes of the GS-GOGAT cycle are again involved, but the cycle operates in a slightly different way in that glutamine itself may be exported from the root nodule or it may transfer its amide group to aspartate, thus generating asparagine, which is exported (Figure 2.38).

Alternatively, the fixed nitrogen may be exported in the form of **ureides**, especially allantoin and allantoic acid (see Glossary), as occurs especially in legumes of warmer regions. In a complicated pathway that occurs in the plastid, glutamate and/or glutamine participate in the synthesis of purines. One of these purines, xanthine, exits the plastid and is converted to uric acid, which is transferred to non-infected root cells. In those cells, uric acid is converted in the peroxisomes to allantoin, which is exported to the phloem for transport to the leaves. Alternatively, the allantoin may be converted in the ER to allantoic acid, which is also translocated via the phloem

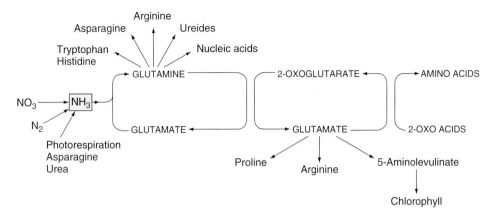

Figure 2.38 The Glutamine Synthetase-GOGAT cycle.
Based on Lea, P and Leegood, R (1999) Plant Biochemistry and Molecular Biology, Wiley, Chichester, page 168.

to the leaves. In the leaves, the ureides are degraded, thus releasing ammonia, which is re-assimilated by the GS-GOGAT cycle.

Selected references and suggestions for further reading

Abrahams, JP. Leslie, AGW, Lutter, R and Walker J.E (1994) Structure at 2.8-Angstrom resolution of F1-ATPase from bovine heart mitochondria. *Nature* **370**, 621–628.

Bowsher, C., Steer, M. & Tobin, A. (2008) *Plant Biochemistry* Garland Science, London.

Bryant, J.A. (2010) Replication of nuclear DNA. *Progress in Botany* **71**, 25–60.

Bryant, J.A. & Aves, S.J. (2011) Initiation of DNA replication: functional and evolutionary aspects. *Annals of Botany* **107**, 1119–1126.

Buchnan, B. *et al.* (2002) *Biochemistry and Molecular Biology of Plants.* ASPB, Rockville, MD, 81.

Caroll, C. (2011) The vacuole-not just an empty hole! https//aobblog.com/2011/12/The-vacuole-not-just-an-empty-hole/

Dambrauskas, G., Aves, S.J., Bryant, J.A., Francis, D. & Rogers, H.J. (2003) Genes encoding two essential DNA replication activation proteins, Cdc6 and Mcm3, exhibit very different patterns of expression in the tobacco BY2 cell cycle. *Journal of Experimental Botany* **54**, 699–706.

Duchesne, L.C. & Larson, D.W. (1989) Cellulose and the evolution of plant life. *Bioscience* **39**, 238–241.

Evans, D.E., Bryant, J.A. & Hutchison, C.J. (2004) The nuclear envelope: a comparative overview. In Evans, D.E., Hutchison, C.J. & Bryant, J.A. (eds), *The Nuclear Envelope*, BIOS, Oxford.

Evans, D.E., Shvedunova, M. & Graumann, K. (2011) The nuclear envelope in the plant cell cycle: structure, function and regulation. *Annals of Botany* **107**, 1111–1118.

Fiserova, J., Kiseleva, E. & Goldberg, M.W. (2009) Nuclear envelope and nuclear pore complex structure and organization in tobacco BY-2 cells. *Plant Journal* **59**, 243–255.

Francis, D. (2007) The plant cell cycle – 15 years on. *New Phytologist* **174**, 261–278.

Francis, D. (2011) A commentary on the G_2/M transition of the plant cell cycle. *Annals of Botany* **107**, 1065–1070.

Fry, S.C., Franková, L. & Chormova, D. (2011) Setting the boundaries: primary cell wall synthesis and extension. *The Biochemist* **33**, 14–19.

Hussey, P.J., Ketelaar, T. & Deeks, M.J. (2006) Control of the actin cytoskeleton in plant cell growth. *Annual Review of Plant Biology* **57**, 109–125.

Jurgens, G. (2005) Cytokinesis in higher plants. *Annual Review of Plant Biology* **56**, 281–299.

Lea, P.J. & Leegood, R.C. (1999) *Plant Biochemistry and Molecular Biology.* Wiley, Chichester.

Raven, P.H., Evert, R.F. & Eichchorn, S.E. (2005) *Biology of Plants* 7th ed. W.H. Freeman, New York.

Ulvskov, P. (ed.) (2010) Plant Polysaccharides – Biosynthesis and Bioengineering. *Annual Plant Reviews, Volume 41,* Wiley-Blackwell, Oxford.

Wheeler, GL, Jones MA and Smirnoff (1998) The biosynthetic pathway of vitamin C in higher plants. *Nature*, **393**, 385–389.

Wightman, R. & Turner, S. (2011) Digesting the indigestible – biosynthesis of the plant secondary wall. *The Biochemist* **33**, 24–28.

CHAPTER 3

Genes, Gene Expression and Development

3.1 Genes

3.1.1 Introduction

The purpose of this chapter is to give an overview of gene expression, of the main ways in which it is regulated and some of the factors involved in that regulation. Over recent decades, increasing knowledge has revealed that these are very complex topics, but at the centre there remains a simple concept, conceived initially, by two French scientists, Boivin and Vendreley as early as 1947. 'DNA makes RNA and RNA makes protein'[i].

This has become known as the *central dogma* of molecular biology and it describes information flow in gene expression. Of course, it is shorthand; DNA does not *make* RNA and neither does RNA *make* protein. In longhand we would say that the coded information in DNA is transcribed into RNA and the coded information in RNA is translated (decoded) to provide the instructions to make proteins.

We now know that the picture is much more complex than this. Genetic information flows from DNA to DNA in the process of replication (as Watson and Crick pointed out in one of their two famous 1953 papers (Watson & Crick, 1953)). It can flow from RNA to DNA, as in retroviruses, and from RNA to RNA in viruses with RNA genomes. We also know that most of the RNA in a cell does not have an information transfer (coding) function. We know, too, that the overall process can be controlled in many different ways. Nevertheless, the simple concept

represented by the central dogma remains at the heart of our understanding. So we now look at how these processes work in plants, starting with a consideration of the genetic material itself, DNA.

3.1.2 DNA, genes and chromatin

In prokaryotes, nearly all the DNA consists of genes and of the regulatory sequences by which gene transcription is controlled. However, this is not so in multicellular eukaryotes. Consider the data in Table 3.1.

It is clear that there is some relation between genome size and the complexity of the organism. Prokaryotes have less DNA than unicellular eukaryotes and, among the eukaryotes in general, complex organisms have more DNA than simple organisms. We might expect this intuitively. We could not envisage an angiosperm plant functioning with the same number of genes as a

Table 3.1 Genome sizes in a range of organisms.

Species	Taxonomic group	Genome size (base-pairs of DNA)
Escherichia coli	Bacteria	4.2×10^6
Saccharomyces cerevisiae, yeast	Fungi	1.3×10^7
Drosophila melanogaster, fruit fly	Insects	1.8×10^8
Fugu rubripes, puffer fish	Fish	3.7×10^8
Homo sapiens, human	Mammals	3.2×10^9
Arabidopsis thaliana, thale cress	Plants	1.2×10^8
Vicia sativa, field bean	Plants	2.25×10^9
Vicia faba, broad bean	Plants	1.3×10^{10}
Fritillaria assyriaca, fritillary	Plants	1.2×10^{11}

[i]Boivin, A and Vendreley, R (1947). Experentia 3, 32–34. Note that the phrase is often attributed mistakenly to James Watson and/or Francis Crick.

Functional Biology of Plants, First Edition. Martin J. Hodson and John A. Bryant.
© 2012 John Wiley & Sons, Ltd. Published 2012 by John Wiley & Sons, Ltd.

bacterium. It seems obvious that very complex organisms need more genes than very simple organisms. In general this expectation is correct but, when the detail is examined, the picture is not quite so clear.

The first point to make here concerns the actual amounts of DNA. How much DNA does an organism actually need to fulfil its genetic requirements? Certainly not as much as many of those listed in the table actually possess. Indeed, with the exception of the simplest organisms, the DNA amounts are far in excess of the genetic needs, even when allowance has been made for the complexity of the genes themselves and for the DNA sequences involved in regulating genes. This excess of DNA over the likely coding requirements in a genome has been confirmed very clearly for those plants and animals whose genomes have been sequenced. This is known as the *C-value paradox* (the amount of DNA in the single copy of a genome is the C-value). But it is actually even more paradoxical. Within some groups, the C-value varies little between species, as is seen in birds and mammals. In plants (and amphibians), however, there is very extensive variation amongst the species in the group.

Why does one plant species 'need' thousands of times more DNA than another? Why does a particular plant species need so much more DNA than a mammal? It is a paradox indeed. However, three major points are clear. The first is that, within a group (e.g. flowering plants), many of the larger genomes have arisen by a doubling during evolution of the whole genome. Second, gene duplication has occurred during evolution. Many genes are present as multigene families, often with differences in expression patterns between different family members. In other instances, gene duplication has led to subtle divergences in gene function, exemplified by the genes that encode proteins involved in the initiation of DNA replication[ii] (see Chapter 2, section 2.13). Further, certain genes are present as many copies; for example, most angiosperm genomes contain several thousand copies of the genes encoding the ribosomal RNAs.

However, most of the variation in genome size is based not on genes but on sequences that have no direct coding function. Furthermore, many of these non-coding sequences are repeated within genomes and, for

some sequences, this amounts to millions of repeats. So, although the number of potential genes has increased by gene duplication and genetic variation as more complex organisms have evolved, much of the variation between organisms has been caused by extensive amplification of non-coding sequences. This is beautifully illustrated by comparing two *Vicia* species, between which there is a nearly six-fold difference in DNA amounts (Table 3.1). These closely related species have similar numbers of genes; the difference between their genomes is almost entirely based on non-coding repetitive DNA sequences, as discussed below.

The excess of DNA over the apparent coding needs is vast in many angiosperms, amounting to several orders of magnitude. What is its function? It is not uncommon to hear the excess DNA referred to as 'junk DNA'. However, this is a misleading term. The location of a particular type of extensively repeated sequence (of about 180 base pairs), known as *satellite DNA* at the centromeres and telomeres, for example, suggests a structural function. The telomeres are the ends of chromosomes at which a modified replication mechanism exists to prevent chromosome shortening, and the centromeres are the structures by which chromosomes are pulled apart during cell division (see Chapter 2, section 2.13).

The involvement of particular DNA sequences in the structural features of chromosomes is another example of the 'many-sidedness' of this remarkable molecule. However, for much of the excess DNA in a genome, there is still no clear idea as to its function. This becomes more apparent on more detailed analysis of differences in the genomes of closely related species.

Taking the genus *Vicia* as an example, there is a greater than ten-fold difference between the sizes of the smallest and largest genomes, whereas the haploid chromosome number varies only between $n = 5$ and $n = 7$. Variation in genome sizes arises almost entirely from variations in repetitive DNA sequences, including genes encoding ribosomal RNA, 'satellite' DNA, simple-sequence repeats, **transposons** (mobile DNA sequences) and **retroelements**.

Some of these repetitive DNAs, for example the rRNA genes and satellite DNA, occur as long tracts containing many sequence repeats next to each other. Others, especially the retroelements, are scattered through the genome. These retroelements appear to have arisen by reverse transcription of RNA, including mRNA

[ii] As discussed by one of us in Bryant, J. & Aves, S. (2011) Initiation of DNA replication: functional and evolutionary aspects. *Annals of Botany* **107**, 1119–1126.

molecules and retroviruses, followed by movement and amplification. Many of them are mobile within the genome and are thus known as **retrotransposons**[iii].

In *Vicia*, retrotransposons comprise between 20 and 45 per cent of the genomes[iv] in different species and the copy numbers of different types vary extensively between species. Thus, comparisons between *V. faba, V. melanops* and *V. sativa* show that retrotransposons of the Ty3-gypsy[v] type are very abundant and comprise 18–35 per cent of the genomes of these three species. Ty1-copia group retrotransposons are less numerous, while LINEs (long-interspersed nuclear elements) are the least abundant. At the other end of the range, the very large (22–25,000 base pairs) *Ogre* retroelement[vi], which is a plant-specific member of the Ty3-gypsy group, exhibits hugely variable copy numbers between species. For example, in *V. pannonica* ($1C = 6.6 \times 10^9$ bp), there are 1×10^5 copies, comprising 38 per cent of the genome, while in *V. faba* ($1C = 1.33 \times 10^{10}$ bp), there are between 100 and 500 copies, comprising on average 0.1 per cent of the genome.

Why plant genomes have accumulated so many retrotransposons in their genomes and why the copy numbers are so variable is not clear. It is a real challenge to understand these evolutionary changes in genomes since the emergence of angiosperms back in the late Jurassic or early Cretaceous (Chapter 1, section 1.7). Some of the mysteries of DNA still remain unsolved.

The size of and composition of plant genomes is thus a matter for continuing amazement, not least when the lengths of the DNA molecules themselves are considered. In *Vicia faba*, for example, the mean length of the DNA double helix per chromosome is about 750 mm, yet 12 chromosomes are packed into a diploid nucleus of 10–15 μm diameter. In order to achieve this, the DNA is (as in all eukaryotes) coiled around barrel-shaped structures called *nucleosomes*.

First discovered in 1974, a nucleosome consists of a complex of small basic proteins called histones, namely two molecules each of histones H2A, H2B, H3 and H4. About 150 base pairs of DNA are wrapped round each nucleosome, and there is between 50 and 70 bp of linker DNA between nucleosomes, giving a structural repeat length of 200–220 base pairs (Figure 3.1).

On the outside of this structure there is one molecule of histone H1, the linker histone that holds the nucleosomes closely together. Although nucleosome-associated DNA, with its characteristic 'beads on a string' structure (Figure 3.1), is relatively easily extractable from cells, it probably does not often exist in this form in the nucleus. The nucleosome-associated DNA is coiled into a 'solenoid' structure with six nucleosomes per turn to form the *30 nm chromatin fibre*. This structure is further 'tightened' when the chromosomes condense for cell division (Chapter 2, section 2.13).

In the nucleus, some regions of chromatin are more compacted than others. These are the regions of *heterochromatin*. Some heterochromatin is a permanent feature of cells and is often referred to as constitutive heterochromatin. It includes the centromeric and telomeric chromatin (both of which are, as already mentioned, rich in the repetitive DNA known as satellite DNA). It contains no genes and was long assumed to be transcriptionally inactive. However, we now know that some transcription occurs in constitutive heterochromatin, and it appears that the RNA molecules are involved in the epigenetic modification of chromatin (see sections 3.4.2 and 3.4.4). Other regions of chromatin may be facultative heterochromatin, compacted in particular cell types and presumed to involve gene domains that are not active in those cells. The remainder of the chromatin, *euchromatin*, is less compacted and it is assumed that this less compacted structure allows access of the enzymes and other proteins involved in gene expression, as discussed more fully in the next section.

Examination of the distribution of chromatin within the nucleus shows that it is anchored to the nuclear

---

[iii]Transposons are genetic elements that are mobile within the genome; their presence was first deduced by Barbara McClintock, working on maize genetics in the 1950s. Retrotransposons are copied from an RNA molecule whereas the 'classical' DNA transposons, such as the *Ac/Ds* elements in maize, can amplify and move without going via an RNA copy.

[iv]In maize, about 50 per cent of the genome consists of retrotransposons, making up the majority of genome-wide repeated sequences.

[v]The nomenclature is a little confusing. *Ty* stands for transposon-yeast; *gypsy* and *copia* are Ty-type elements first described in *Drosophila*. LINES – long interspersed nuclear elements – and SINES – short interspersed nuclear elements – are sometimes called *retroposons* rather than retrotransposons because they lack the terminal repeat sequence characteristic of the latter. Instead they have a poly(A) sequence.

[vi]Transcripts of *Ogre* retroelements have been detected in several plants. However, it is not clear how often they are translated, if at all.

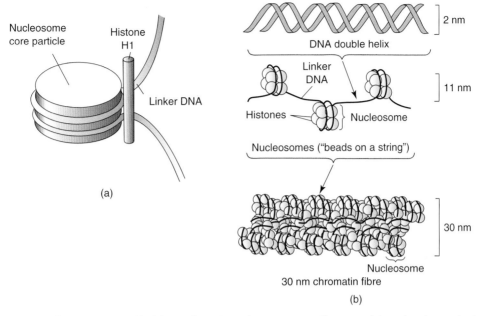

Figure 3.1 Structure of chromatin. (a) Simplified diagram illustrating nucleosome structure. The core particle consists of two molecules each of histones H2A, H2B, H3 and H4. Histone H1 is on the outside of the particle. This is based on a diagram in Raven *et al.* (2003). (b) DNA is wrapped round nucleosomes to form the 'beads on a string' structure (note: histone H1 is omitted from this diagram for the sake of clarity). The beads on a string structure is further coiled to make the 30 nm 'solenoid' fibre.

matrix (Chapter 2, section 2.7) by a protein that binds to AT-rich tracts of DNA known as matrix-attachment regions (MARs). Between the MARs, the chromatin is looped out, each loop containing between 5 and 200 thousand base pairs (kbp) of DNA.

The loops contain genes that are potentially active and are thus known as functional domains, with the borders of each functional domain marked by the MARs. Within domains, the chromatin structure of active genes is much more open than in the inactive genes. This can be shown experimentally because these open regions are vulnerable to DNase I (they are called DNase I hypersensitive regions). The more open structure is caused first by *remodelling* of the nucleosomes. This involves changes in the interactions between the histone proteins, so that the structure becomes much looser (observable as a marked increase in the size of the nucleosome) and the interaction with DNA is also looser. The second is *repositioning* of the nucleosomes by sliding or displacement so that short tracts of DNA are not associated with histones. These mechanisms then allow access to the DNA by the regulatory and transcriptional proteins.

3.2 Gene expression

3.2.1 The basics

Genes are transcribed into RNA by DNA-dependent RNA polymerases, of which plants contain several types. Detailed discussion of the biochemistry of the

Table 3.2 DNA-dependent RNA polymerases.

Type of RNA polymerase	RNA product(s)
Polymerase I*	25S, 18S and 5.8S ribosomal RNAs[vii]
Polymerase II	mRNA, some small nuclear RNAs
Polymerase III.1	5S ribosomal RNA
Polymerase III.2	tRNAs
Polymerase III.3	Some small nuclear RNAs and small cytoplasmic RNAs

*Located in the nucleolus

[vii]Note that the ribosomal RNAs are delineated here by their sedimentation coefficients in Svedberg units (S). The molecular weights of these molecules are about $1.3 \times 10^6, 0.7 \times 10^6$ and 5.4×10^4 respectively. The molecular weight of the 5S rRNA is about 4.0×10^4.

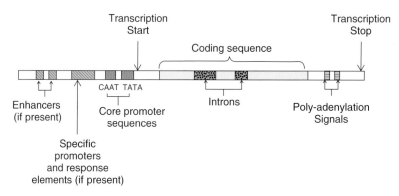

Figure 3.2 General structure of a plant gene.

polymerases lies outside the scope of this chapter, but we should note that each is specific for particular types of RNA, as shown in Table 3.2.

Plant RNA polymerases are unable to interact directly with their DNA template as they require the presence of transcription factors already bound to the DNA. We can illustrate this by reference to a 'housekeeping' gene (i.e. a gene whose product is required in all cells). In all genes, there is a **promoter** which lies 'upstream' from the transcription start site (Figure 3.2).

The key element of the promoter in many house-keeping genes is the *TATA box*, often associated with a *CAAT box* a few nucleotides further upstream; this is recognized by the TATA-binding protein, which is part of a *general transcription factor*. The transcription factor actually consists of about 12 different subunits, which are assembled together in stages after the TATA box is bound. During this process, the RNA polymerase (in this example, RNA polymerase II), which itself consists of several subunits, is recruited to the template. The largest subunit of RNA polymerase II is then phosphorylated by a protein kinase which is one of the subunits of the transcription factor. This allows the polymerase to move away from the transcription factor and thus initiate transcription (this is sometimes referred to as *promoter escape*). More proteins – the transcription elongation factors – are then recruited.

In general, plant genes, like those of all eukaryotes, are long because of the presence of introns (see next section), and one of the roles of the elongation factors is to assist processivity (i.e. to make sure that polymerase completes the job!). Elongation factors are also involved in the modifications of nucleosome structure (see section 3.1.2), ensuring that the polymerase has access to the template.

Soon after transcription is under way, the free 5' end of the nascent mRNA (pre-mRNA) is modified by the addition of methyl-GTP via an unusual 5'$^-$5'triphosphate linkage. This is the **cap**, which will later be involved in the binding of mRNA to the ribosome. The other end of the pre-mRNA molecule is also modified. As the polymerase complex approaches the end of the gene, it copies into the mRNA the *polyadenylation signal sequence*. This is recognized and bound by another group of proteins, which first displace the polymerase from the template, thus terminating transcription, and also recruit poly(A) polymerase, which adds a *poly(A) tail* to the pre-mRNA.

Poly(A) tails, which vary considerably in length between different mRNAs, occur on all eukaryotic mRNA molecules except those encoding histones. Although the existence of poly(A) tails has been known for many years, their function is still not clear. Some have proposed that they are involved in the initiation of translation, while other scientists point out the short-lived nature of histone mRNAs, with no poly(A), compared with the long tails on very stable mammalian globin mRNA. This suggests that the poly(A) may be involved in the stability of the mRNA, even though some longer-lived messages have short poly(A) tails. Opinions are thus divided.

However, the pre-mRNA is not yet ready to be exported to the cytoplasm; discoveries made with mRNA in the mid and late 1970s revealed what was then a very surprising feature. The sequences of genes and of the pre-mRNA

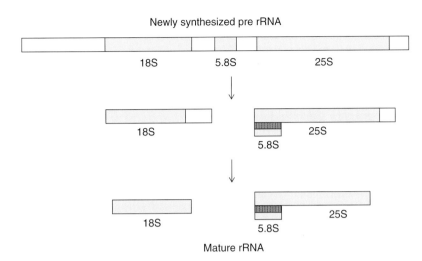

Newly synthesized pre rRNA

Mature rRNA

Figure 3.3 Diagram of the processing of ribosomal RNA. Note that in the mature rRNA, the 5.8S molecule is hydrogen-bonded (base-paired) to the 25S molecule (indicated in the diagram by upright lines between the two).

molecules transcribed from them (which are immediately capped and poly-adenylated, as described earlier) are longer, sometimes very much longer, than the coding sequences of the equivalent mRNA molecules. The excess length in the sequence of a gene (and the corresponding pre-mRNA) is arranged in discrete sections along the gene; the coding sequence is interrupted.

The interrupting sequences are called *introns* and the sequences represented in the final mRNA are called *exons*. Introns are a feature of nearly all eukaryotic genes, and in some genes the total length of introns far exceeds the length of joined-up exons[viii]. Thus the longest plant gene intron known is about 70,000 bases long, although most are between 80 and 140 bases in length.

In plant pre-mRNA, the introns tend to be AU-rich and, as in all eukaryotes, there are conserved sequences at the intron-exon junctions. The removal of the introns and the rejoining of the RNA is catalysed by the **spliceosome**, a complex consisting of four different small nuclear RNAs (snRNAs) and associated proteins. We note in passing that introns also occur in pre-transfer RNA; the ribonuclease that is involved in intron removal from

[viii]Actually, the existence of introns and the extent to which they interrupt some genes is, if we stop to think about it, still very surprising. It is just that we have become, over the years, familiar with this strange feature of eukaryotic genes.

pre-tRNA, RNase-P, contains an RNA molecule as an essential component.

Finally, the mRNA is exported as a ribonucleoprotein complex via the nuclear pores to the cytoplasm, where it directs the synthesis of protein on the ribosomes. The mRNA first binds to the small ribosome subunit to form the translation initiation complex, and from there the complete ribosome is assembled.

3.2.2 Synthesis of ribosomal RNA

Although a discussion of RNA synthesis in gene expression tends to focus on mRNA, most of the RNA in a plant cell is non-coding, and the bulk of that, by mass at least, is ribosomal RNA. Ribosomal RNA is also processed after synthesis and, indeed, this was discovered before the discovery of introns. The DNA coding sequences for the 25S, 18S and 5.8S RNAs form a single transcription unit – the rRNA gene – which is extensively repeated in a tandem array in the nucleolus. Each gene is copied by RNA polymerase I into the long precursor RNA molecule. This is processed in a stepwise manner by a specific ribonuclease, RNase MRP, cutting at specific sites (Figure 3.3). As with the nuclease involved in processing pre-tRNA, this RNase contains an RNA molecule. Other small nuclear RNAs are also involved. The RNA that is trimmed from the precursor is rapidly degraded; it does not accumulate in the cell.

The genes encoding the 5S ribosomal RNA are also extensively repeated, but they are not located in the nucleolus. They are transcribed by RNA polymerase III.1 and the product is 5S RNA itself. This is very unusual among eukaryotic RNAs, in that no post-transcriptional processing occurs.

Before leaving the nucleus, the ribosomal RNAs are assembled, together with the relevant proteins, into the two subunits of the ribosome. 25S RNA (with 5.8S RNA hydrogen-bonded to it by base-pairing) and 5S RNA are the RNA components of the large subunit, and 18S RNA is the RNA component of the small subunit. The subunits leave the nucleus separately and do not combine until after mRNA has bound to the small subunit.

3.2.3 Many sorts of RNA

As noted in the previous section, most of the RNA in a cell is not directly involved in the transfer of genetic information – it is non-coding RNA. The following are the major classes:

• *Ribosomal RNA*: 25S, 18S, 5.8S and 5S – the RNA components of ribosomes.
• *Transfer RNA*: transfer RNAs are the 'adapter' molecules that form an informational bridge between the order of nucleotides in mRNA and the order of amino acids in the corresponding protein. Each type can carry a specific amino acid and has the 'anti-codon' sequence that corresponds to one of the codons for that amino acid.
• *Small nuclear RNAs* (snRNAs): involved in processing the ribosomal RNA precursor and pre-mRNA (some writers divide this group into two classes: small nucleolar RNA, involved in rRNA processing; and small nuclear RNA, involved in mRNA processing).
• *Nuclease-associated RNAs*: the RNA components of the ribonucleases involved in processing pre-rRNA and pre-tRNA.
• *Small cytoplasmic RNA*: part of the signal-recognition particle involved in uptake of nascent proteins into the ER.
• *Micro RNAs*: a very heterogeneous class of small RNA molecules, each of which is complementary to a sequence in a particular mRNA species (technically then, they carry some genetic code, but it is not used in protein synthesis). Similar molecules are synthesized in the cytoplasm in a process that uses mRNA as a template. All are involved in gene regulation (see section 3.4).
• *Intronic RNAs*: some introns that are spliced out of pre-mRNA molecules have regulatory functions in the cell. An example is the RNA known as 'COLDAIR', which is involved in the vernalization response (see Chapter 8, section 8.4.3).

3.2.4 Protein targeting to the endomembrane system

Details of the mechanisms of protein synthesis itself lie outside the scope of this chapter. However, following the discussion of organelles and membranes in the previous chapter, it is important to understand how proteins reach the right cell compartment. We start by considering the ER and associated components of the endomembrane system (Figure 3.4).

Proteins destined for the lumen of the ER (with possible onward transport to the Golgi, tonoplast or plasma membrane) contain at their N-terminus (the first part of the protein to be made) a 15–30 amino acid sequence called the **signal peptide**. When the growing protein consists of about 70 amino acids, it sticks out from the ribosome and the signal peptide can then be recognized and bound by the *signal recognition particle* (SRP). This is an interesting structure consisting of a small RNA molecule[ix], to which are bound six different protein molecules.

Binding of the SRP has two effects: first, translation is temporarily halted; and second, the ribosome is moved to the surface of the ER (thus creating 'rough' ER), where the SRP is recognized by the *SRP receptor*. This releases the SRP and the nascent protein is inserted into the channel of the translocation complex. Translation is re-started and the signal peptide is removed. The growing protein chain eventually reaches the lumen of the ER, and finally the whole protein is extruded into the lumen. It is fascinating to note that, although this pathway as described here relates to plants and other eukaryotes, a very similar pathway for signal recognition – including the use of an RNA molecule in a signal recognition particle – exists in bacteria. In these prokaryotic organisms, the pathway is used for the export of proteins.

3.2.5 Protein targeting to the nucleus

As was noted earlier, the ER is continuous with the outer envelope of the nucleus and ribosomes are often seen on the outer envelope. However, this does not mean that

---

[ix]It is actually about 300 nucleotides long, so, although classified as a small cytoplasmic RNA, it is of moderate length.

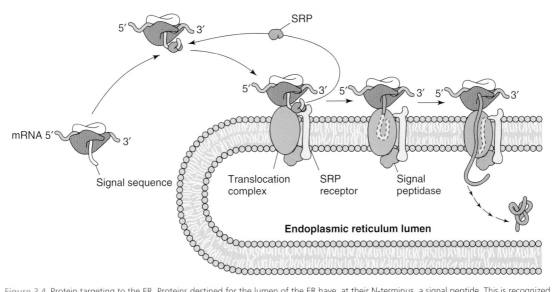

Figure 3.4 Protein targeting to the ER. Proteins destined for the lumen of the ER have, at their N-terminus, a signal peptide. This is recognized and bound by the signal-recognition particle (SRP), temporarily halting synthesis of the protein. The SRP docks onto the SRP receptor in the ER membrane and protein synthesis is resumed. As it is being synthesized, the protein is threaded through the translocation complex into the lumen of the ER. During this process, the signal peptide is removed by signal peptidase.
From Buchanan B. *et al.* (2002) Biochemistry and Molecular Biology of Plants, ASPB, Rockville, MD, p 181.

proteins are targeted to the nucleus in the same way as they are to the ER. The mechanism is very different, but it does include the use of a particular amino acid sequence, the *nuclear localization signal* (NLS) in the structure of the protein. NLSs are not necessarily located at or near the N-terminus of protein; indeed, their position in the protein chain varies considerably between proteins.

There are three types of NLS in plants, of which the commonest is the bi-partite type. These consist of two short regions containing basic amino acids (i.e. arginine, histidine or lysine), separated by a spacer of up to 10 amino acids (Figure 3.5).

The other two types are the SV-40 virus-type monopartite NLS, which consists of a run of five basic amino acids, and the yeast Matα2 type, in which a run of four basic amino acids is interrupted by three hydrophobic amino acids.

MATKRSVGLKEADLKGKRV
MAVKKSVGSLKEADLKGKR

Figure 3.5 Two N-terminally located plant bi-partite nuclear localization signals. The basic amino acids arginine (R), histidine (K) and lysine (L) are shown in red*. Note that both these NLSs also have basic amino acids in the spacer region. * The full list of single letter abbreviations for amino acids is given in the Glossary.

Transport of NLS-containing proteins into the nucleus first requires recognition of the NLS and the binding of the NLS-containing protein by two cytosolic proteins, *importin-α* and *importin-β*. Importin-α binds the NLS itself, while it is importin-β that docks with the nuclear pore complex (see Chapter 2, section 2.7). The cargo is only carried through the pore in the presence of GTP and a small GTPase, *Ran*. Ran hydrolyses the GTP, remaining bound to the resultant GDP, which enables the *Ran*GDP plus the importins, carrying their cargo, to traverse the pore complex (Figure 3.6).

Inside the nucleus, *Ran* exchanges its GDP for a GTP and is now able to bind to importin-β. This binding causes dissociation of the importins from the cargo, which is now safely inside the nucleus. Importin-β-*Ran*GTP and importin-α then leave the nucleus by a mechanism that is still not entirely clear, although for importin-α it appears that an export protein may be involved.

Although we now understand at least the essential features of targeting to the nucleus, some puzzles remain. First, there are nuclear-located proteins that do not possess an NLS. One of the proteins involved in the initiation of DNA replication is an example. Without an NLS, they cannot be recognized by importin-α. It is currently

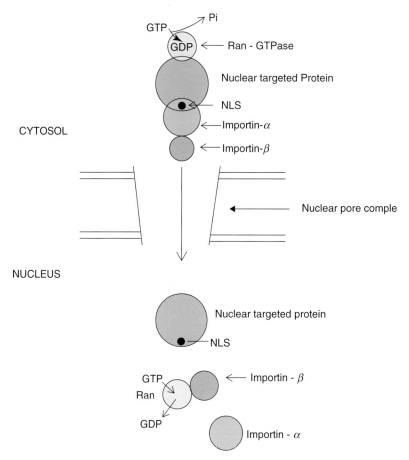

Figure 3.6 Diagram illustrating transport of proteins through the nuclear pore complex. A protein carrying a nuclear localization signal (NLS) is recognized by importin-α (which binds to the NLS) and importin-β (which docks at the nuclear pore complex (NPC)). Transport through the pore can only occur in the presence of the *Ran*-GTPase, which hydrolyses its bound GTP to GDP and Pi. In this form, the *Ran* facilitates the transport of the transport complex through the pore. Inside the nucleus, *Ran* swaps its bound GDP for a GTP. This enables it to bind directly to importin-β, leading to the dissociation of the transport complex. Note: the exact mode of association of *Ran* with the transport complex prior to transit of the NPC is not clear. In this diagram, its positioning is for the sake of clarity, not a specific indication of actual *in vivo* position.

thought that they must be 'piggy-backed' into the nucleus by an NLS-containing protein.

Second, there are many cytosolic proteins that possess an NLS; why is the NLS not functional in these proteins? Third, there are '*moonlighting proteins*' – proteins that have more than role. Phosphoglycerate kinase (PGK) is an example, with a major role in the cytosol in respiration and a minor role in the nucleus as an accessory protein in DNA replication or repair. PGK carries a bi-partite NLS at the N-terminus (Figure 3.7), but only about 10 per cent of the protein enters the nucleus.

A clue to this type of partitioning comes from study of hormone responses. In responses to several hormones,

phosphorylation status of NLS-containing proteins determines whether or not they actually enter the nucleus (see section 3.6).

3.2.6 Peroxisomes

Peroxisomes, as described in Chapter 2 (section 2.10), are organelles bounded by a single membrane (contrasting with the double membranes of chloroplasts and mitochondria). They do not contain DNA, RNA nor a protein-synthesizing system. All their proteins (whether for leaf peroxisomes, 'general' peroxisomes or glyoxysomes) are synthesized on 'free' ribosomes in the cytosol and transferred to the peroxisome. This requires a

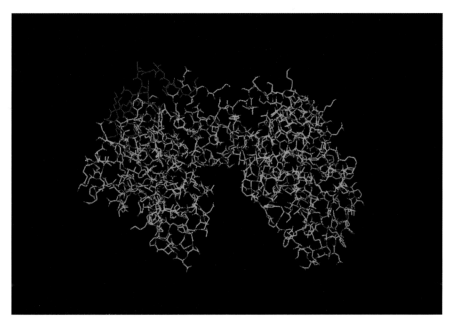

Figure 3.7 Computer-generated model of phosphoglycerate kinase (PGK) from pea. PGK is a 'moonlighting' protein with a primary role in the cytosol and a secondary role in the nucleus. The protein possesses a nuclear localization signal, shown in red on the computer model. Photograph by Kirsty Line, Richard Kaschula, Jennifer Littlechild and JAB.

peroxisome-targeting signal, of which two types have been discovered.

The commoner of the two (PTS1) consists of three amino acids – Ser-Lys-Leu – located at the C-terminus of the protein immediately downstream of a group of mainly basic amino acids. It is not removed from the protein during transport into the organelle.

The second type (PTS2) is a short sequence located at the N-terminus; this signal is removed on entry into the organelle. At the time of writing, the details are yet to be worked out, but it is clear that several of the *peroxin* (PEX) proteins are involved in the recognition of the targeting signals and the uptake of the proteins into the organelle. One protein – PEX14 – appears to be involved in overall regulation of import via both types of PTS, while other PEX proteins are involved specifically in the import of either PTS1 or PTS2 proteins.

3.3 Chloroplasts and mitochondria

3.3.1 Introduction

It has already been noted in Chapter 2 that, although chloroplasts and mitochondria possess their own genomes and protein-synthesizing machinery, they do not have anything like the genetic capacity to code for all the proteins that they need. This situation has arisen because of the transfer of genes from the initial symbionts to the nucleus of the host cell. The expression of the nuclear and organellar genomes, and the activities of the cytosolic and organellar protein-synthesizing systems, must therefore be tightly coordinated in regulating the development and metabolism of the organelles. In the next two sections, we describe the main features of the gene expression systems in chloroplasts and mitochondria, followed by discussion of protein-targeting mechanisms.

3.3.2 Chloroplasts

Chloroplasts possess circular genomes, similar in organization to those in bacteria but much smaller, typically between 130 and 160 kbp. The coding capacity is thus limited, especially when we consider that a significant proportion of the plastid genome of most angiosperms consists of two inverted repeats, which contain the same subset of genes but in opposite orientation. There are about 100 protein-coding genes (most of these proteins are involved in photosynthesis, thylakoid membrane organization or ribosome structure), plus the genes encoding four different ribosomal RNAs (similar to the RNAs of cytoplasmic ribosomes but 'bacterial' in size: 23S, 16S, 4.5S, 5S) and about 30 tRNA molecules.

Both strands of the DNA are transcribed, and there it at least one instance of trans-splicing of transcripts to make the final RNA product (a tRNA). A number of the genes are organized as *polycistrons* (another bacterial feature) that are transcribed to give a polycistronic mRNA. There are also about 20 introns, some of which are in genes located within polycistrons.

From all that we know about chloroplasts, it is not surprising that the DNA-dependent RNA polymerase responsible for gene transcription resembles closely the RNA polymerase of *Escherichia coli*. It recognizes bacterial-type promoter sequences and, like the bacterial polymerase, it consists of four subunits, three of which are encoded in the chloroplast genome and the fourth in the nuclear genome. But that is not the whole story. Rather surprisingly, chloroplasts contain a second RNA polymerase, consisting of just one protein chain, similar to the polymerases encoded in some bacteriophage[x] genomes. The gene is located in the nucleus and the function of the polymerase seems to be the maintenance of background levels of transcription, especially in non-green plastids.

The features of genome and gene organization mentioned above inevitably means that RNA processing is a feature of chloroplast gene expression, as it is of nuclear gene expression. For rRNA, the gene encodes all four rRNA species and the initial transcript is cut and trimmed to produce the mature rRNA.

However, there is another complication: between the 16S and 23S coding sequences are two tRNA genes, both of which are interrupted by introns of about 1 kbp in length. These are much longer than the introns that occur in pre-tRNA encoded in the nucleus and, indeed, are much longer than the tRNAs themselves. So, processing of the rRNA precursor releases the mature rRNAs and also two interrupted tRNA sequences which require intron removal. For mRNA transcripts, the processing may be equally complex. Unlike their bacterial counterparts,

chloroplast ribosomes cannot translate polycistronic messages. Polycistronic transcripts must therefore be processed to release the individual mRNAs and, again, this may require intron removal.

Messenger RNA is translated on the chloroplast ribosomes, which, notwithstanding their inability to use polycistronic mRNA, resemble closely the ribosomes of bacteria, both in size and in many of the biochemical details of protein synthesis. They are also inhibited by different compounds from those that inhibit cytoplasmic ribosomes. Thus, in the past, indications of the sites of synthesis of particular proteins were obtained by applying inhibitors specific for either cytoplasmic or chloroplastic ribosomes.

It has already been mentioned in this chapter, and in Chapter 2, that most of the proteins required for a fully functional chloroplast are encoded in the nucleus and synthesized on cytoplasmic ribosomes. Further, some of the structurally more complex proteins, including RNA polymerase, Rubisco (ribulose bisphosphate carboxylase-oxygenase, the CO_2-fixing enzyme – see Chapter 7, section 7.4) and the F_0F_1 ATP synthase, possess both nuclear and chloroplast encoded subunits. Thus, Rubisco consists of eight large (56 kDa) sub-units encoded in the chloroplast genome and eight small (14 kDa) sub-units encoded in the nuclear genome. This emphasizes that not only is transfer of proteins from the cytoplasm into the chloroplast required, but also that the two systems are coordinated.

Signal peptides (section 3.2.4), nuclear localization signals (section 3.2.5) and peroxisome-targeting signals have already introduced the concept of 'address labels' incorporated into the amino acid sequences of proteins. The same concept is applicable to nuclear-encoded proteins destined for the chloroplasts (Figure 3.8).

As with nuclear proteins, proteins destined for the chloroplast are made on 'free' ribosomes in the cytoplasm. They possess at their N-terminus a **transit peptide** which is between 40 and 50 amino acids in length. This transit peptide remains 'exposed' because the protein is held in an unfolded state by a '**chaperone**' protein (in this instance, chaperone Hsp70).

Uptake into the chloroplast occurs at sites where the outer and inner chloroplast envelopes are tightly pressed together; at these sites, proteins embedded in the outer envelope and proteins embedded in the inner envelope come together to form pores. These pore-forming

---

[x]Bacteriophages are viruses that infect bacteria.

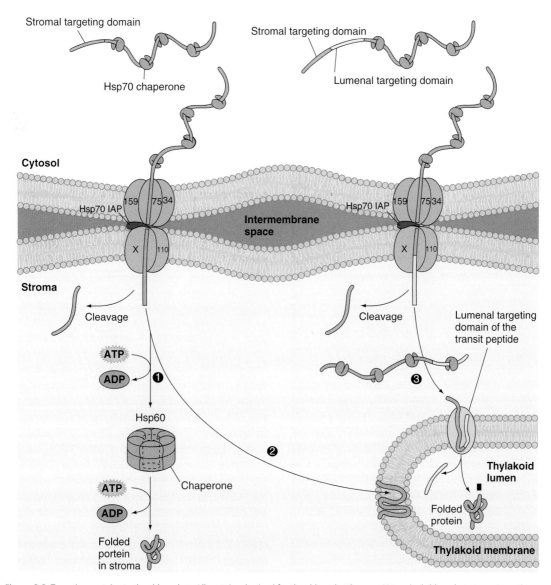

Figure 3.8 Targeting proteins to the chloroplast. All proteins destined for the chloroplast have an N-terminal chloroplast stroma targeting domain. The newly synthesized proteins are bound by the chaperone Hsp70 and 'escorted' to the proteinaceous pore complexes (which bring the outer and inner envelopes into contact). Once inside, the stromal targeting domain is removed. On the left side of the diagram, the protein possesses no further targeting domains and either folds to assume its active configuration or is inserted into the thylakoid membrane. The right side of the diagram illustrates a protein that has a second targeting sequence for insertion into the thylakoid lumen. These are escorted by Hsp70 to the thylakoid surface and then imported into the lumen by an import complex. The second targeting domain is removed when the protein reaches the lumen. Note: more details may be found in the text. From Buchanan, B. *et al* (2002) Biochemistry and Molecular Biology of Plants, ASPB, Rockville, MD, p 167.

proteins are known as the 'toc' and 'tic' proteins (toc – *translocon of outer envelope*; tic – *translocon of inner envelope*). The transit peptide of the newly-synthesized protein engages with the pore proteins in the outer envelope and the whole protein is taken through the pore. On entry into the chloroplast stroma, the signal peptide is removed and, from here, three different destinations are possible.

First, proteins that remain in the stroma are folded to give their correct final configuration (note: this may involve combination with chloroplast-encoded polypeptides already inside the chloroplast, as with Rubisco and RNA polymerase). Correct folding – and the formation of multi-subunit proteins – also requires chaperone proteins and is dependent on ATP hydrolysis.

Second, proteins that are destined for the thylakoid membrane are recognized by a *signal-recognition particle* (unlike its ER counterpart, this SRP does not contain an RNA molecule). Protein recognition depends on a sequence of about 18 amino acids in length, within a hydrophobic region of the protein. The SRP facilitates uptake, via an SRP receptor, into the membrane in a process that requires GTP hydrolysis. This is clearly a post-translational mechanism, but the same particle can also work co-translationally (as does the ER system) with proteins synthesized on chloroplast ribosomes. Assembly of complex proteins such as ATP synthase takes place within the membrane.

Third, proteins destined for the thylakoid lumen possess a second transit peptide which is exposed by removal of the main chloroplast transit peptide (as described immediately above). This thylakoid transit peptide is recognized by a protein complex that moves the protein through the thylakoid membrane into the lumen, where the second transit peptide is removed. The uptake process is dependent on a pH gradient across the thylakoid membrane and, for some proteins, it also involves ATP hydrolysis.

The fact that some complex proteins require subunits encoded in the nucleus and subunits encoded in the chloroplast reminds us again of the problem of coordination between the two systems. They must keep in step with each other. The extent of the problem becomes even clearer when we consider the two genomes. Within the one diploid nucleus possessed by the cell, the relevant genes are present only at low copy number – for many, only one copy per haploid genome. Chloroplasts, on the other hand, possess multiple copies of their genomes (the number varies at different stages of leaf development) and leaf cells contain many chloroplasts.

The imbalance is obvious. For the proteins containing both nuclear-encoded and chloroplast-encoded subunits, it is now clear that the nuclear-encoded polypeptides have a key regulatory role, working via a post-transcriptional control mechanism. The details vary between proteins but, in general, the presence in the chloroplast of the nuclear-encoded sub-unit(s) allows or stimulates translation of the mRNA(s) of the chloroplast-encoded sub-unit(s).

There are also more subtle controls. For example, if a situation arises in which the Rubisco large subunit starts to accumulate without being assembled into the mature enzyme, then it inhibits the translation of its mRNA – a process known as *autoregulation*.

However, the problem is wider than this. In considering chloroplast development, the expression of the nuclear genome is coordinated with the state of chloroplast development, whether that is the full, light-dependent development of a photosynthetic organelle or the development of non-green plastids such as amyloplasts.

Taking chloroplasts as a specific example, expression of nuclear genes encoding chloroplast proteins is responsive to the accumulation of chloroplast-encoded proteins, to the development of chloroplast membranes, to the accumulation of chloroplast pigments and to the rate of import of proteins into the chloroplast. This effect of chloroplasts on expression of nuclear genes is called *retrograde signalling*; details are still being worked out, but it is at least partly based on the activities of the nuclear *GENOMES UNCOUPLED* (*gun*) genes encoding the GUN proteins.

GUN1 (and possibly also GUN2 and GUN3) inhibits the activity of at least two transcription factors, thus down-regulating the genes controlled by those transcription factors. GUN4 and GUN5, however, simulate chlorophyll synthesis by interacting with one of the enzymes in the biosynthesis pathway, which tends to ameliorate the negative regulation imposed by GUN1. There is also evidence that the GUN pathways interact with cryptochrome signalling (see section 3.7.4) and with sucrose signalling. Furthermore, because different types of plastid occur in different types of cell or organ, or at different developmental stages, there must also be some interaction with these higher levels of regulation (see Chapter 4, section 4.12).

3.3.3 Mitochondria

Much of what has been said about the genetic systems of chloroplasts is also true of mitochondria, although, as will become apparent, there are some important differences between the two organelles. Mitochondrial genomes are larger than chloroplast genomes and are extremely variable in size between plant species (there is an approximately 13-fold range in size amongst the angiosperms). However, the larger size does not mean a greater genetic capacity.

Indeed, plant mitochondria contain fewer genes than chloroplasts. There are very few protein-coding genes, and most of these are involved in respiration – subunits of ATP synthase, of NADH dehydrogenase, of cytochrome oxidase and cytochrome b. Some of these genes contain introns, which are spliced out of the pre-mRNA molecule. Two ribosomal proteins are also encoded in the chloroplasts. There are genes encoding the three ribosomal RNAs (see below) and genes encoding tRNAs. The number of tRNA genes present appears variable; in some species there are just enough to support mitochondrial protein synthesis, but in others there are too few. Maize, for example, has just three tRNA genes in its mitochondrial genome, which means that tRNAs must be imported from the cytoplasm.

With so few genes, why are the genomes so large (and variable between species)? This is because of the presence of a large amount of non-coding DNA amongst the coding sequences. For example, in the 367 kbp mitochondrial genome of *Arabidopsis thaliana*, only 10 per cent is coding DNA. Indeed, the information density is less than in the nuclear genome. The amounts, sequences, genomic locations and arrangements of the non-coding DNA vary considerably between species. In addition to this, there has been re-arrangement of the order of genes during the evolution of different species, together with the acquisition of segments of chloroplast DNA (albeit non-functional).

Finally, in several plant species, including maize, tobacco and *Brassica campestris*, it has been shown that the mitochondrial genome exists as subgenomic circles. These may be assembled to give the full circular genome – the 'master circle'.

The ribosomes of mitochondria are larger than those of chloroplasts, approaching the size of cytoplasmic ribosomes. They contain only three types of rRNA: 26S, 18S and 5S. Although the sizes of these RNAs, and of the ribosomes, resemble those of the cytoplasm, the ribosomes are of the prokaryotic type, as discussed in the previous section.

Organization of the genes coding for the rRNAs varies between species. In maize, the genes for all three are located close together, but the 26S RNA is transcribed separately while the 18S and 5S coding sequences form a 'di-cistron' which is transcribed as one piece, followed by processing of the transcript. In rice, the 26S and the 18S-5S genes are separated by a gene encoding an ATP synthase subunit and a piece of chloroplast DNA (see above).

Although mitochondria have far fewer metabolic functions than chloroplasts, there is still a need to import several hundred proteins. Further, as with the chloroplast, several large multi-subunit proteins consist of both nuclear and organelle-encoded subunits. Import of proteins into plant mitochondria also relies on an 'address label', in this instance known as a *pre-sequence*.

The mechanism for uptake into the organelle is very similar to that already described for the chloroplast. The chaperone Hsp70 keeps the protein unfolded so that it can engage with the tom and tim proteins (*translocons of the outer and inner membranes*). After entry into the matrix (which requires an electrochemical gradient), the pre-sequence is removed and the protein is folded with the help of chaperone proteins. This again requires hydrolysis of ATP. Proteins that are destined for the mitochondrial inner membrane or the inter-membrane space have a second pre-sequence which is rich in hydrophobic amino acids. This second sequence is removed within the inter-membrane space once the protein has either embedded in the inner membrane or has crossed completely into the space.

3.4 Control of gene expression – switching genes on and off

3.4.1 Transcriptional control – general features

In the patterns of gene expression that we see in the development and life of a plant, there are many variants of the processes described above. Many genes have more specialized promoter sequences in addition to the basic promoters discussed in section 3.2.1. Specialized promoter sequences are recognized and bound by transcription factors which are specific for particular organs or phases of life, or which are involved in responses to hormones or to environmental factors. Some of these factors are negative transcription factors, effectively blocking access to the promoter. Several hormone-response pathways involve the removal of negative transcription factors.

The activities of many genes are also up-regulated by **enhancer** sequences, which may be adjacent to the promoter or may actually be some distance away (some enhancers are even located within introns). These are also recognized by transcription factors which, by bridging between the enhancer and the promoter, bring about a conformational change in DNA which increases the ability of the general transcription factor and/or RNA polymerase II to bind. Many enhancers are tissue or organ-specific or are involved in responses to hormones or environmental factors. As well as this type of up-regulation, genes may be down-regulated by sequences known as **silencers**, which are bound by negative transcription factors to prevent or reduce binding at the general promoter site.

3.4.2 Chromatin and transcription

We have already noted the existence in the nucleus of regions of condensed transcriptionally inactive chromatin known as heterochromatin. This shows that there is a relationship between chromatin structure and gene expression, and it introduces us to the concept of **epigenetics**. The term 'epigenetics' refers to a range of phenomena in which genes are kept switched on or off by modifications of DNA or of chromatin proteins. Some of these DNA modifications may be preserved through cell divisions, and those that are preserved through meiosis are effectively heritable.

The best known epigenetic modification is *DNA methylation*, which is a gene silencing mechanism. Cytosines are methylated in CpG dinucleotides and CpNpG trinucleotides. When DNA is replicated, one strand of each daughter molecule is methylated and one is not. The 'hemi-methylated' DNA is strongly recognized by the DNA methyltransferase enzyme, which thus methylates the other strand. In this way, methylation patterns can be carried through mitotic cell division. They can also

be carried through meiosis, but actually epigenetic patterns are mostly lost during meiosis, so that gametes have (mostly) unmethylated DNA.

Histone H3 may also be methylated, and the effect of this varies according to the site of methylation. Methylation of lysine at position 9, for example, provides a binding site for a protein called HP1, which induces tighter packaging of chromatin and hence repression of gene activity (see Chapter 8, section 8.4.3). Methylation of the lysine at position 4, on the other hand, is often correlated with *histone acetylation* in the same region of chromatin. Histone acetylation is well known as a mechanism for promoting gene expression. The effects of acetylation are to reduce the affinity of histones for DNA and to loosen the nucleosome structure, thus promoting more open and accessible tracts of chromatin.

3.4.3 Other levels of regulation

The picture that has been built up thus far is of gene regulation at the level of transcription – switching genes on or off or modulating the activity of genes. It might be expected that, in a complex multicellular organism, control at this level would be the most frequent way of controlling gene expression. That assumption is probably correct, but there are several other mechanisms, which vary between fine tuning and much coarser control.

A quick overview of the basic processes outlined in this section suggests several points at which the expression of a particular coding sequence may be modified:

• Alternative transcription start sites[xi], exemplified by the two forms of glutathione-S-transferase in *Arabidopsis*. Genomic analysis suggests that about 30 per cent of *Arabidopsis* genes have possible alternative transcription start sites.

• Alternative intron splice sites: examples include the genes in *Arabidopsis* that encode a group of serine and arginine-rich proteins, which are themselves involved in splice site selection.

• Alternative poly-adenylation sites giving different 3'UTRs (untranslated regions) in mRNA, exemplified by the gene encoding the Rubisco small subunit (section 3.3) in *Phaseolus vulgaris* (common bean). Sequences in UTRs are often involved in regulating mRNA degradation, so

---

[xi]Alternative transcription sites, alternative splicing and leaky ribosome scanning are all means of obtaining more than one protein from a single gene. This raises interesting questions about how a gene should be defined.

it is possible that this mechanism generates mRNAs of different stabilities.

• Alternative translation start sites (also known as *leaky ribosome scanning*). There are two examples here from chloroplasts and mitochondria. These organelles both possess a second type of RNA polymerase which is similar to those encoded by bacteriophages and which is encoded in the nucleus (see section 3.3.2). Leaky ribosome scanning produces the mitochondrial and the chloroplastic versions from the same mRNA. The same is true of the organellar DNA polymerases. This feature raises questions about gene transfer and/or loss during the evolution of the organelles.

• Some proteins require post-translational modification, including:

  • Processing to release the active protein from a precursor, as with proteases active during the mobilization of stored proteins during germination (see Chapter 4, section 4.11.2).
  • Glycosylation, as in cell wall proteins (Chapter 2, section 2.2.1).
  • Association with other proteins in order to become active, as exemplified by the cyclin-dependent kinases in cell cycle regulation (Chapter 2, section 2.13) and by assembly of Rubisco (section 3.3.2 and Chapter 7, section 7.4.5).
  • Reversible phosphorylation, exemplified by several proteins involved in regulation of the cell cycle (Chapter 2, section 2.13).
  • Reversible **sumo**ylation, in which a **s**mall **u**biquitin-like **mo**difier protein is attached via its C-terminal glycine residue to lysine in the target protein. Unlike the ubiquitin attachment which is involved in protein degradation (section 3.4.5), sumoylation increases protein stability and enhances particular protein-protein interactions, especially those involved in transport between cytoplasm and nucleus and those involved in the formation of transcription complexes (sumoylation of transcription factors usually leads to a down-regulation of transcription).

These mechanisms, coupled with those controlling transcription itself, thus provide for very tight control over gene expression and of the protein end-products. However, the question arises as to what happens to mRNA and protein molecules that are no longer needed after a gene has been switched off. We therefore now consider mRNA and protein degradation.

3.4.4 mRNA degradation

The stability of mRNAs is determined by a number of factors, including the length of the poly(A) tail (see section 3.2.1) and the sequence of the 3'UTR (see previous section). The role of the poly(A) tail probably involves a *poly(A) binding protein* (PABP), which requires a certain minimum length of Poly(A) for binding. It is suggested that, when binding by PABP is not possible, the mRNA becomes vulnerable to de-capping (see section 3.2.1). De-capped mRNA cannot be translated; it is no longer useful and is degraded. In respect of the 3'UTR, particular sequences mark the mRNA for degradation; one such sequence is the 40-base DST ('downstream element'), which is present in the 3'UTRs of some very unstable mRNAs. The presence of DST leads to accelerated de-polyadenylation and hence degradation of the mRNA via removal of the cap.

The mechanism of degradation just described might be termed a general mechanism. However, there exist mechanisms that are much more specific. The discovery of these started in the late 1970s and early 1980s, when several scientists, including one of us (JAB), found that plants possess an *RNA-dependent RNA polymerase* enzyme. Plants that are infected with RNA viruses (such as tobacco mosaic virus) may also possess such an enzyme, encoded in the virus genome. The plant's own enzyme is different from the virus-encoded enzyme and is present in plants that have never been exposed to a virus. The function of this polymerase, and hence the significance of the finding, remained completely unknown, so the topic was put to one side.[xii]

However, it was picked up again following some interesting results obtained in genetic modification experiments. It is possible to down-regulate expression of a particular gene by introducing a DNA molecule that encodes an mRNA which is antisense to the mRNA encoded by the target gene (see Box 3.1). The theory is that the antisense RNA base pairs with the mRNA, preventing it from being translated (and in fact the double-stranded RNA is actually degraded, as later became apparent). For many genes, the theory works in practice and, for several years, antisense techniques were used to

---

[xii]One of the reasons for lack of progress was that the techniques needed to make progress were not then available – an illustration of the interaction between technical advances and advances in knowledge.

Box 3.1 Genetic modification

The first successful plant genetic modification (genetic engineering) experiments were carried out in the early 1980s. They made use of a naturally occurring process in which *Agrobacterium tumefaciens* transfers a piece of DNA, T-DNA, from its Ti-plasmid (Ti = tumour-inducing) to the genome of a host plant. This normally causes crown gall disease (see Chapter 11.2, section 11.2.2) but, in order to use the Ti-plasmid for genetic modification, the disease-causing genes are removed. Indeed, the only T-DNA sequences needed for integration into host DNA are the short repeated border sequences, between which scientists can splice any gene that they wish to transfer.

From those first successful experiments, things moved on very fast. The use of specific promoters allowed the inserted gene to be expressed only in particular cells, tissues or organs, or in response to specific internal or external signals. It is even possible to send the protein encoded by the inserted gene into an organelle. New methods were developed for delivering the engineered T-DNA to the plant, including shooting the DNA into plant cells (although, it needs to be said, the bacterium is still widely used as a means of delivering the DNA to the recipient plant).

As well as adding new genes to plants or plant cells, it is also possible to switch genes off. This was originally done by 'antisense' technology, where a reverse version of the target gene is inserted; this leads to the synthesis of antisense mRNA, which hydrogen-bonds with the real mRNA, forming a double-stranded molecule which cannot be translated (and, indeed, is degraded). However, with the discovery of siRNA, it is now possible to switch off any gene by insertion into the plant of the appropriate siRNA sequence (this is often known as RNAi technology).

The first products from GM plants came on the market in the USA in the mid-1990s and, since then, the commercial use of GM crops has spread to nearly 30 countries – although, at the time of writing, those countries do not include the UK (see Chapter 12, section 12.4).

Although it is the agricultural/commercial applications of plant genetic modification that often come to mind first, these are actually the 'tip of the iceberg'. By far the most extensive use of GM, and of the many techniques that have spun off from GM, coupled with rapid advances in sequencing technology, in transcription analysis with microarrays, in protein analysis ('proteomics') and in the applications of information technology ('bio-informatics'), has been in research on genes and on gene function. Much of what is written about genes in this book would have been unknown, had GM and associated technologies not been developed.

investigate and modify gene function (e.g. in the work on tomato ripening in Donald Grierson's lab in Nottingham: Chapter 4, section 4.9).

It is also possible to introduce a second *bona fide* copy of an endogenous gene in order to increase the amount of gene product. However, for some genes it was noted that introducing a second copy of an endogenous gene (for example in an attempt to increase the intensity of flower colour) actually led to down-regulation of that gene. A gene could apparently be switched off by the presence of another active copy of the same gene. This is **co-suppression**, and it shows that it *is* possible to have too much of a good thing.

Investigation of the mechanism(s) behind co-suppression led to the discovery of a previously unknown phenomenon – **RNA silencing**. The first point to note is that double-stranded (ds) RNA molecules are degraded very efficiently in plant cells. The dsRNA may be viral dsRNA, or antisense RNA base-paired with mRNA or mRNA that has been copied by the plant's own RNA-dependent RNA polymerase (so it does have a function). The latter reaction, the copying of mRNA, is triggered by the over-expression of some, but no means all, genes.

Double-stranded RNAs are cut into small (21 nucleotides) pieces by a ribonuclease called *Dicer* (Figure 3.9). These short pieces of dsRNA are very powerful silencers and are known collectively as siRNA (small inhibitory or interfering RNAs). The dsRNA pieces associate with *Argonaute* (AGO) proteins in the RISC (RNA-induced silencing complex); the dsRNA pieces are denatured, and one strand base-pairs with a complementary region of the original viral or mRNA, which is then either degraded by the *Slicer* ribonuclease in the RISC or is blocked from being translated (Figure 3.9) Either way, the original RNA target is silenced. This discovery has led to development of a new technique in genetic modification: RNA interference is widely used to study the effects of silencing individual genes.

As mentioned earlier, an 'orphan' observation on RNA-dependent RNA polymerase might have provided an early clue as to the existence of this phenomenon. However, David Baulcombe at the University of Cambridge, undoubtedly one of the leaders in this exciting field, states that: '*the first RNA silencing paper may have been published as long ago as 1928*'.

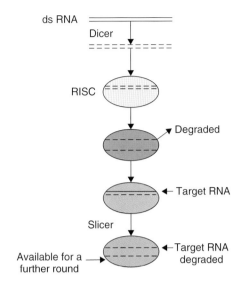

Figure 3.9 Diagram of an RNA silencing mechanism. Double-stranded RNA is a target for Dicer ribonuclease, which cuts the RNA into 21 bp pieces. The pieces associate with the AGO protein (not shown) in the RNA-induced silencing complex (RISC). Denaturation of the small dsRNAs releases the small inhibitory RNA molecules to base-pair with the target RNA (which, in the example shown here, will be the same as the first strand of the original dsRNA at the top of the diagram) to form a partially double-stranded molecule. This is a target for Slicer ribonuclease and is thus degraded.

The paper in question describes tobacco plants infected with tobacco ringspot virus, where the lower leaves showed extensive development of symptoms but the upper leaves remained healthy. Baulcombe suggests that the viral RNA evoked the Dicer-Slicer system in the upper leaves and thus inhibited viral gene expression. Furthermore, there are now indications that siRNA is mobile in the phloem and/or via plasmodesmata, providing a degree of systemic protection.

The silencing of excess mRNA or of viral RNA is not the whole story. Plant (and animal) cells contain small RNA molecules called *microRNAs* (miRNAs: see section 3.2.3) which are encoded in the genome. So many are now known that extensive miRNA databases have been established. They are complementary to short tracts of mRNAs encoded in the same genome and thus can silence endogenous genes, providing yet another level of genetic regulation. MicroRNAs are initially synthesized as an inverted repeat RNA which is able to form base-paired regions. These precursor molecules are cleaved by Dicer and the resulting 21–24 nucleotide-long RNAs then work as already described.

Finally, some miRNAs are involved in epigenetic silencing of genes because they can guide DNA methylation enzymes (see section 3.4.2) to specific coding sequences. Other RNAs, such as the intronic RNA, *COLDAIR*, may also be involved in gene silencing (Chapter 8, section 8.4.3).

3.4.5 Protein degradation

There are several types of protein degradation in plant cells. First, processing proteases, involved in protein transport and/or maturation, occur in several cell compartments. Second, several proteases are involved in the mobilization of stored proteins during germination (Chapter 4, section 4.11.2) and during senescence. In the latter, the proteases are often located in the vacuole, and their release exposes cellular proteins to their activity. A similar process occurs during necrotic cell death which may occur as a result of damage caused by herbivores, pathogens or environmental factors (see Chapters 9 and 11). Third, a specific set of caspase-like proteases is involved in programmed cell death (section 3.5).

However, except for those situations in which large-scale protein degradation occurs (germination, senescence), most proteins are degraded by large multi-protein complexes called **proteasomes**, which are located in the cytoplasm and the nucleus. Degradation of specific proteins *via* the proteasome pathway is a further mechanism involved in the control of gene expression. Indeed, Kate Dreher and Judy Callis at the University of California, Davis, state that: '*The ubiquitin-proteasome system has been implicated in the regulation of almost every developmental process in plants, from embryogenesis to floral organ production, probably through its central role in many hormone pathways.*'

Protein molecules destined for proteasome-mediated degradation are first 'tagged' by the transfer of a small protein, **ubiquitin**, to a lysine residue in the target protein. This is a complex, multi-step, ATP-dependent process which culminates in the formation of a peptide bond between the side-chain amino group of the lysine and the C-terminus of the ubiquitin (Figure 3.10). Any accessible lysine residue in the target protein can be ubiquinated and, once the first ubiquitin has been transferred to a particular lysine, more can be added to form a short oligo-ubiquitin chain. It is presumed that the more ubiquitin residues a protein carries, the stronger is targeting for degradation.

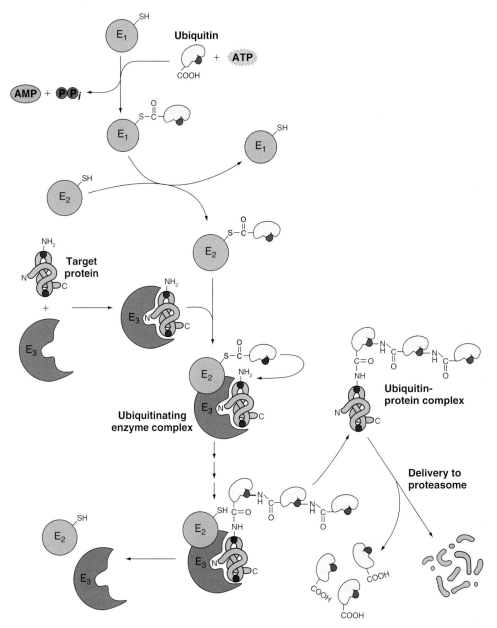

Figure 3.10 Diagram of the ubiquitin-proteasome system for protein degradation. Proteins that are destined for degradation are 'marked' by ubiquitin, a small (76 amino acids) protein. Ubiquitin is transferred by a ubiquitination complex to lysine residues in the target protein by a five-step process. Further ubiquitin residues may then be added in the same way. The marked protein is then delivered to the proteasome for degradation and the ubiquitin is recycled. Key E_1 = ubiquitin-activating enzyme; E_2 = ubiquitin-conjugating enzyme; E_3 = ubiquitin-ligating enzyme. These are the main members of the ubiquitination complex.
From Buchanan B. *et al*. (2002) Biochemistry and Molecular Biology of Plants, ASPB, Rockville, MD. p 451.

The ubiquinated protein is delivered to the proteasome, where it is unfolded and fed into a central channel in the proteasome to be degraded. The ubiquitin residues do not enter the channel, but are removed and thus recycled.

3.5 Molecular aspects of development

3.5.1 General points

It hardly needs to be said that the growth and development of a plant involves the differential expression of genes in different cells, tissues and organs. That process is controlled by a diverse range of interacting signals, including hormones, as discussed in this and subsequent chapters. One feature that emerges is that some genes are responsible for establishing and maintaining the identity of particular organs. Under the 'umbrella' activities of such genes, the genes whose products characterize a particular organ are active. We also note the control and balancing of activities at cell, tissue, organ and whole plant levels. The balance between apical and lateral growth is one example; the balance between division and expansion is another.

Another general feature of plant development is its plasticity. Most plants can recover routinely from herbivory (Chapter 11, section 11.2.1) or other forms of damage, provided this is not extremely severe. The loss of a shoot apex need not be disaster if lateral buds are available to take over. Plasticity is also seen in the ability of plants to generate a whole organism from one part of the whole. Gardeners and horticulturalists make use of this in propagation from cuttings, or even from cultured undifferentiated cells. Some plants have evolved extreme versions of this phenomenon. Anyone who has tried to remove an *Acanthus* plant from a garden knows that whole plants may spring up from seemingly insignificant fragments of root.

These general statements provide a very tidy picture, but at least one very large puzzle remains. Study of genomics shows that each angiosperm uses essentially the same set of genes, albeit with some sequence divergence between species, to regulate its growth and development. However, despite this, and despite the plasticity of plant growth and development, different species look very different from each other; the basic body plan shows a huge amount of variation (e.g. shoot organizational

forms in Chapter 6, section 6.4). Understanding how this works is going to be one of the challenges of the early 21st century, and it is good to note that some clues are beginning to emerge.

3.5.2 Programmed cell death

In animals, one route by which cells die is the highly controlled process of *apoptosis*. This involves a regulatory cascade that eventually leads to an ordered dismantling of the major cell components and, finally, the engulfment by phagocytosis of the dead cell, usually by a macrophage. This last phase obviously cannot happen in plants, and there has been much discussion about whether plants exhibit a cell death mechanism equivalent to apoptosis.

What is clear is that there are several ways in which a plant cell may die; furthermore, as in animals, programmed cell death (PCD) often occurs as part of normal development. This form of PCD has many similarities to apoptosis. Proteins are degraded by cysteine proteases which, although they have only very limited sequence similarity to the *caspases* involved in apoptosis, clearly have the same function. Chromatin is degraded in an ordered manner, producing a characteristic 'ladder' of DNA fragments whose sizes correspond to multiples of the amount of DNA associated with nucleosomes. Organelle breakdown starts with deformation of membranes or envelopes, and the whole process moves in a regulated sequence to the final stage of autophagy ('self-eating'), in which the cell contents are completely broken down.

Whether or not this can be called apoptosis (as some plant scientists do call it) is a matter for discussion. The importance for this chapter is that, whatever term we use to describe it, this form of PCD is an important feature of the control of gene expression during plant development.

3.6 Plant hormones

3.6.1 Introduction

Like all complex organisms, plants need to respond to developmental cues and to endogenous and exogenous signals (e.g. environmental factors) in a regulated and coordinated manner. At the heart of this coordination are signalling molecules, many of which are hormones. By the late 1960s, auxin, cytokinins, abscisic acid, gibberellic acid and ethylene were considered to be the 'big five' hormonal regulators, possibly with auxin taking a particularly important role. However, there were already clues that other regulatory molecules existed, and we now include brassinosteroids, salicylic acid, jasmonic acid and strigolactone as hormones. We also recognize that compounds such as polyamines, sucrose (and other oligosaccharides) and nitrate have a role in signalling which goes beyond their role in metabolism. It is also now clear that calcium and nitric oxide are signalling molecules, with calcium in particular being involved in a wide variety of pathways and responses.

In general, modes of action of hormones involve the recognition of the hormone either at the cell surface or inside the cell. This recognition leads to a signal transduction pathway, culminating in specific cellular effects (e.g. stomatal closure) and/or to changes in gene expression (using the latter term in the widest possible sense, i.e. not just switching genes on and off).

In considering both the main functions and the modes of action of each hormone, there is a danger that we will think of each one in isolation. Instead, we must note that the functioning plant contains a network of signalling systems and that different parts of the network interact with each other. Furthermore, it is equally important to note that sensitivity to a hormone is as important as the presence and concentration of the hormone itself. This is nicely illustrated by the different responses of root and shoot growth to auxin.

3.6.2 Auxin

The discovery of auxin came about through studying the effects of illumination, and especially unilateral illumination, on plant growth. The fact that growing plants bend towards the light (Figure 3.11a) has been known for well over 2000 years. For example, it forms the basis of the Greek story in which Venus, the goddess of love, transforms Clytie, a water nymph, into a green plant. Thereafter, Clytie is condemned to follow the daily movements of the sun god, Apollo. However, it was Charles Darwin who first provided some insight into the underlying mechanisms (as described in his book, *The Power of Movement in Plants*, 1880). In particular, Darwin suggested that the perception of the light signal by the shoot tip can be separated from the bending response, and that 'an influence' passes from tip to the elongation zone where the bending occurs.

In the 20th century, Darwin's theory was further developed. In the late 1920s, a Russian scientist, Nicolai Cholodny, and a Dutch-American scientist, Frits Went, independently developed the theory that growth in response to a one-sided stimulus[xiii] causes an asymmetric redistribution of a plant growth hormone. This leads, in turn, to differential growth and thus curvature.

Discovery in 1931 of the hormone itself, named auxin[xiv] and identified as indole-acetic acid, helped to strengthen this hypothesis. Further confirmatory evidence came first from the demonstration that the shoot tip produces auxin, and second from experiments in which auxin was applied asymmetrically to the cut surface of decapitated shoots (Figure 3.11b). The Cholodny-Went hypothesis has been challenged a number of times, but nevertheless it remains at the core of our understanding of the role of auxin in phototropism, albeit that it has been modified to incorporate the rapid and apparently direct inhibition of growth on the illuminated side of the stem (see also section 3.7.5 and Chapter 6, section 6.11).

As well as phototropism (and other tropic responses), auxin is involved in growth in general, apical dominance (and therefore phyllotaxis), embryogenesis, morphogenesis, cell division and differentiation. Because of the breadth of its influence, it is sometimes known as the 'master hormone'.

The main target of auxin action is the SCF complex which ubiquitinates proteins (see section 3.4.5). The main auxin-binding protein is a nuclear-located F-box protein[xv], TIR (although at least three other F-box proteins, AFB1-3, also have auxin-binding activity). TIR (and presumably the AFBs) is associated with the SCF complex and, when auxin is bound to TIR, the latter stimulates the interaction of the SCF complex with the IAA and Aux transcriptional repressors, leading to their ubiquitination and subsequent hydrolysis by the proteasome. This releases the auxin-response factors (ARFs), transcription factors which are already bound to auxin-response elements (AREs; canonical sequence is TGTCTC), thus up-regulating the transcription of auxin-responsive genes. This pathway is shown diagrammatically in Figure 3.12.

[xiii]Their research involved responses to light and to gravity.
[xiv]From the Greek word *auxein* (to grow).
[xv]F-box proteins contain an essential 50-amino-acid tract that is involved in protein-protein interaction.

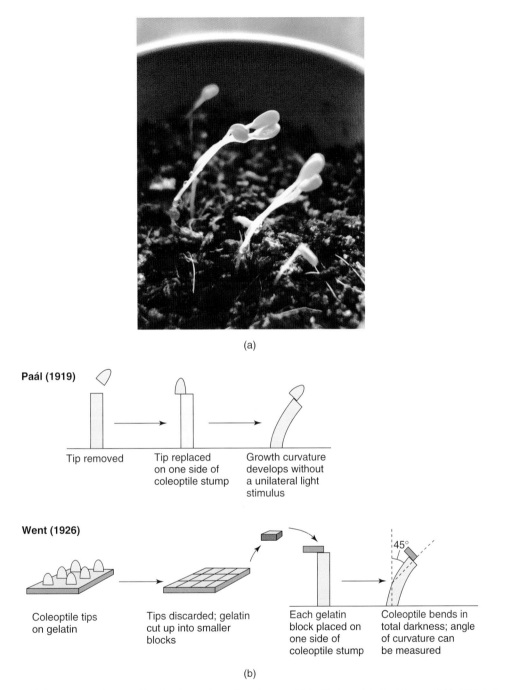

(a)

(b)

Figure 3.11 (a) Phototropism: Clover seedlings bend towards a unilateral source of light. Photograph by Brad Bowman (b) In 1919, Paál showed that asymmetric replacement of a coleoptiles tip caused bending similar to that seen in phototropism. This suggested that phototropic bending was caused by the asymmetric distribution of a diffusible substance. In 1926, Went collected that substance, namely auxin, in gelatine blocks; placing blocks asymmetrically on decapitated coleoptiles also induced bending. From Taiz, L and Zeiger, E (2002) Plant Physiology, Sinauer, Sunderland, MA.

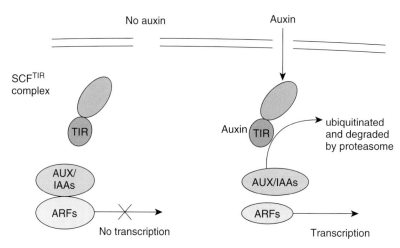

Figure 3.12 Auxin signalling pathway. In the absence of auxin, the SCF^{TIR} ubiquitination complex is inactive. AUX and IAA proteins which are transcriptional repressors prevent ARFs (auxin-response factors) from transcribing auxin-responsive genes. When auxin is bound to its receptor TIR, AUX/IAA proteins are ubiquitinated by the SCF^{TIR} complex and degraded by the proteasome. ARFs are then able to transcribe the auxin-response genes.

There is some discussion as whether the destruction of IAA/Aux proteins is involved in all auxin responses. In particular, it is suggested that proton pumping in auxin-induced cell expansion (Chapter 2, section 2.2.3) is mediated via the binding of auxin to ABP1, an auxin-binding protein located on the cell surface.

3.6.3 Cytokinins

Cytokinins are modified adenine molecules that influence a diverse range of processes, including cell division, meristem activity in relation to root and shoot development, differentiation of vascular tissue, acclimation of leaves to the light environment, leaf senescence and some stress responses. In several of these processes, it interacts with auxin.

Thus, cytokinins promote the formation of lateral shoots by activating axillary buds. This is partly achieved by inducing cell cycle activity in the buds. Auxin, on the other hand suppresses lateral shoot formation, at least partly because it can inhibit cytokinin synthesis, an effect mediated via the AXR1 signalling pathway (see above). The situation is further complicated by the role of recently discovered hormones called strigolactones (see section 3.6.10) in repressing shoot branching. Finally, auxin and cytokinin(s) act together in maintaining cell division, and both are usually needed to maintain dividing

cells in culture[xvi]. The synthesis of both is induced when infection by *Agrobacterium tumefaciens* (see Box 3.1) leads to the re-initiation of cell division.

Cytokinin receptors, of which three (CRE1, AHK2 and AHK3) have been identified in *Arabidopsis*, are histidine kinases that span the plasma membrane. Cytokinin binding causes dimerization of the receptor and auto-phosphorylation of histidine residues. The phosphate groups are transferred to aspartate residues on the receiver domain and a phospho-relay then occurs, in which the histidine phospho-transfer protein, AHP is phosphorylated. AHP can then enter the nucleus and transfer the phosphate group to a response regulator protein, ARR, leading to transcription of cytokinin-responsive genes (Figure 3.13). Although this pathway appears straightforward, there are several possibilities for fine tuning, with two active cytokinins and three receptors providing the means of varying the pathway in relation to different cytokinin responses.

3.6.4 Gibberellins

Gibberellins, of which the best known is gibberellic acid (GA), are complex diterpenoids. They were first isolated from the fungus *Gibberella fujikuroi*, the causative agent

---

[xvi]The famous tobacco BY2 cells are an exception: they can make their own cytokinins.

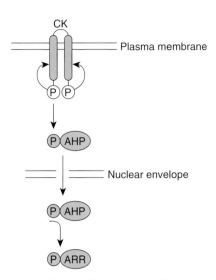

Figure 3.13 Cytokinin signalling pathway. Cytokinin receptors span the plasma membrane. They have an external receptor domain and an internal histidine kinase domain. Binding of cytokinin (CK) causes dimerization of the receptor, followed by autophosphorylation and phosphate transfer. The histidine kinase domain also phosphorylates AHP, a histidine phospho-transfer protein. This enters the nucleus and transfers the phosphate group to a response-regulator protein (ARR), leading to the transcription of cytokinin-responsive genes.

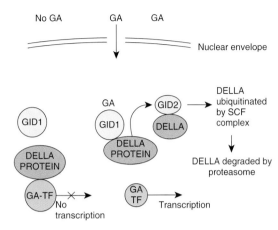

Figure 3.14 GA signalling pathway. GA signalling is very similar to IAA signalling. In the absence of GA, DELLA proteins inhibit the up-regulation by the GA-transcription factors (GA-TFs) of the GA-responsive genes. When GA is present, it is bound by GID1, which then interacts with the DELLA protein (or proteins), bringing them into contact with GID2, which is part of the SCF^{GID2} ubiquitination complex. The DELLA protein(s) is/are ubiquitinated and then degraded by the proteasome. This allows transcription of the GA-responsive genes. Note: different DELLA proteins are involved in different situations. In control of rice stem elongation, the relevant DELLA protein is SLR1.

of 'foolish seedling' disease in rice, the symptoms of which include greatly elongated stems. In addition to regulating stem elongation, GA is involved in retardation of senescence, in the breakage of seed dormancy, in germination, in the mobilization of nutrient reserves in seeds, in the regulation of the juvenile-to-adult phase change and in sex determination. In seeds with light-dependent germination, GA synthesis is an early phytochrome-mediated response to light. Another known effect of GA is that, in several plants, it has been shown to bypass day-length signals in the induction of flowering.

GA responses are initiated by degradation of a group of transcriptional repressors, the DELLA proteins. The major GA-binding protein, GID1 (located in the nucleus), is able, when GA is bound to it, to interact with the DELLA proteins. The DELLA proteins are brought to the SCF (a ubiquitination complex) via GID2 (which, in GA-responsive cells, is associated with the SCF). Ubiquitination and proteolysis then follow (Figure 3.14). It is possible that different DELLA proteins may be involved in different situations. For example, a DELLA protein known as SLR1 is involved in the control of stem

growth in rice. Nevertheless, overall GA signalling is very similar to the main auxin-signalling pathway. We also note in passing that this model is *very* different from that prevalent in the 1990s, in which it was proposed that GA acted via a membrane-spanning G-protein[xvii].

3.6.5 Abscisic acid

The existence of a growth-inhibiting substance had been discovered in the 1950s. The actual compound, abscisic acid (an isoprenoid), was isolated in the early 1960s and its structure was worked out later in the same decade. Its name, unfortunately, is misleading; it does not regulate abscission, but has a role in bud dormancy. In seed development, it prevents precocious germination by imposing a temporary dormancy. Genes involved in seed maturation are regulated by ABA. It has a role in imposing post-harvest dormancy in seeds, and antagonizes GA in the breakage of dormancy. It also antagonizes cytokinins in some aspects of the control of cell division. It is involved in root-shoot signalling in the drought response and it also causes stomatal closure (see Chapter 9, section 9.4).

[xvii] G-proteins bind GDP or GTP and are usually activated when bound GDP is replaced by GTP.

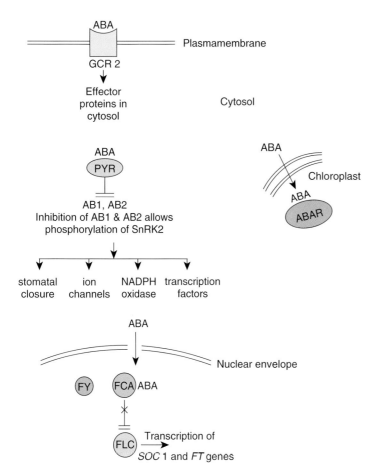

Figure 3.15 ABA signalling pathways. Proteins with ABA receptor properties occur in four different locations. At the plasma membrane, ABA binds to and activates a membrane-spanning G-protein. The downstream effects of this are not yet clear. In the cytoplasm, ABA binds to the PYR family of receptors. This leads to inhibition of the protein phosphatases AB1 and AB2, allowing phosphorylation of the regulatory protein snRK2. Downstream of this are the effects of ABA on stomatal closure, ion channels, NADPH oxidase activity and transcription of some ABA responsive genes. In the chloroplast, ABA binds to ABAR, leading to effects on chlorophyll synthesis and plastid-nuclear signalling. In the nucleus, ABA binds to FCA (an RNA-binding protein), dissociating it from its partner, FY. This, in turn, prevents FCA-FY from inhibiting the floral repressor protein FLC. When FLC is not inhibited, the activity of genes leading to flowering, including *FT* and *SOC1*, is inhibited. ABA thus prevents, or at least delays, flowering.

There are at least three different ABA receptor proteins (Figure 3.15). The first to be identified, FCA (Flowering Time Control Protein A), is associated with the inhibition of flowering by ABA. FCA is a nucleus-located RNA-binding protein. This regulates the 3′ processing of mRNA molecules and also combines with another mRNA processing enzyme, FY (Flowering Locus Y). A major target of FCA-FY is the pre-mRNA encoding the FLC protein; disruption of the processing of pre-mRNA leads to down-regulation of *FLC* gene expression. Since FLC is a negative transcription factor that inhibits flowering, down-regulation effectively promotes flowering. When ABA is bound to FCA the latter can no longer associate with FY, *FLC* gene expression is not down-regulated and thus flowering is inhibited (Figure 3.15).

The second ABA-binding protein is located in the chloroplast. Initially named ABAR (ABA Receptor) it was later shown to be a protein known as CHLH, one of the sub-units of Mg-chelatase (involved in chlorophyll synthesis). The immediate downstream events following ABA binding are not yet clear, but the final effects include inhibition of germination and/or early seedling growth, and possibly modification of chloroplast-to-nucleus retrograde signalling (see section 3.3.2).

Among the more recently discovered ABA receptors are the PYR/PYL/RCAR family of proteins, which occur in the cytosol. In the presence of ABA, these proteins act as powerful inhibitors of two members of the protein phosphatase 2C (PP2C) family of proteins, namely ABI1 and ABI2. From here, it appears several pathways branch out. Inhibition of AB1 and AB2 allows phosphorylation and hence activation of regulatory proteins. These include SnRK2, which in turn phosphorylates a range of proteins that includes transcription factors that initiate transcription of ABA-responsive genes, CPK21, involved in regulation of stomatal closure, ion channels and NADPH oxidase (Figure 3.15).

While the ABA story already seems complex, there is a further twist in the tale. ABA is also bound specifically by two plasma membrane G-protein-coupled receptors (see footnote on page 79). Since G-proteins are generally involved in signal transduction, it has been suggested that these receptors represent a further signalling pathway for ABA. Currently, the available data do not give a clear enough picture to support or reject this suggestion, but further research should resolve the issues.

3.6.6 Ethylene

Ethylene is a gaseous hormone with a simple chemical structure. In 1901, Dimitry Neljubow, a student in St Petersburg, showed that ethylene given off by the coal-gas lamps in the lab caused aberrant growth in pea stems. Out in the streets, leaf abscission often occurred prematurely in trees growing near lamps. Thus, Neljubow should be credited with the first demonstration that ethylene can affect plant growth and development.

The better-known examples of ethylene (e.g. from kerosene heaters) causing fruit ripening came over the next 20 years or so and, by 1934, ethylene had been identified as a naturally occurring plant hormone. However, research on ethylene did not really get going until the 1950s, and it was only in the 1980s that its biosynthetic pathway (The Yang cycle) was finally worked out. The last steps in its synthesis, catalysed by ACC synthase and ACC oxidase (Figure 3.16), are major control points in ethylene action. For example, the gene encoding the ACC oxidase is subject to transcriptional regulation by a variety of factors, including ethylene itself.

Ethylene has a major role in fruit ripening and in leaf abscission. It is involved in some responses to plant pathogens and to waterlogging (in the latter, it has role in aerenchyma formation – see Chapter 10, section 10.3.1) and also in cell death.

As with several hormones, negative regulation is at the heart of ethylene signalling. Immediately 'downstream' of the initial ethylene receptor is the CTR1 (constitutive triple response) protein. Mutations in its gene lead to constitutive expression of genes that are normally ethylene-regulated. CTR1 normally represses those genes, and ethylene itself removes that repression.

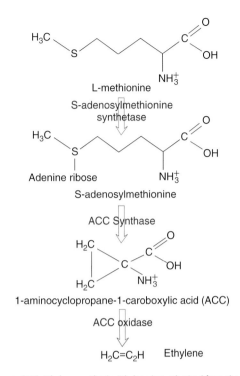

Figure 3.16 Ethylene synthesis. Ethylene is synthesized from the amino acid, methionine via the cell's methyl donor, S-adenosyl methionine. Control over ethylene synthesis is exerted at the last two steps.

How, then, does it work? The ethylene receptor ETR1, which spans the ER membrane, was the first plant hormone-binding protein to be unequivocally identified (in 1993). Since then, several other similar proteins have been identified, including ETR2, ERS1, ERS2 and EIN4. Indeed, *Arabidopsis* possesses 11 ethylene receptors, grouped into two sub-families. These proteins are classic two-component response regulators with an external hormone-binding domain which propagates a signal via an internal histidine kinase domain. Binding of ethylene causes the receptor to dimerize. A histidine residue on the kinase domain is auto-phosphorylated and the phosphate is transferred to an aspartate residue on the receiver domain (Figure 3.17). The internal domain of the receptor then initiates ethylene action by inhibiting the repressive activity of the next protein in the pathway, CTR1.

The exact mode of interaction with CTR1, which is also located in the ER membrane, is not clear. The histidine kinase domain of the receptor is essential for the interaction, but kinase activity itself appears not to

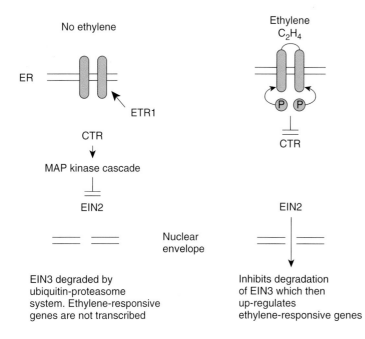

be. The phosphorylation of histidine and the subsequent phosphorelay appear to modulate, rather than mediate, the interaction. The interaction inhibits CTR1 and this event leads to the activation of a previously inactivated protein, EIN2.

The transmission of signal from CTR1 to EIN2 is not yet fully understood, but it may involve a sequence of kinases similar to the MAP kinase cascade in animals. This suggestion receives support from the similarity between CTR1 and MAP kinase kinase kinase (MAPKKK). Activated EIN2 enters the nucleus and prevents the ubiquitin-proteasome-mediated degradation of the transcription factor EIN3. EIN3 is then able to initiate transcription of the ethylene-responsive genes (Figure 3.17). These include genes encoding a family of transcription factors, the ethylene-response factors (ERFs). Interestingly, some of these ERFs are up-regulated by other signals, indicating again the existence of cross-talk between signalling pathways.

3.6.7 Brassinosteroids

The structure of brassinosteroids clearly reveals them as steroid in nature (Figure 3.18). They were discovered following the observation that pollen tube elongation is regulated by a compound that is distinct from GA. In the late 1970s, this was shown to be a steroid and

was named brassinolide (it was extracted from *Brassica* pollen). Nearly 50 brassinosteroids are now known, but brassinolide is by far the most hormonally active. Others may act as antifeedants, causing premature moult in herbivorous insect larvae. Brassinosteroids are involved in a very wide range of responses, including pollen tube growth, the unrolling of grass leaves, increased stem elongation (where, alongside auxin, they have a role in proton pumping and in orientation of cellulose microfibrils), differentiation of xylem (again alongside auxin) and stimulation of ethylene biosynthesis.

Figure 3.18 Structure of brassinolide.

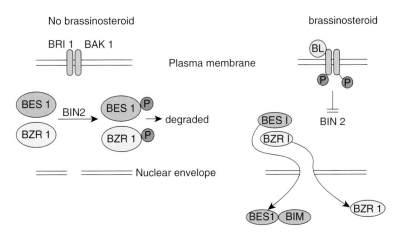

No brassinosteroid

brassinosteroid

Figure 3.19 Brassinosteroid signalling pathway. Brassinosteroid receptors such as BRI1 are large, complex proteins that span the plasma membrane. Also required for activity is another membrane-spanning protein, BAK1, and the two form a hetero-dimer. In the absence of brassinosteroid (BL), a protein kinase, BIN2 phosphorylates two transcription factors, preventing them from entering the nucleus and marking them for degradation by the ubiquitin-proteasome system. When BL binds to BRI1, its internal domain and that of BAK1 are phosphorylated. This leads to the inhibition of BIN2, preventing the destruction of BES1 and BZR1. These enter the nucleus. BES1 combines with BIM to up-regulate the BL-responsive genes, while BZR1 inhibits the transcription of genes involved in BL biosynthesis.

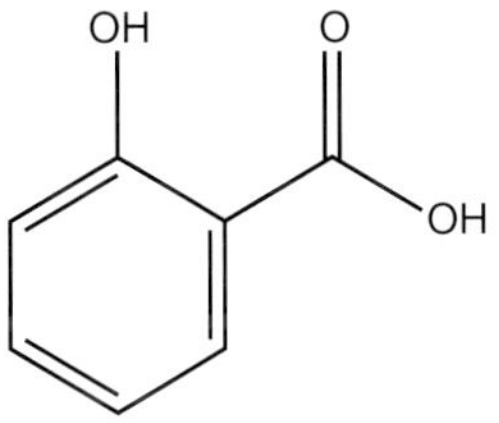

Figure 3.20 Salicylic acid structure.

Brassinosteroids do not cross the plasma membrane but bind to a membrane-spanning receptor-kinase, of which three have been described (BRI1, BRL1 and BRL3). Also essential for brassinosteroid signalling is a second plasma membrane spanning protein, BAK1, which forms a heterodimer with the receptor (Figure 3.19). Binding of a brassinosteroid causes phosphorylation of both proteins, which leads to an interaction with the cytosolic protein kinase, BIN2, thereby inhibiting it.

When active, BIN2 phosphorylates two transcription factors – BES1 and BZR1. In the phosphorylated state, these transcription factors are degraded by the ubiquitin-proteasome system but, in the non-phosphorylated state when BIN2 is inactivated, they can enter the nucleus. There, BES1 associates with another transcription factor, BIM, to up-regulate brassinosteroid-responsive genes. At the same time, BZR1 down-regulates the genes involved

in brassinosteroid synthesis, thereby mediating negative feedback control.

3.6.8 Salicylic acid

Salicylic acid is a relatively simple phenolic compound (Figure 3.20), used in the pharmaceutical industry for making aspirin. It is involved in responses to plant pathogens and, in these responses, it may interact with auxin responses (as described below). It regulates the switch from the normal mitochondrial electron transport chain to the alternative oxidase, for example during thermogenesis in heat-generating flowers (see Chapter 2, section 2.14.2 and Chapter 8, section 8.6.2). It also inhibits ethylene synthesis and, in doing so, it may slow senescence. This effect is made use of in extending the life of cut flowers by putting aspirin in the water.

In the response to pathogens, SA synthesis and some of the effects of SA are dependent on cytoplasmic receptor-like protein kinase (see Chapter 11, sections 11.3.8 and 11.3.9). One of the effects of SA is to counter the auxin over-production that is stimulated by many pathogens. This achieved by preventing access of auxin to the TIR protein. TIR is then unable to interact with IAA and Aux proteins, which are now not susceptible to degradation by the ubiquitin-proteasome system; auxin action thus is prevented (see section 3.6.2).

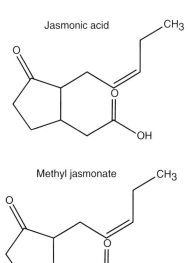

Jasmonic acid

Methyl jasmonate

Figure 3.21 Structure of jasmonic acid and its close derivative, methyl jasmonate.

3.6.9 Jasmonic acid

Jasmonic acid (JA: see Figure 3.21) is derived from the unsaturated fatty acid, linolenic acid (18:3). JA and Me-JA were both identified in the 1970s and 1980s as inhibitors of plant growth and are now regarded as *bona fide* hormones. JA certainly inhibits some facets of plant growth but also has a positive effect, together with ABA, on the accumulation of storage proteins, with GA on pollen maturation and with ethylene and ABA on pollen grain separation and anther dehiscence. JA also stimulates some ethylene-regulated events such as fruit ripening, because it induces the ethylene synthesis enzyme, ACC oxidase (see section 3.6.6).

JA accumulates in response to herbivory (Chapter 11, section 11.2.1) or to pathogen infection (Chapter 11, section 11.3.9); it is converted to jasmonyl-L-isoleucine, the hormonally active form, which then induces the synthesis of a range of defence chemicals. Local wounding leads to the processing of a peptide, *pro-systemin*, to yield a smaller peptide, the 18-amino acid *systemin*. This peptide activates the enzymes of JA synthesis.

As implied by its name, systemin is mobile within the plant, so that JA synthesis and the downstream effects of JA (including expression of genes encoding proteinase inhibitors) can occur in parts of the plant that have not yet been exposed, for example, to herbivore attack.

The system is further amplified in that JA stimulates prosystemin expression and also activates the ORA47 transcription factor that regulates positively several genes involved in JA synthesis.

Indeed, the JA signalling pathway initially appears as widening network of positive and negative transcription factors, and it shows evidence of cross-talk with signalling from other hormones, including salicylic acid and ethylene. However, this network is explained when we consider the initial event in JA signalling. JA binds to jasmonate-ZIM-domain (JAZ) proteins; these are negative transcription factors, which repress the positive transcription factor MYC2. Binding of JA to the JAZ proteins promotes their interaction with the SCF complex and, as we have noted already, this marks the proteins for degradation via the ubiquitin-proteasome system. MYC2 then initiates transcription of the JA-responsive genes, some of which also encode transcription factors – hence, the widening network mentioned at the start of this paragraph. Again, we note, that, as with IAA and GA, an early event in JA signalling is to remove an inhibitory protein by sending it to the ubiquitin-proteasome degradation system.

3.6.10 Strigolactones

The existence of a substance (or substances) that stimulate(s) the germination of parasitic plants in the genera *Striga* and *Orobanche* (Chapter 11, section 11.1.2) has been known for many years. In the late 20th century, these were identified as strigolactones – unusual molecules, namely sesquiterpene lactones, carotenoid derivatives (Figure 3.22). However, early in the 21st century it was realized that strigolactones are a group of much more widely distributed plant hormones, with roles in root development (especially the growth of root hairs). In shoots, they suppress lateral branching and appear to be involved in the coordination of root and shoot growth, possibly via an effect on auxin transport. Further research on strigolactones as hormones is set to reveal some very interesting new data.

3.6.11 Hormone responses – general points

Although we have dealt with the hormones one by one, it is essential to remember that they do not work on their own in the plant. Indeed, attention has already been

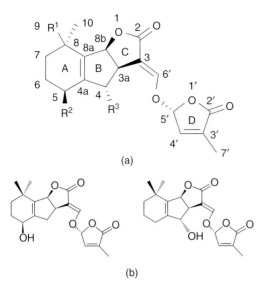

(a)

(b)

Figure 3.22 (a) General structure of strigolactones. (b) Two naturally occurring strigolactones: on the left, (+)-strigol; and on the right, orobanchol.

drawn to processes that are regulated by several hormones working together. Some of these 'collaborations' are quite subtle, with one hormone modulating the effect of another, or with multiple effects arising because of the overlap or even integration of signalling pathways. As well as acting together, hormones may work against each other, with the net effect being the result of relative concentrations of the active hormones in a cell or tissue. Some of the interactions occur because some hormones can influence, positively or negatively, the synthesis of other hormones. Hormones can also regulate, both negatively and positively, their own synthesis, as seen here, for example with ethylene and jasmonic acid.

Subtle variations in effect can arise when a hormone exists in more than one form, as exemplified by the cytokinins. Other variations may occur because of the existence of multiple isoforms of receptor proteins, or of other components in signalling pathways. Furthermore, the pathways themselves are likely to be subject to feedback inhibition when the effect of the hormone has been fully 'achieved'. This is nicely illustrated in jasmonate signalling.

We also have to consider the active concentration of a hormone at its site of perception. This depends on many factors, including rates of synthesis and degradation, the possibility of sequestration or conjugation as an inactive

ester, transport through the plant (where relevant), uptake into cells, the abundance of carrier proteins or of receptors (and hence the sensitivity of a particular cell to the hormone) and so on. Thus, in respect of just *one* hormone, auxin, Ottoline Leyser at the University of Cambridge writes:

'*Auxin action is driven by the interplay between context-sensitive transcriptional readouts of auxin concentration and active positioning of auxin within tissues by localized and oriented transporters, with likely additional contributions from localized synthesis and catabolism . . . There are obvious advantages to such a system in maintaining the environmental responsiveness that is a touchstone of plant development, as well as providing triggers for genetically regulated developmental progression . . . The possibilities for communication in this system are . . . extraordinary in their complexity and diversity. There are effects driven by auxin concentration, with the strong possibility that intracellular and extracellular auxin are measured independently. In addition, there are effects based on competition between tissues for auxin sources or auxin sinks, which can be driven by the capacity for auxin flow. On top of this there is the as yet unsupported possibility that auxin flux through transporters into and/or out of cells could also be directly monitored. All these systems are interdependent and regulate one another. There is no doubt that a serious effort in mathematical modelling will be required, closely linked to experiments, to have any chance of understanding auxin biology.*'

3.6.12 Calcium and nitric oxide

Although they are not hormones, calcium and nitric oxide must be included here because of their roles in signalling pathways. The idea that calcium (Ca) is a signalling molecule, equivalent in some ways to a plant hormone, has been with us for several decades. However, in the past 20 years or so, we have been able to obtain a much clearer understanding of intracellular calcium fluxes through the use, in living cells, of fluorescent calcium-binding proteins such as *aequorin*.

It is now apparent that cells are able to change very quickly the concentration of free Ca in the cytosol by release from locations such as the ER, vacuole and even the cell wall. This operates by stimulation of specific gated calcium channels. Thus, calcium fluxes are involved in establishing polarity (e.g. in the tip growth of pollen tubes), in aspects of circadian regulation, in some tropisms (e.g. thigmotropism: see Chapter 6, section 6.13) in the action of some hormones (see, for example, section 3.6.4) and in response to biotic and abiotic stresses (see, for example, Chapter 9, sections 9.3.2 and 9.5; and Chapter 10, section 10.2.2) where again an interaction with hormone signalling may occur. More specifically, in relation to stress, calcium is regarded as a key player in the linkage of stress responses to hormone signalling pathways.

At least five types of intracellular calcium receptors or sensors have been recognized, namely **calmodulins** (CaM), CaM-like proteins, calcium-dependent protein kinases (CDPK), Ca-CaM-dependent protein kinases and calcineurin-B-like proteins (CBL proteins). CBL proteins are the regulatory subunits of a class of protein kinases called CBL-interacting protein kinases or CIPKS. These are, in effect, a subgroup of CDPKs, but the dependence on Ca is mediated by the binding of Ca by the CBL proteins.

This array of sensors/receptors gives some sense of the complexity of calcium signalling, but that sense is compounded when we consider that plants contain multiple isoforms of all these proteins, giving a huge flexibility in the generation of signalling pathways. Indeed, current estimates suggest that calcium can affect the activity of well over 1,000 genes, a detailed discussion of which is well beyond the scope of this chapter. In some instances, there is a direct interaction between the calcium receptor (with its bound calcium) and the relevant transcription factors, as happens with some CaM-regulated genes. In other instances, there are signalling cascades between the calcium receptor and the target transcription factors; some of these cascades involve mitogen-activated kinases (MAPKs).

As with other signalling molecules, the end effect on gene expression may be up- or down-regulation, according to the gene in question. Furthermore, the interaction with hormones is again seen, in that calcium may regulate the activity of transcription factors that bind to hormone-response elements in gene promoters, for example the ABA response element (ABRE).

Finally, we must emphasize that it is not only transcription factors that are targets of Ca signalling. Ca-CaM, for example, directly interacts with about 20 proteins that are not transcription factors, including kinases, enzymes involved in metabolism, several membrane-located, ATPase-linked ion carriers and also proteins in the cytoskeleton. The influence of calcium is thus very extensive.

Study of calcium as a signalling molecule has a relatively long history, but it is only recently that attention has been focused on nitric oxide (NO). It was noted in the 1970s that plants sometimes produce NO, and in the past decade or so it has been implicated as having roles in several developmental processes and in responses to stress. It has, in effect, 'joined the club' of plant signalling molecules.

NO is a small, gaseous uncharged free radical, with a short half-life, and it is highly diffusible across biological membranes. It is capable of direct modification of proteins by nitrosylation and by nitration of tyrosine. It possibly acts as an anti-oxidant, and it has been suggested that it activates enzymes (possibly by the modifications mentioned) that scavenge reactive oxygen species. By contrast, in the hypersensitive reaction to plant pathogens (see Chapter 11, section 11.3.8), NO and jasmonate together reduce the activity of the ROS scavenging enzymes ascorbate peroxidase and catalase, leading to increases in ROS concentration and eventually to programmed cell death. NO may also activate Ca channels, thus stimulating the release of Ca in response to stress. However, Ca, when bound to an appropriate receptor, can also antagonize some aspects of NO activity, for example in the regulation of specific transcription factors.

At the time of writing, it is still not entirely clear how plants make NO; seven different mechanisms have been proposed, including the direct formation by NO synthase and as a partial nitrite reductase reaction. Current evidence suggests that most NO is probably produced by NO synthase, with the concomitant conversion of arginine to citrulline. There is equal uncertainty about signalling pathways, although, as indicated already, some of the effects of NO are mediated by direct effects on proteins. Doubtless, this will be a fruitful line of research over the next few years.

3.7 Light receptors

3.7.1 Introduction

Green plants are *par excellence* users of energy from sunlight. The role of the light receptors involved in photosynthesis is discussed in Chapter 7 (section 7.4.2).

Because of the importance of photosynthesis, it is not surprising that plants have developed ways of optimizing their interception of sunlight in whichever habitat they are growing. It is equally unsurprising that plants use light as a developmental cue and that they exhibit circadian rhythms entrained to the daily changes in the light environment, as night succeeds day with a 24-hour periodicity (Chapter 6, section 6.10). The non-photosynthetic light receptors involved in these phenomena fall into three groups: *phytochromes, cryptochromes* and *phototropins*. Phytochromes were the first to be discovered.

3.7.2 Phytochromes

Many plant species that quickly colonize open ground (a large number of which may be classified as weeds or ruderals) have a light requirement for germination. This maximizes the chances of seeds germinating only in appropriate habitats. It is noted in passing that some of the species with this open ground requirement can form 'seed banks', in which seeds lie buried and dormant until the ground is disturbed in some way and the seeds are brought to the surface, at which time they are able to germinate (Chapter 11, section 11.1.3).

It was this requirement for light in germination that was involved in the discovery of phytochromes. The plant used was not, in fact, a wild species, but a variety of lettuce, Grand Rapids (*Lactuca sativa*). Somewhat unusually for a highly selected horticultural crop, this variety still requires light for germination.

It was shown by Harry Borthwick and Sterling Hendricks and their colleagues at the USDA Agricultural Research Station at Beltsville (Maryland) that seeds germinate after exposure to a short flash of red light (650–680 nm); however, if the red light flash is followed by a flash of far-red light (710–740 nm), then the seeds do not germinate. The effect of red light on the seed is thus reversible by far-red. Further, the far-red light inhibition can itself be reversed by a further flash of red light and, in theory, this induction-inhibition reversibility can be carried on *ad infinitum*.

More detailed analysis showed that the peak wavelength for the red light effect is at 666 nm and, for far-red light, 730 nm. It was noted that the optimum red light wavelength was the same as that involved in the flowering response of plants that are photoperiod-sensitive (see Chapter 8, section 8.4.1).

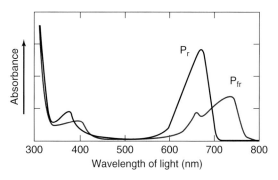

Figure 3.23 Phytochrome spectra. The inactive form of phytochrome, P_R, is converted to the active form, P_{FR}, by exposure to red light (peak absorbance at 666 nm). P_{FR} is converted back to P_R by exposure to far-red light (peak absorbance at 730 nm).

These data may be interpreted in two ways. First, it is possible that two antagonistic photoreceptors with different action spectra are responsible for the effects of red and far-red light. Secondly, it is equally possible that a single photo-reversible receptor is involved. Identification and characterisation of the blue pigment *phytochrome*[xviii] showed that the second interpretation is correct. Phytochrome was shown to exist in two forms – P_R, with an absorption peak in the red region of the spectrum, and P_{FR}, with an absorption peak in the far red region of the spectrum (Figure 3.23).

Absorption of red light converts P_R to P_{FR}, in which form the phytochrome is biologically active. Absorption of far-red light converts P_{FR} back to P_R, the inactive form. There is also a slow, dark reversion of P_{FR} to P_R. The phytochrome protein exists naturally as a dimer, with each of the pair carrying the actual light receptor or **chromophore**. This is a linear tetrapyrrole[xix] linked to the protein via a thio-ether linkage (Figure 3.24). Absorption of red light converts one of the chemical bonds from the *cis* to the *trans* configuration. This change initiates the various series of events that lead to the biological processes that phytochrome regulates.

Some of the roles of phytochrome are listed in Table 3.3. The molecular mechanisms involved in performing some of these roles are discussed in more detail in section 3.7.3.

[xviii]The name *phytochrome*, simply meaning 'plant colour', was coined by biophysicist Warren Butler, who, with Harold Siegelman, is credited with the first identification of the pigment.
[xix]Chlorophyll and haem are also tetrapyrroles but the four porphyrin rings are in a cyclic configuration (see Chapter 7, Figure 7.6).

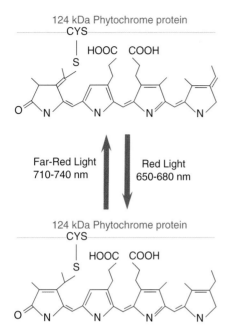

124 kDa Phytochrome protein

Far-Red Light
710-740 nm

Red Light
650-680 nm

124 kDa Phytochrome protein

Figure 3.24 Phytochrome chromophore structure. The chromophore is joined to the phytochrome protein via a thio-ether linkage (S-S bridge) between the chromophore and a cysteine residue. The red/far-red reversibility lies in the 'flipping' of a double bond in the pyrrole ring nearest to the protein attachment point.

Table 3.3 Examples of processes mediated by phytochrome.

Light intensity	Biological activity
Very low fluence	Seed germination[xx]
Low fluence	Seed germination[xxi]
	Circadian clock
	Flowering
	De-etiolation and greening of seedlings
	Chloroplast orientation in cells
High irradiance	Stem elongation[xxii]

Meanwhile, we consider some of the ecological aspects of phytochrome action.

Observation of the two absorption spectra shows that phytochrome can never exist as pure P_R or pure P_{FR}. The

[xx] Note that very low fluence processes are not reversible by far-red light.
[xxi] As in the experiments with *Grand Rapids* lettuce, and subsequently with many other species.
[xxii] Longer-term responses, not far-red reversible.

Table 3.4 Variations in light intensity and quality.

	Light intensity, μmol quanta m^{-2}s^{-1}	Light quality: R/FR ratio
Bright daylight	2000	1.15–1.20
Sunset	20	0.98
Under leaf canopy	20	0.1–0.7

absorption spectrum of P_R extends slightly into the far-red region, while that of P_{FR} extends markedly into the red region, such that the maximum possible proportion of P_{FR} is 85 per cent and that of P_R is 97 per cent. In real light environments, of course, these percentages are never attained, and the ratio of P_R to P_{FR} at any one moment enables the plant to sense the light environment. Consider the data in Table 3.4.

Although the light intensity at sunset is much lower than in the middle of the day, the red/far-red ratio is altered only slightly. However, at an equally low light intensity under a leaf canopy, the red/far-red ratio is dramatically reduced. Light that is transmitted through leaves or reflected by leaves has a much greater proportion of far-red than the light that reaches the earth's (or leaf's) surface out in the open. The increased proportion of far-red transmitted through the canopy, or reflected from close neighbours, leads to internode elongation and thus the plant escapes, as far as is possible, from being shaded by its neighbours. Internode elongation, unlike germination of lettuce seeds, is stimulated by far-red light.

How, then, can the inactive form of phytochrome, P_R, promote internode elongation? The answer is that it does not directly do so. It is the *lowered concentration of P_{FR}* that is the key feature here; this is discussed more fully in section 3.7.4.

Nevertheless, many plants *do* grow in the shade of others. The effects of this include, among other things, changes in leaf morphology and in the response of photosynthesis to light (see Chapter 9, section 9.7.1; and Chapter 10, section 10.7.1). In two plants of the same species, one grown in shade and one unshaded, the shaded plant will form leaves that are thinner and larger than those on the unshaded plant (Figure 3.25). These differences in leaf growth are mediated by phytochrome's effect on the distribution of cytokinins (see section 3.6.3).

The differences in surface area and in leaf thickness effectively mean that shaded plants synthesize lower

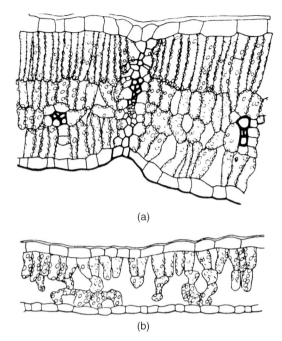

(a)

(b)

Figure 3.25 Transverse sections of leaves of sugar maple (*Acer saccharum*): (a) leaf from sunny side of the tree; (b) leaf from centre of the canopy.
From Fig. 6.16 of Hart JW (1988) Light and Plant Growth, reproduced by permission of Springer, Heidelberg.

amounts of photosynthetic enzymes per surface area of leaf. At very low light intensities, the shade leaves photosynthesize more effectively than non-shade ('sun') leaves. This advantage is quickly lost as the light intensity increases, and photosynthesis in shade leaves saturates at much lower light intensities than sun leaves. If a shade leaf is brought into light of a higher intensity, it will suffer photobleaching and oxidative damage, eventually causing inhibition of photosystem II (see Chapter 9, section 9.7.2).

Study of gene expression and of mutants in *Arabidopsis thaliana* has given us extensive information on how phytochrome works at the molecular level (section 3.7.3). This research has also revealed that there are five different forms of phytochrome – phytochromes A to E – encoded by five different genes. All angiosperms possess phytochromes A, B and C (although there has been some doubt as to whether peas possess phyC). PhyD and phyE are only known in eudicots and, among those, phyD is currently known only from the Brassicaceae, including *Arabidopsis*. Most of the major phytochrome effects are, indeed, mediated by phyA and phyB, with the relative

contributions of each varying according to the biological activity that is being regulated.

Phytochrome A is unusual in that it accumulates in darkness but is broken down in the light, whereas the synthesis of other phytochrome proteins is not affected by the light environment. It also functions in high far-red environments, in which it may act as a far-red sensor. The behaviour of phyA therefore does not fit the pattern discussed earlier (in retrospect it is probable that our general understanding of phytochrome action was developed around phyB).

The particular properties of phyA make it suited to participating in the very low fluence reactions involved in the germination of some seeds, and this role has been confirmed in subsequent experiments. However, as already indicated, it also participates in other light-regulated phenomena and may be a sensor of light fluence levels. Overlap and redundancy of function is also seen in phytochromes C, D and E (in those plants that possess them). These phytochromes modulate the activities of phyB (and, to a much lesser extent, of phyA), with particular phytochromes being especially associated with particular light-regulated phenomena. In *Arabidopsis*, PhyD and PhyE, for example, interact with PhyA and PhyB in entraining the circadian clock, while all five phytochromes participate in regulation of leaf architecture.

3.7.3 Phytochrome signalling pathways

It is already clear that phytochrome action is complex. Phytochromes regulate a number of different physiological activities and developmental changes. Individual phytochromes collaborate with each other to differing degrees in different situations. A further complication is the participation of cryptochrome alongside phytochrome in several light-stimulated processes. Inevitably, then, there is a complicated array of signalling pathways that, because of space limitations, cannot be described here in detail. We thus present an overview but include some detail of specific pathways (Figure 3.26).

Early experiments with *Arabidopsis* mutants that grow in the dark as if they were in the light (i.e. do not show the etiolation response) revealed the existence of the *COP/DET/FUS* group of genes, which encode negative regulators of transcription. In response to the appropriate light signal, the activated form of phytochrome migrates into the nucleus, leading to the inactivation

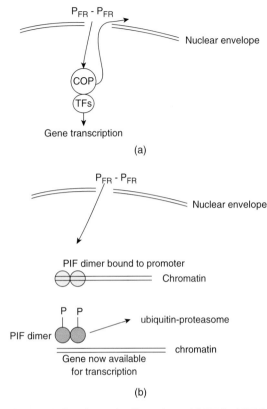

Figure 3.26 Phytochrome signalling pathways: (a) COP (and DET & FUS) proteins inhibit phytochrome-responsive transcription factors. These proteins are inactivated by the active form of phytochrome (P_{FR}), probably acting as a dimer. The inactivated proteins leave the nucleus and the transcription factors are freed to interact with the relevant genes. (b) Dimers of the phytochrome-interacting factors (PIFs) bind to promoters of phytochrome-responsive genes. Active phytochrome dimers bring about phosphorylation of the PIFs. This leads to their degradation by the ubiquitin-proteasome system. The gene promoters are then available for binding by positive transcription factors.

of the COP/DET/FUS proteins (and their export from the nucleus). This, in turn, causes the release of HY5 and other transcription factors which up-regulate the light-responsive genes.

However, there are other nuclear proteins with which the active form of phytochrome may interact, namely the *phytochrome-interacting factors* (PIFs). These are helix-loop-helix transcription factors which inhibit light-dependent responses by binding with G-box regions in the promoters of relevant genes. PIFs exist in the nucleus as homo- and hetero-dimers.

Active phytochrome dimers enter the nucleus and interact with the PIF dimers; this interaction leads to the phosphorylation of at least the PIF3 homo-dimer, and possibly also of other PIF dimers. In general, this marks the PIF proteins for ubiqitination and subsequent proteolysis (see section 3.4.5), thus activating phytochrome-dependent gene expression. It is noted that genes involved in GA synthesis are among those up-regulated in the germination response of lettuce seeds, leading to rapid increases in GA, a known promoter of germination. The identity of the kinase that carries out the phosphorylation of the PIF proteins is not entirely clear. At present it is thought to be phytochrome itself.

There is one further aspect of PIF activity to consider: *pif3* mutants exhibit some deficiencies in phytochrome-mediated responses. This suggests that PIF3 also has a positive regulatory role. Further, phyB dimers have been shown to bind to PIF3 dimers already associated with the G-box regions in the promoters of, among others, the *CCA1* and *LHY* genes, leading to an activation of gene expression. Both of the latter genes encode MYB-like transcription factors which are especially involved in regulating the circadian clock.

3.7.4 Cryptochromes

Although red light plays a major role in the processes discussed in section 3.7.2, it became clear that blue light is also involved in several of them. This led to painstaking research to identify the relevant photoreceptors, culminating in the discovery of cryptochromes[xxiii], nuclear-located proteins with strong sequence homology to the DNA repair enzymes, DNA photolyases (Chapter 10, section 10.7.2). The action spectra of the cryptochromes suggest that their chromophores are flavin compounds and, indeed, each protein molecule possesses a non-covalently linked molecule of flavin adenine dinucleotide (FAD) which can undergo photo-reversible redox reactions.

However, that is not the whole story, because a molecule of *N5,N10-methenyl-5,6,7,8-tetrahydrofolate* (MTHF)[xxiv],

[xxiii]Perhaps the name *cryptochrome* ('hidden colour') reflects the difficulty experienced in the early days of the search! Cryptochromes are, in fact, found in a very wide range of organisms, including animals, and they have a role in regulating the circadian clock.
[xxiv]Note that this compound is sometimes referred by the generic name *pterin*.

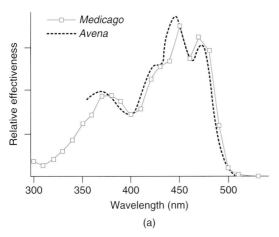

(a)

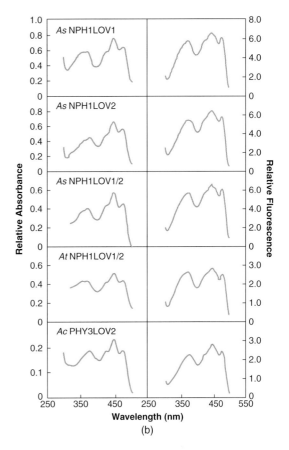

(b)

Figure 3.27 (a) Action spectrum for phototropism. (b) LOV domains are the domains in phototropin proteins that bind the FMN photoreceptors. Absorption spectra (left) and fluorescence spectra (right) for LOV domain proteins from Arabidopsis (top four panels in each column; the bottom panels show data from a phytochrome-related protein associated with two LOV domains). Part of Figure 1 in Christie *et al.* (1999), Proceedings of the National Academy of Sciences, USA 96, 8779–8783, © National Academy of Sciences, USA; reproduced with permission.

with an absorption peak at 380 nm is also bound to the protein and like FAD, this exhibits photo-reversible redox reactions. It is currently thought that the MTFH acts an antenna, directing the absorbed light energy towards the FAD.

Cryptochrome proteins form dimers by association of their N-terminal domains, and this dimerization is essential for activity. Cryptochromes that are activated by light of an appropriate wavelength interact directly with the negative regulator COP1 (see section 3.7.3), thereby causing it to release the transcription factor HY5 and leading to activation of the relevant genes.

3.7.5 Phototropins

Charles Darwin's work on phototropism has already been discussed in section 3.6.2. Here we need to add that he also observed blue light to be the most effective for induction of phototropism (see also Chapter 6, section 6.11). More recent detailed analysis of the action spectrum reveals two peaks in the blue region of the spectrum and one in the ultraviolet (Figure 3.27). This type of spectrum is characteristic of flavins (see above), leading to the idea that a flavin-based compound may be the light receptor involved in phototropism.

Study of mutants in *Arabidopsis thaliana* reveals that there are two proteins at the centre of the phototropic response. These are called phototropin 1 and phototropin 2, encoded by *phot1* and *phot2* respectively (see also Chapter 6, section 6.11). In responding to blue light, phot1 and phot2 have different but overlapping roles, as shown in Table 3.5.

The two proteins are very similar in structure, each possessing a photoreceptor region and C-terminal protein kinase domain involved in signalling. The photoreceptor region contains two LOV domains, to each of which is bound one molecule of flavin adenine mononucleotide (FMN – Figure 3.28). FMN is the photoreceptor, thus confirming the deduction made from the action spectrum that the photoreceptor is a flavin (see above). In the phototropins, FMN is excited by blue light, thereby enabling it to activate the protein kinase domain at the C-terminal end of the protein. From this point, the signalling pathway that leads to auxin re-distribution

Table 3.5 Roles of phototropins.

Response	Which phototropin?
Phototropism	Mostly phot1; phot2 only involved at high fluence rates
Stomatal aperture	Phot1 and phot2 across a wide range of fluence rates; cryptochromes may also have a role here
Chloroplast migration to illuminated side of cell in low light	Phot1 and phot2
Retreat of chloroplasts away from illuminated side of cell in very bright light (avoidance of photobleaching)	Phot2

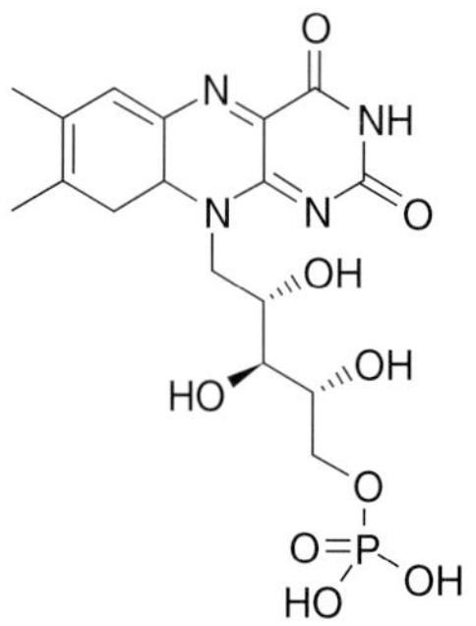

Figure 3.28 FMN – flavin mononucleotide, the phototropin photoreceptor

(and for that matter, to the very rapid inhibition of growth on the illuminated side – see section 3.6.2) is not understood. One event that follows is the phosphorylation of the phototropin protein itself by the kinase domain – but how, or indeed whether, this is involved in subsequent signalling is not clear.

3.8 Concluding comments

In this and the previous chapter, we have focused mainly on structures, events and processes at the molecular, organellar and cellular levels. However, we cannot leave it there. For growth, development and maintenance of the whole plant, it is vital that cellular and molecular events and processes are integrated and coordinated throughout the plant. Thus, the next few chapters consider development and function at higher levels, starting with the

earliest phases of plant life, from the single-celled zygote to the functioning multi-cellular green plant.

Selected references and suggestions for further reading

Baulcombe, D. (2004) RNA silencing in plants. *Nature* **431**, 356–363.

Binder, B. M. (2008) The ethylene receptors: complex perception for a simple gas. *Plant Science* **175**, 8–17.

Bishopp, A., Mahonen, A.P. & Helariutta, Y. (2006) Signs of change: hormone receptors that regulate plant development. *Development* **133**, 1857–1869.

Black, D. & Newbury, S. (2004) RNA interference and what it does. *The Biochemist* **26**, 7–10.

Brown, T. A. (2006) *Genomes 3*. Garland Science, London.

Dreher, K. & Callis, J. (2007) Ubiquitin, hormones and biotic stress in plants. *Annals of Botany* **99**, 787–822.

Evans, D. M. A., Bryant, J. A. & Fraser, R. S. S. (1984) Characterization of RNA-dependent RNA polymerase activities in healthy and TMV-infected tomato plants. *Annals of Botany* **54**, 271–281.

Franklin, K. A. & Quail, P. H. (2009) Phytochrome functions in *Arabidopsis* development. *Journal of Experimental Botany* **61**, 11–24.

Galon, Y., Finkler, A. & Fromm, H. (2010) Calcium-regulated transcription in plants. *Molecular Plant* **3**, 653–669.

Hart, J. W. (1988) *Light and Plant Growth*. Unwin Hyman, London.

Heslop-Harrison, J. S. & Schwarzacher, T. (2011) Organisation of the plant genome in chromosomes. *Plant Journal* **66**, 18–33.

Holland, J. J., Roberts, D. & Liscum, E. (2009) Understanding phototropism: from Darwin to today. *Journal of Experimental Botany* **60**, 1969–1978.

Inaba T. & Schnell D. J. (2008) Protein trafficking to plastids: one theme, many variations. *Biochemical Journal* **413**, 15–28.

Jones, R., Thomas, H., Ougham, H. & Waaland, S. (2012) *The Molecular Life of Plants*. Wiley-Blackwell, Chichester and Oxford.

Jorgensen, R. A. (2011) Epigenetics: biology's quantum mechanics. *Frontiers in Plant Science* **2**, article 10, doi:10.3389/fpls.2011.00010.

Kaufmann, K., Pajoro, A. & Angenent, G. C. (2010) Regulation of transcription in plants: mechanisms controlling developmental switches. *Nature Reviews Genetics* **11**, 830–842.

Kendrick, M.D. & Chang, C. (2008) Ethylene signalling: new levels of complexity and regulation. *Current Opinion in Plant Biology* **11**, 479–485.

Lee, Y., Lee, H-S., Lee, J-S., Kim, S-K. & Kim, S-H. (2008) Hormone- and light-regulated nucleocytoplasmic transport in plants: current status. *Journal of Experimental Botany* **59**, 3229–3245.

Leyser, O. (2006) Dynamic integration of auxin transport and signalling. *Current Biology* **16**, R424–R433.

Müller, D. & Leyser, O. (2011) Auxin, cytokinin and the control of shoot branching. *Annals of Botany* **107**, 1203–1212.

Nature News and Comment/Mello CC (2011) Video animation: RNA interference http://www.nature.com/news/video-animation-rna-interference-1.9673

Schwenkert, S., Soll, J. & Bolter, B. (2011) Protein import into chloroplasts: how chaperones feature into the game. *Biochimica et Biophysica Acta – Biomembranes* **1808**, 901–911.

Shinshi, H. (2008) Ethylene-regulated transcription and cross-talk with jasmonic acid. *Plant Science* **175**, 18–23.

Sullivan, J. A. & Deng, X. W. (2003) From seed to seed: the role of photoreceptors in *Arabidopsis* development. *Developmental Biology* **260**, 289–297.

Swartz, T. E., Corchnoy, S. B., Christie, J. M., Lewis, J. W., Szundi, I., Briggs, W. R. & Bogomolni, R. A. (2001) The photocycle of a flavin-binding domain of the blue light photoreceptor phototropin. *Journal of Biological Chemistry* **276**, 36493–36500.

Tsukaya, H. (2006) Mechanism of leaf shape determination. *Annual Review of Plant Biology* **58**, 249–266.

Wang, X-F. & Zhang, D-P. (2008) Abscisic acid receptors: multiple signal-preception sites. *Annals of Botany* **101**, 311–317.

Wasternack, C. (2007) Jasmonates: An update on biosynthesis, signal transduction and action in plant stress response, growth and development. *Annals of Botany* **100**, 681–697.

Watson, J. D. & Crick, F.H.C. (1953) Genetical implications of the structure of deoxyribonucleic acid. *Nature* **171**, 964–967.

Whippo, C.W. & Hangarter, R.P. (2006) Phototropism: Bending towards Enlightenment *Plant Cell* **18**, 1110–1119.

Woldermariam, M. G., Baldwin, I. T. & Galis, I. (2011) Transcriptional regulation of plant inducible defences against herbivores: a mini-review. *Journal of Plant Interactions* **6**, 113–119.

CHAPTER 4

From Embryo to Establishment

4.1 Introduction

In Chapter 1 we introduced the concept of alternation of generations between the gametophyte and the sporophyte (and it will be again discussed in Chapter 8). This alternation is very obvious in simple land plants, but it is far from obvious in flowering plants themselves. This is because, as indicated in Chapter 1 (section 1.8) and Chapter 8 (section 8.5), the gametophyte generation has, during the course of evolution, been reduced to a few cells contained within the sporophyte. The organisms that we know as flowering plants thus represent the extreme in development of the sporophyte. The first cell in each new sporophyte generation is the zygote, resulting from the fusion of a sperm cell with an egg cell (Chapter 8, section 8.6.3). It is at this point where we 'break into' the plant's life-cycle, looking at the earliest stages in the life of the new sporophyte. The initial focus is therefore on the development of the embryo from the zygote.

4.2 Embryogenesis

The general pattern of embryogenesis is similar throughout the flowering plants, although there are variations in detail both between monocots and dicots and within each group. In the general eudicot pattern, the zygote elongates about three-fold in the direction of the future apical-basal axis. Positional information is generated; the embryo, albeit one cell, now has an established polarity, though there has been no cell division. Subsequently, the first division of the zygote, which occurs at right angles to the long axis of the cell, is asymmetric, and this division reinforces the up-down polarity of the embryo. This first cell division is accompanied by the deposition of a cell wall, and recent evidence suggests that the orientation of particular glycoproteins in the cell wall reinforces the polarity of the embryo.

The smaller, uppermost of the two cells arising from the asymmetric division, will give rise to most of the embryo, while the larger basal cell undergoes a limited series of divisions to become the suspensor. This establishment of polarity, starting with the elongation of the zygote, which is always the same way round with respect to the structure of the embryo sac, almost certainly relies on positional information. Exactly how that information is signalled to the zygote to set up the asymmetric division giving rise to cells with such different subsequent fates is not clear.

However, it *is* clear that the asymmetric division involves asymmetric expression of genes that will later be involved in further development of the embryo and suspensor. For example, the zygote, prior to division, expresses two *WOX*[i] genes, *WOX2* and *WOX8*. These encode homeodomain transcription factors (which are discussed later). After the asymmetric division, *WOX2* is only expressed in the apical (embryonic) cell and *WOX8* in the basal (suspensor) cell. The latter cell also expresses *WOX9*. Furthermore, after the first division, as illustrated in Figure 4.1, the transcriptional regulators MONOPTEROS (MP) and BODENLOS (BDL) are expressed only in the apical (embryonic) cell, while the auxin-response protein, PIN7 is present only in the basal (suspensor) cell, where it localized to the apical end of the cell. This is consistent with the observation that there is a unidirectional transport of auxin from the basal to the apical cell.

This directional transport is, however reversed, after a few embryonic cell divisions, prior to the establishment of the root apical meristem (see below). At about the

[i] *WUSCHEL-related homeobox* genes.

Functional Biology of Plants, First Edition. Martin J. Hodson and John A. Bryant.
© 2012 John Wiley & Sons, Ltd. Published 2012 by John Wiley & Sons, Ltd.

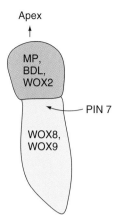

Apex

MP,
BDL,
WOX2

PIN 7

WOX8,
WOX9

Figure 4.1 Diagram showing asymmetric expression of genes in the two-celled embryo. MONOPTEROS (MP), BODENLOS (BDL) and WOX 2 are expressed only in the smaller apical cell (which will become the embryo proper), while WOX8, WOX9 and PIN7 are expressed only in the larger basal cell. Within that cell, the PIN7 protein is localized at the apical end.

same time, the expression of *WOX8* is down-regulated in the suspensor and is now expressed in the embryo cells adjacent to the suspensor.

The suspensor anchors the embryo at the micropyle (see Chapter 8) and provides a route for the passage of nutrients, hormones and other growth regulators from maternal tissues to the embryo. Indeed, the suspensor has been called the angiosperm equivalent of the mammalian placenta (although, unlike the placenta, it has only a short period of existence during embryonic development). The cells of the suspensor become enlarged and undergo extensive endoreduplication of their DNA, i.e. the replication of DNA in the absence of cell division (as described in Chapter 2, section 2.13.5).

At this point, we note that suspensor cells of some species are the 'record-holders' in respect of the number of rounds of DNA replication that occur in an undivided cell. For example, in suspensors of runner bean (*Phaseolus coccineus*), cells with 16,384C amounts of DNA have been recorded, representing 13 rounds of DNA replication from the original 2C value! The function of endoreduplication remains a matter for debate, but it is relevant here that it leads to nuclear and cellular enlargement. In the suspensor, this provides a larger area of cell surface via which nutrients may be transferred.

Meanwhile, the embryo proper has started to undergo a series of divisions, initially forming a globe-shaped

structure in eudicots such as *Arabidopsis thaliana*[ii] and an elongated globe in monocots such as rice (*Oryza sativa*). Even at this very early stage, it has proved possible to determine the subsequent developmental programmes of specific cells or groups of cells. A cell layer on the surface of the embryo, formed by cell divisions parallel to the surface (periclinal divisions) becomes the protoderm. Its position on the outside leads to its later role of the epidermis. The transcription factor WOX2, possibly 'aided' by WOX1 and WOX3, is important in the regulation of planes of division at this stage and thus has a direct role in subsequent patterning.

Also clear is that there is now differential expression of several genes between the protoderm and the inner cells. Further divisions, internal to the protoderm, set up a pro-cambium which will later give rise to the vascular tissue; surrounding it is the ground meristem, the precursor of the less specialized tissues between the vascular strands and the epidermis.

In the vertical axis, the base of the embryo becomes the root apical meristem (RAM)[iii]. Very interestingly, the cell at the top of the suspensor (i.e. adjacent to the embryo), the hypophysis or hypophyseal cell, is incorporated into this region of the embryo and actually becomes the founder cell of the embryonic root meristem. It that role, it divides asymmetrically, leading to the formation of the quiescent centre, which we may regard as a group of stem cells or organizer cells within the meristem. Such incorporation of a specific cell in the suspensor into the embryo proper is reliant on positional information, which involves signalling from the embryo to the hypophysis. This is dealt with here in detail as an example of the way in which the transport of signalling molecules can mediate position-dependent development. It has been known for several years that auxin is involved in this process, and we are now beginning to understand how this hormone works.

[ii]We need to note that, in some respects, *A. thaliana* may give us a slightly atypical view of embryogenesis. Its cell division patterns are much more regular, and thus subsequent developmental programmes may be more clearly localized than in other angiosperms. It has been suggested that this situation arises because the plant, and hence its embryo, is small – there is no room for wastage.

[iii]Note that whatever type of root meristem is possessed by the adult plant (see Chapter 5, section 5.3.1), all embryonic RAMs are of the closed configuration. Changes to other configurations (open, intermediate) occur after the resumption of growth.

Referring again to Figure 4.1, the original basal (suspensor) cell has, meanwhile, divided transversely to give a total of six to nine cells. PIN7 is still localized to the apical end of the two uppermost cells of the suspensor, while another protein involved in auxin signalling – PIN1 – is now present in specific locations in the embryo cells.

However, by the time the embryo reaches the early globe stage, things have changed. In the uppermost cell of the suspensor, the hypophysis and in the adjacent cell, the localization of PIN7 has changed from the apical end to the basal end. The membrane-trafficking control protein GNOM[iv] is essential for this change of location. In the embryo cells adjacent to the hypophysis, auxin activates the auxin-dependent transcription factor, MP. This in turn depends on the auxin-dependent degradation of BDL, which is a negative regulator of MP. MP then activates the transcription of several genes, which results in, among other effects, the transport of auxin from the embryo to the suspensor. Although other cells in the suspensor receive and respond to the auxin signal, only the hypophysis reacts in this very specific way.

What controls this cellular specificity is not clear, but a recent finding from Weijers' laboratory (Figure 4.2) in Wageningen (The Netherlands) suggests that MP can influence gene activity 'at a distance'. Among the genes activated by MP are two transcription factors which are expressed in the embryo cells adjacent to the hypophysis. One of these proteins actually moves from where it is made in the embryo into the hypophysis (and not to other suspensor cells); it thus a *mobile* transcription factor. Its exact role in the hypophysis is not clear, but based upon its structure, it is suggested that it combines with another protein, with which possibly it acts to regulate cell cycle genes (see Chapter 2.13.3).

A further indication of the difference between the hypophysis and the other suspensor cells is that the latter synthesize a specific protein kinase, encoded by the *SHAGGY* gene. Its role is not clear but it appears to be associated with the specific functions of the suspensor.

Auxin signalling is thus important in setting up the quiescent centre in the new RAM. However, this is not the whole story. Recent research on *Arabidopsis thaliana* indicates that cytokinin is also involved, and there may be a critical and dynamic balance between auxin and

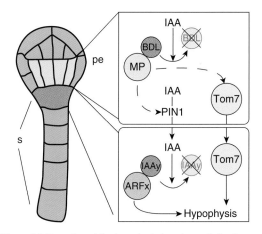

Figure 4.2 Formation of the hypophysis, based on activity of a mobile transcription factor. In the pro-embryo (pe), IAA induces the degradation of BDL protein. This releases MP, which activates auxin transport via PIN1; thus, auxin is moved to the apical cell of the suspensor(s). This transport of auxin is necessary, but not sufficient for hypophysis formation; another mobile signal is also needed. That signal is the transcription factor TOM7, synthesis of which is up-regulated in the pro-embryo by MP. Reproduced by permission from Schlereth, A. *et al.*(2010) Nature 464, 913–916

cytokinin. Further, mutations affecting brassinosteroid signalling also affect specification of the hypophysis, and there is evidence for an interaction between brassinosteroids and auxin in this process. Finally, it must be pointed out (as made clear in Chapter 5, section 5.3.1) that the quiescent centre is just one part of a larger but nevertheless defined region, the root apical meristem (RAM), containing dividing cells. Overall specification of the RAM requires the activity of another member of the WOX family of homeodomain transcription factors – WOX5.

The developmental pathway exhibited by particular cells depends at least partly on what is happening in adjacent cells. A further example of this is seen in the relationship between embryo and suspensor. Some of the many mutations known to affect embryonic development (and discussed above) result in the formation of an embryo that is extremely rudimentary; this affects signalling to the suspensor. Thus, in *RASPBERRY1* and *SUS* mutants, suspensor cells proliferate to produce embryo-like structures. It appears, then, that normal embryos inhibit the proliferation and embryogenic potential of the suspensor cells, but that in these mutants there is no transmission of inhibitory signals.

Another mutant in which the inhibitory signalling breaks down is *TWN1* ('twin'). In these mutants, the

---

[iv]GNOM is involved in many situations in which membrane trafficking is involved, for example in root hydrotropism.

suspensor actually forms a second fully viable embryo in the presence of an initially normal embryo (although cotyledon formation may be affected at a later stage). The *TWN1* gene is thus thought to be involved in cell-to-cell signalling, but it is not known whether this involves a failure of embryo cells to transmit signals to the suspensor or failure of the suspensor to respond to the signal. The occurrence, in some *TWN1* mutants of aberrant cotyledon development, suggests, but by no means proves, that the lesion is in embryonic signalling rather than in suspensor perception.

As cell division continues, the cotyledons are initiated (leading to a heart shape in eudicots and a cylindrical shape in monocots) and the shoot apical meristem (SAM) is also laid down (see Figure 4.3). The actual relative timings of the establishment of the SAM and the cotyledon(s) vary between species. In *Arabidopsis*, the events occur at more or less the same time but they are under the control of different genes. Two main genes, *WUSCHEL* (*WUS*) and *STEMLESS* (*STM*) are essential for specifying the SAM, with the activity of *STM* being dependent on the activity of several other genes.

The cotyledonary primordia, by contrast, are laid down mainly under the influence of the *LATERNE* and *PID* genes but maintenance of cotyledonary identity is dependent on *LEC1*[v], *LEC2 and FUS3* all of which encode transcription factors. Further growth of the region behind the root apical meristem forms the radicle (the embryo's 'root'). Auxin is involved here in regulating cell number and size in specific zones of the root apex, as shown by the effects of mutations in the *AUX-LAX* gene family. The cotyledons also continue to expand and, in many species, become the major storage organs; there is evidence for the involvement of auxin here, too.

As already indicated, the role of the suspensor is to channel nutrients and growth factors to the growing embryo. The suspensor is thus the final link in the supply chain, while the funiculus (the 'stem' of the ovary) provides the first part of the route from the ovary wall into the ovule. The whole supply chain obviously has a limited life, because the embryo stops growing as the seed approaches maturity (see below). The suspensor's role is complete by the time the cotyledons fill out, and most of the suspensor cells undergo programmed cell death (PCD – see Chapter 3, section 3.5.2 and Chapter 11,

section 11.3.8). Thus, in Figure 4.2, the only visible sign of the suspensor that remains is its large basal cell. The specific nature of the signal that initiates PCD in the suspensor is not known.

Overall, then, the development of the embryo involves a number of different processes. Cell division is regulated in time and space to establish form. The embryo is patterned, both radially and in the vertical axis; the developmental programmes of different parts of are specified in a sequential manner. The whole process is under genetic control within a context of positional information and cell-to-cell signalling.

We have discussed some aspects of the genetic regulation already but, at the most recent count, some 250 genes are known to affect embryogenesis in the 'model' eudicot *Arabidopsis thaliana*, and over 50 of these are involved in the patterning of the embryo. Constraints on space (and on our readers' patience) do not allow us to discuss the function of all these genes, but the more significant ones have been mentioned where appropriate. Three further points may be made in summary.

The first point is that the expression of only a few of these genes is specific to the embryo. This becomes obvious when we think about the genes that operate at various different levels of regulation. Thus, maintenance of meristem identity is as much part of, for example, a root in an established plant as it is in the embryo. Some cell signalling processes occur in many phases of a plant's life; essential 'housekeeping' genes are expressed throughout the plant, and so on.

It is thus not surprising that mutations affecting more generally expressed genes affect embryo development. Taking our examples from DNA replication in *Arabidopsis thaliana*, insertional inactivation of the *PROLIFERA* gene or of the *TILTED* gene results in embryo lethality. The former encodes MCM3, involved in the initiation of DNA replication, while the latter encodes a sub-unit of one of the replicative DNA polymerases, polymerase-ε. Interestingly, in plants with a weaker, non-lethal mutation in *TILTED*, cell-cycle length is increased by 35 per cent throughout embryogenesis and there are changes in cell patterning. This implies that there is an interaction between specification of cell type and the regulation of the cell cycle.

A similar situation is seen later in the life of the plant. Weak mutant alleles of the *INCURVATA2* gene, which codes for the catalytic sub-unit of DNA polymerase-α,

---

[v] *LEC: leafy cotyledon.*

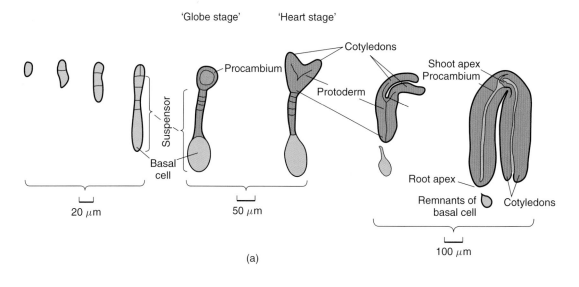

(a)

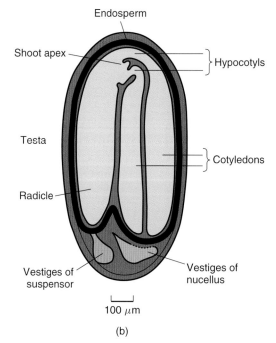

(b)

Figure 4.3 (a) Diagram showing the main stages in a developing eudicots embryo. (b) The mature embryo *in situ* in the seed. Reproduced with permission from Bryant, J (1985), Seed Physiology, Edward Arnold. London.

cause, amongst other things, disruption of the homeotic patterning of floral organs (Chapter 8). Strong mutant alleles of the same gene are embryo-lethal.

The second point is that mention of homeotic patterning takes us right back to the early stages of embryo development. **Homeotic genes** were discovered in research on development in *Drosophila*, the fruit fly. Mutations in certain genes have very specific effects on the developmental patterning of the segmented body of the fly, such that, for example, feet grow where antennae should be.

These genes are, in effect, the master regulators of the tissue patterning associated with segmentation. They encode transcription factors which contain a particular DNA binding region, the *homeodomain*, a tract of 60 amino acids (the corresponding coding sequence in the gene is the *homeobox* – the genes are therefore called the *Hox* genes).

It was originally thought that *Hox* genes were confined to segmented animals, but it is now clear that they are present in plants and their appearance in evolution thus pre-dates the plant-animal split (see Chapter 1, section 1.4). Plants do not have a segmented body plan, but nevertheless these genes are involved in the patterning and identity of specific regions and organs. Several have already been mentioned in this chapter, including the *WOX* gene family, different members of which are involved in specification of the suspensor and root apical meristem and in patterning at the globe stage of embryo development. At the opposite end of the embryo, *WUSCHEL* and *STEMLESS* are among the homeobox genes involved in the initiation, establishment and activity of the shoot apical meristem. Maintenance and continuing activity of the SAM require the activity of KNOX genes that are members of the KNOTTED family of homeobox transcription factors (see Chapter 6, section 6.3.1). This list of homeobox genes/homeodomain proteins involved in embryo development is not complete[vi], but nevertheless the examples give an indication of the integrated and sequential expression of key regulatory genes in the formation of embryonic organs.

The third point is that the discussion on gene expression has been focused on the new sporophyte. However, in at least the early stages of embryogenesis, much of the mRNA actually comes from gene activity in the gametophyte and, more specifically, in the gametes themselves. Since the ovum is larger than the sperm cell, transcripts from the maternal genome are more abundant in the zygote than are transcripts from the paternal genome. There is thus a real possibility of maternal effects in the development of early phenotype[vii].

4.3 Endosperm

The remarkable triple fusion event between one of the pollen's sperm nuclei and the diploid central cell in the embryo sac[viii] (see Chapter 8, section 8.6.3) leads to the development of a tissue called the endosperm. Because the triple fusion event is spatially separate from the conventional sexual fusion that gives rise to the embryo, the endosperm is not part of the embryo. However, in the early stages of embryo growth, the embryo becomes embedded or partially embedded in the endosperm and, in some species, endosperm still surrounds or almost surrounds the embryo in mature seeds.

The role of the endosperm in seed development and in germination varies considerably from species to species. In many eudicots, typified by *Arabidopsis thaliana*, the endosperm undergoes development and expansion early in seed development, but at a later stage its contents are absorbed by the embryo so that the endosperm is reduced to just one cell layer. In other, non-endospermous eudicots, the endosperm may indeed disappear totally. In seeds with reduced or non-existent endosperm, the cotyledons become the major storage organs. In other eudicots, such as the castor bean, the endosperm remains the major storage tissue of the seed, while some species exhibit intermediate situations in which the endosperm remains as a few layers of cells at maturity. In monocots, the best characterized endosperms are those of cereals, in which the endosperm mainly consists of dead cells at maturity, in contrast to endospermous eudicots, where the endosperm remains alive in the mature seed.

The first division of the triploid endosperm founder cell usually occurs before the first division of the zygote. In the majority of species, this first division is followed

---

[vi]For a fuller account the reader is referred to De Smet *et al.* (2010).

[vii]This should not be confused with maternal imprinting, in which only the maternal allele at a particular locus is active.

[viii]In very primitive angiosperms, including water lilies, the second sperm nucleus fuses with a single female haploid cell. Thus, in these species, the endosperm is diploid.

by a very rapid series of partially truncated cell cycles, in which cytokinesis (cell division) does not keep pace with karyokinesis (nuclear division). This results in a multinuclear **syncytium**. Remarkably, the nuclear division cycles during this phase are synchronous with each other. Endosperms that exhibit this phenomenon are known as *nuclear-type* endosperms. Cellularization then occurs as cell division and cell wall synthesis catch up with nuclear divisions.

In other species, mitosis and cytokinesis in the endosperm are 'conventional'. Such endosperms are called *cellular-type* endosperms. At the cellularization stage (or its equivalent in cellular-type endosperms), two zones of transfer cells become apparent. One zone is adjacent to a region of the ovule known as the chalaza, and it is involved in the transfer of nutrients from maternal tissues to the endosperm. The other zone is adjacent to the embryo and is involved in the transfer from endosperm to embryo. In those species in which the endosperm disappears during seed development, this transfer is initiated soon after the cellularization stage, so that the newly established cellular endosperm is gradually emptied and the cells atrophy and die.

In persistent endosperms which remain in the seed at maturity, typified by the cereals, the cellularization phase is followed by DNA endoreduplication; DNA is replicated, but mitosis does not occur. This can lead to cells with up to 384C amounts of DNA (seven rounds of DNA replication of the original 3C genome). The switch to endoreduplication occurs in response to an increase in the amount of auxin, so that the auxin to cytokinin ratio increases markedly.

The function of this process is still the subject of discussion. Is it to increase the dosage of genes involved in the synthesis of storage products? Is it to provide large cells in which to deposit those products? Or is it a mixture of both?

The balance of parental genomes in the endosperm of most angiosperm species is obviously unusual, namely two maternal (M) and one paternal (P). This balance is very important in the correct development of endosperms of both the persistent and transient types. If plants of the same species but with different ploidy levels (Chapter 2, section 2.13.5) are crossed, then variants of the normal genome balance are created. M : P genomes in a ratio greater than 2 lead to underdevelopment of the endosperm; ratios of less than 2 lead to excessive development.

Following a normal fertilization between plants of the same ploidy, imprinting mechanisms cause particular genes to be expressed from one genome rather than another. Mi-RNAs (see Chapter 3, sections 3.2.3 and 3.4.4) are abundant in endosperm and are thought to have a role in this imprinting. Changing the genome ratios disturbs this imprinting, as do mutations in genes in a small family of *POLYCOMB*[ix] genes. The group involved in flowering plant endosperm formation is known as the *FIS* (Fertilization-Independent Seed) gene complex. The proteins encoded by the *FIS* gene complex bring about histone modifications that specifically silence target alleles in the maternal or paternal genomes, thus causing imprinting. Mutations in these genes upset the imprinting and can even, as the gene name implies, cause endosperm to start developing in the absence of fertilization. However, the seeds do not progress to maturity and are aborted.

4.4 Perisperm

As was indicated in Chapter 1, section 1.7, the second fertilization event leading to endosperm formation is confined to angiosperms. In gymnosperms, the main nutritive tissue is the perisperm, derived from the nucellus which surrounds the embryo sac within the ovule. This also occurs in some angiosperms, mostly members of the more primitive groups. However, with one exception (the very primitive angiosperm *Hydatella*), nutrient deposition in angiosperm perisperm follows fertilization, whereas in gymnosperms it begins before fertilization.

4.5 Late embryo growth, storage deposition and desiccation

4.5.1 Non-endospermous eudicots

The overall pattern of embryo growth in a typical eudicot is shown in Figures 4.3 and 4.4. As has been emphasized already, the timescales over which these events take place vary considerably between species, but the pattern remains similar. After the establishment of the general plant body plan (described above), cell division ceases but embryo size increases dramatically, mainly through cell enlargement in the cotyledons. Correlated with this

---

[ix]Named for the effects in *Drosophila* of a mutation in one of these genes.

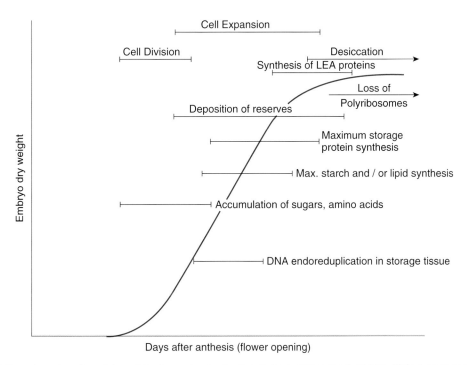

Figure 4.4 General overview of events in seed development and maturation. Notes on DNA endoreduplication: (1) there are some species in which DNA endoreduplication does not occur in the storage tissue (e.g. cotyledons of *Brassica napus,* oil-seed rape or canola). (2) DNA endoreduplication occurs early in embryogenesis in the embryo suspensor (not indicated in this diagram).

transition is a decrease in hexose content and an increase in sucrose content. The decline in the ratio of hexose : sucrose appears to be an important factor in the switch to the seed maturation phase. This is consistent with the role of sucrose as a signalling molecule, almost certainly acting in this instance alongside, or as part of, a signalling network with auxin.

At this stage, the deposition of the storage reserves is initiated in the cotyledons and continues until seed desiccation sets in (see Figure 4.4). In many species, including legumes and *Arabidopsis*, the cotyledon cells undergo DNA endoreduplication. DNA amounts of 16 and 32C are relatively common, while in many legumes, some cotyledon cells undergo five rounds of endoreduplication, giving 64C amount of DNA. As noted above in respect of endosperm cells, the function of DNA endoreduplication remains unknown. It is clear however, that it is not an absolute necessity in the deposition of reserves. For example, DNA endoreduplication does not occur in the cotyledons of *Brassica napus*, which are also storage organs with large cells.

The major storage compounds are proteins, polysaccharides and lipids (mainly tri-acyl glycerols, also known as triglycerides) with the relative proportions of these compounds differing between different species (see Table 4.1). Less widespread or less abundant compounds, including oligosaccharides, may also be deposited.

Some of the less abundant compounds are defence chemicals (anti-nutritional factors), including non-protein amino acids such as canavanine, poisonous glycosides (including cyanogenic glycosides), condensed tannins and the purine derivative, caffeine. Other compounds, including protease inhibitors and phytic acid (inositol hexaphosphate) serve a dual role of storage and defence. The biochemical pathways involved in the synthesis of the major compounds were described in Chapter 2. Here we need to point out that the different compounds may be stored at different locations within the seed and within specific sub-cellular organelles.

Plant storage proteins were originally grouped and classified according to their solubility: **albumins** are water-soluble; **globulins** are soluble in dilute salt

Table 4.1 Composition, as percentage of total dry weight, of storage compounds in seeds of four eudicot crop plants.

	Garden pea, Pisum sativum	Peanut, Arachis hypogaea	Soybean, Glycine max*	Oilseed rape, Brassica napus
Protein	22%	30%	37%	21%
Carbohydrate	50%	15%	30%	20%
Lipid	5%	48%	22%	48%

*Note: In common with several other legumes, soybean stores carbohydrate in the form of structural polysaccharides (mainly arabinogalactans) in enlarged primary walls of the cotyledon cells. It also stores galacto-oligosaccharides such as raffinose and stachyose (Figure 4.5). Only a small proportion of the stored carbohydrate in soybean – up to one third – consists of starch.

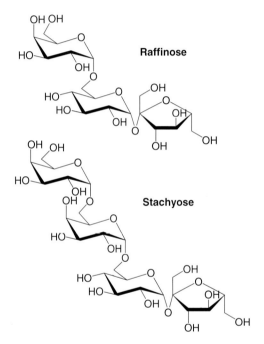

Figure 4.5 Structures of stachyose and raffinose. Images from Yikrazuul, Wikimedia Commons.

solutions; and **prolamins** dissolve in aqueous alcohol. The latter only occur in cereals and other grasses.

There is also a specific sub-group of cereal/grass storage proteins, the **glutelins**, that do not dissolve in aqueous alcohol; dilute acid or alkali is required for their extraction. Originally placed in a group of their own because of the difficulty in dissolving them, they are now recognized as chemically belonging to the prolamin group, but with polypeptide chains that are held together by disulphide bridges.

The synthesis of all storage protein groups is under tight temporal/developmental control in relation to the morphological development of the seed, with control being exerted at the level of transcription. Thus, in genetic modification experiments, it has been possible to confine expression of the inserted gene to seeds simply by linking the gene to a promoter from a gene encoding a seed storage protein.

The major storage proteins in eudicots are firstly the small albumins, usually termed the 2S albumins (based on their speed of sedimentation when subjected to ultra-centrifugation). These proteins are found right across the range of eudicots and, in some species (e.g. soybean), they also act as protease inhibitors (see above). The second and larger group of eudicot storage proteins are the globulins, initially characterized in legumes. There are two main types, the 7S vicilins and the 11S legumins. Despite their 'legume-specific' names, these proteins also occur across the range of eudicots.

Storage proteins are synthesized on polyribosomes attached to the endoplasmic reticulum (ER) and are inserted into the lumen of the ER during synthesis, as described in Chapter 3 (section 3.2.4). Protein-filled vesicles then bud off from the ER and are transported either directly or via the Golgi to the protein-storage vacuoles (protein bodies). These may arise by coalescence of Golgi-derived vesicles or by incorporation of protein into pre-existing lytic vacuoles. After synthesis, but before final deposition, the 7S vicilins undergo complex cleavage patterns (Figure 4.6) to generate polypeptides of several different sizes and are also modified by glycosylation (the addition of carbohydrate residues). The 11S legumins are also processed after synthesis (Figure 4.6). The initial

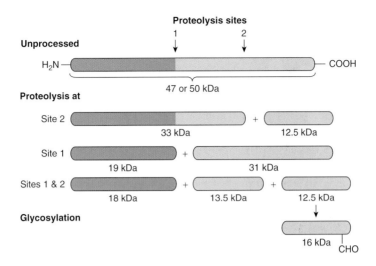

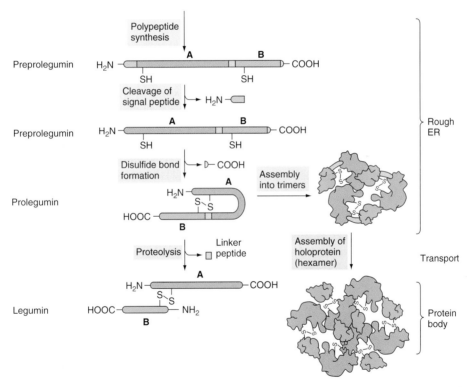

Figure 4.6 Processing of eudicot storage proteins. Top: 7S vicilin-type proteins. Three different-sized polypeptides are generated from one precursor. Each of these polypeptides may be glycosylated, with the extent of glycosylation varying between species. Bottom: 11S legumin-type proteins. After removal of the signal peptide (see text), the protein is folded so that an intramolecular S-S bridge can form. At this stage, the protein assembles into a trimer and then the linking peptide is removed. Each molecule of legumin now consists of two polypeptides held to together by an S-S bridge and assembled into a trimer with two other legumin molecules. Trimers are transported to protein bodies, where they join in pairs to form the final hexameric product.

Redrawn with permission from page 1034 in Buchanan, B. *et al.* (2002), Biochemistry and Molecular Biology of Plants, ASPB, Rockville, MD and Figure 1 in Krochko, JE and Bewley, JD (1988). Electrophoresis 9, 751–763.

polypeptide chain folds and is then cleaved to form a large acidic polypeptide and a smaller basic polypeptide. Six each of the two polypeptides are assembled within the protein bodies to form a large 'holoprotein.'

Although the major storage carbohydrate is starch, others are also stored, depending on species, as already mentioned. Indeed, the relatively widespread occurrence in legumes of the galactose-based oligosaccharides, raffinose and stachyose, is one of the factors that contribute to the digestive discomfort when some types of beans are ingested by non-ruminant animals (including humans). However, at this point we focus on starch. Its deposition is straightforward and does not significantly depart from the pattern discussed in Chapter 2, section 2.5.3. The main point to note is that the starch grains in cotyledons are usually much larger than those in normal leaves. In pea cotyledons, for example, it is not unusual for starch grains to be up to 20 μm diameter (Figure 4.7).

The major storage lipids are the triacylglycerols (TAGs) – three long-chain fatty acids combined with a glycerol molecule (Chapter 2, section 2.14.5). There is variation between species in the lengths of the fatty acid chains and in the degree of fatty acid unsaturation (Table 4.2). These two features are important in the use of plant lipids for human and animal nutrition, and in other applications, because they affect other chemical characters such as susceptibility to oxidation and physical characters such as fluidity.

As discussed in Chapter 2 (section 2.14.5), the site of fatty acid synthesis in plants is the plastid. Within the plastid, the main product of the synthetic pathway is oleic acid (18:1), linked to coenzyme A (i.e. it is a long-chain acyl-CoA) which is exported from the chloroplast and transferred to the ER. Palmitic acid (16:0) and stearic acid (18:0) are minor products of the synthetic pathway and are also transferred to the ER as an acyl-CoA.

Elongation and desaturation (insertion of double bonds) reactions take place within the ER as does esterification with glycerol. The triacylglycerols accumulate between the two layers that form the phospholipid bi-layer of the ER membrane. This accumulation causes the outer layer to swell and, eventually, to break off as a lipid droplet bounded by a single layer of phospholipid. At the same time, a specialized group of proteins, the oleosins, are synthesized on the surface of the ER and coat the lipid droplet before it breaks off (as shown in Figure 4.8), thus forming 'oil bodies' (sometimes referred to as oleosomes). Oleosins are small proteins which have, towards the middle of the molecule, a hydrophobic 'finger' of about 80 amino acids that sticks into the lipid droplet. It is thought that their function is to stabilize the oil bodies at low water potential after desiccation has occurred. Consistent with this view is the observation that they hardly occur in seeds that that are not desiccation-tolerant (such as cocoa) nor in lipid-rich, non-desiccated fruit, including olive and avocado.

The final stage of seed maturation is desiccation (Figure 4.4). In preparation for this, anti-oxidant compounds and protective proteins are synthesized. The latter are the LEA (Late embryogenesis abundant) proteins, of which there five or six different classes found in different plant groups. In general, they are hydrophilic, low-molecular-weight glycine-rich proteins, but the mechanisms by which they protect cell contents from desiccation are not known. Their synthesis is up-regulated by abscisic acid (ABA), the concentration of which increases in late embryogenesis. Class 2 LEA proteins, known as dehydrins, are synthesized in leaves in response to drought; again it is ABA that is the

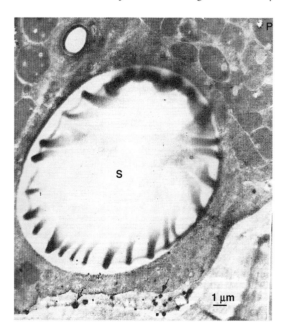

Figure 4.7 Section of a cotyledon of pea (Pisum sativum) after one day of imbibition. S = Starch grain; P = Protein body. The arrows indicate phosphatase activity adjacent to the cell wall. Phosphatases mobilize stored phosphates, e.g. in phytic acid.
Picture is cropped from Figure 1 in Bowen, ID and Bryant, JA (1978). Protoplasma 97, 241–250

Table 4.2 Percentage fatty acid composition of storage tri-acyl glycerols in a range of species. Note that these are 'typical' values; there is some variation in composition between different varieties within a species. The numbers indicate first, the number of carbons in the fatty acid chain and second, the number of double bonds (i.e. the degree of unsaturation). Thus 18:1, oleic acid, has 18 carbons and one double bond.

Plant	16:0 Palmitic acid	18:0 Stearic acid	18:1 Oleic acid	18:2 Linoleic acid	18:3 Linolenic acid	Others*
Oilseed rape, *Brassica napus*	4	2	60	22	10	2
Flax, *Linum usitatissimum*	3	7	21	16	53	
Peanut, *Arachis hypogaea*	11	2	48	32		7
Soybean, *Glycine max*	11	4	24	54	7	
Sunflower, *Helianthus annuus*	7	5	19	68	1	
Castor bean, *Ricinus communis*	1		3	5		91

*Other fatty acids include: Oil-seed rape, 22:1 (Erucic acid). The data are from a low-erucate line, bred for animal and human nutrition. In high-erucate lines up to 40 per cent of total fatty acids may consist of erucic acid; Castor bean (an endospermous eudicot), 12-hydroxy-18:1 (Ricinoleic acid)

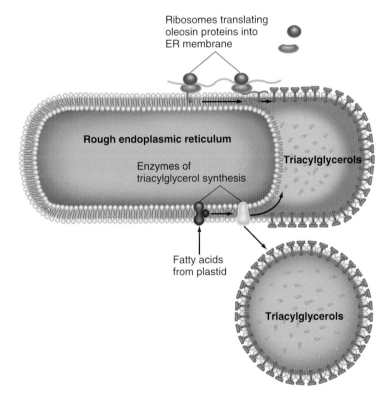

Ribosomes translating oleosin proteins into ER membrane

Rough endoplasmic reticulum

Enzymes of triacylglycerol synthesis

Triacylglycerols

Fatty acids from plastid

Triacylglycerols

Figure 4.8 Formation of oleosomes (oil bodies) during seed maturation. Redrawn, with permission, from p 516 in Buchanan, B. *et al.* (2002), Biochemistry and Molecular Biology of Plants, ASPB, Rockville, MD and Figure 4 in Huang, AHC, (1992) Annual Review of Plant Biology 43, 177–200.

specific regulator. They are also synthesized in response to freezing (Chapter 10, section 10.2.2).

Returning to the seed, ABA is thus involved in specific synthetic events in late embryogenesis, but has an overall regulatory role both in the synthesis of storage compounds and in the process of seed desiccation.

This role is highlighted by experiments in which ABA was prevented from acting in developing bean seeds by complexing it with specific antibodies. Making the ABA unavailable pushed the immature seeds into the germination programme. However, it is also clear that, although the absolute amount of ABA is important, the ratio of

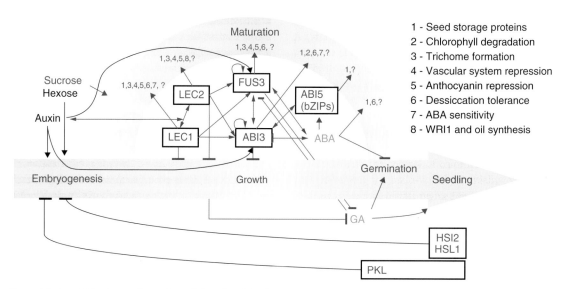

Figure 4.9 A model for the overall genetic regulation of seed development, maturation and early germination in *Arabidopsis thaliana*. Details of ABI, FUS and LEC functions are given in the text. During the early phases of germination, GA represses embryogenesis-type activities. Also involved in this repression are the chromatin-remodelling factor (see Chapter 3), PICKLE (PKL) and the transcriptional repressors HS12 and HSL1. The latter appear to be specific for events in which sugars have a regulatory role.
Reproduced by permission from Santos-Mendoza m. *et al.* (2008) Plant Journal 54, 608–620.

ABA to the growth-promoting hormone GA is also a significant factor. GA is implicated in general in maintaining the growth of the embryo in the earlier stages, and its synthesis is inhibited (by repression of the relevant genes) during the seed maturation phase.

Another clear biochemical feature of late embryogenesis is the changing gene transcription pattern. For example, genes involved in the general background biochemistry of the cells are slowly down-regulated, while genes involved in desiccation-tolerance are up-regulated. The embryo also synthesizes mRNA species that have no role at this stage of development, but which are stored in the dry embryo and are only translated when the seed is rehydrated prior to germination.

However, a large proportion of the mRNA, perhaps as much as 75 per cent of that present in the dry embryo, is actually 'residual' and is broken down in the early stages of germination. Nevertheless, the transcriptional and metabolic switches that occur during the desiccation stage are important for the subsequent germination of the mature seed; thus Ruthie Angelovici and her colleagues refer to seed desiccation as '*a bridge between maturation and germination.*'

The specific mechanisms involved in water loss are not clearly understood. It may well be that desiccation ultimately depends on the detachment of the funiculus, which effectively severs the vascular connection to the whole seed. There is also the possibility that water is actively removed from the seed, to be relocated in parental tissues. Overall, it is very probable that ABA and responsiveness to ABA continue to be involved.

Finally, study of mutants in *Arabidopsis thaliana* reveals that a group of master regulator genes encoding the B3 class of transcription factors exert overall control of seed maturation (Figure 4.9). This includes both the morphological/physiological aspects and the deposition of the storage compounds. The three genes are *LEC2*, *FUS3* and *ABI3*. The first two were mentioned earlier in the context of maintenance of cotyledon identity; *ABI3* is involved in the ABA receptor/signalling pathway. All three have been shown to regulate the synthesis of storage proteins and tri-acyl glycerols and to be involved, along with *LEC1*, in preventing precocious germination.

4.5.2 Endospermous monocots

Cereal endosperm may be regarded as the most important plant tissue in the world, providing the main source of dietary carbohydrate and protein for a large proportion of the human population (see Box 4.1). Endosperm development has therefore been the subject of intensive research, with the aim of maximizing yields and possibly modifying the biochemistry of the storage products and the balance between them.

Box 4.1 Seeds and human nutrition

The total dependence of humans and other animals on the photosynthetic ability of green plants is obvious. In addition to this general dependence, we also rely directly on plants as food, much of it based on seeds. The storage reserves of seeds laid down to provide nutrients to the growing seedling have been very widely exploited by humans, either in direct consumption or after processing (e.g. in bread making). Cereals are especially important, with rice, wheat and maize (corn) providing directly or indirectly up to 90 per cent of the world's food. Indeed, rice is the staple foodstuff of about 50 per cent of the world's population, contributing up to 75 per cent of their food intake (see Chapter 10, textbox 10.1). Pulses (peas, beans and their relatives) are also important.

Cereals, especially but not exclusively barley, are also used in making beer. In a process known as **malting**, grains are set to germinate, which results in the initiation of starch degradation (section 4.11.2). The grain is then heated to stop germination, and the partially broken-down starch, mostly in the form of the disaccharide maltose, is available for further breakdown and fermentation by yeast.

We note in passing that specific products, including lipids (oils) and starch are also extracted from seeds for both food and non-food uses.

The main stages of endosperm formation in cereals conform to the general pattern described earlier for nuclear-type endosperms: syncytium formation; followed by cellularization; followed by 'conventional' cell division; followed by DNA endoreduplication. The timing varies slightly between the endosperm itself and the boundary layer of cells, the aleurone. Indeed, there is also variation between species in respect of aleurone cell behaviour. In barley, for example, aleurone cells show a limited extent of DNA endoreduplication, whereas this does not occur in *Sorghum bicolor* aleurone.

Storage reserves are deposited rapidly during the endoreduplication phase, adding further to the expansion of the endosperm cells. At the same time, the embryo draws nutrients from the endosperm, contributing to embryonic growth and, especially, to growth of the single cotyledon or scutellum[x]. This will act as a transfer organ between the endosperm and the growing parts

---

[x]Named from the Latin *scutella* (a small shield) because of the shape of the cotyledon at maturity.

Table 4.3 Composition, as percentage of total dry weight, of storage compounds in cereals.

	Barley, *Hordeum vulgare*	Maize (Corn), *Zea mays*	Oats, *Avena sativa*	Wheat, *Triticum aestivum*
Protein	12%	10%	13%	12%
Carbohydrate	76%	80%	66%	75%
Lipid	3%	5%	8%	2%

of the embryo when the reserves are mobilized during germination.

As shown in Table 4.3, the most abundant reserves in cereals are carbohydrates. The bulk of the stored carbohydrates is starch. Amyloplasts show two patterns of deposition – several small (1-4 µm) starch grains or one large (20 – 40 µm) per amyloplast. In addition to starch, there are some soluble carbohydrates, including glucose, fructose and sucrose (and in some species, raffinose).

Sweetcorn is a form of maize which is homozygous for a mutant form of the *sugary-1* gene which encodes isoamylase, involved in the control of branching in amylopectin. A side-effect of the mutation is the accumulation of higher levels of sucrose. Interestingly there are records going back to the 18th century indicating that the Iroquois, a native American tribe, were familiar with sweetcorn and cultivated it separately from the wild-type plant. More recently, plant breeders have utilized mutations in another gene, *shrunken-1* in sweetcorn production. This encodes sucrose synthase, involved in the conversion of sucrose to starch (Chapter 2, section 2.5.3).

Thus the 'supersweet' varieties carry mutations in both *sugary-1* and *shrunken-1*; the sucrose content of the grains is much higher and their starch content is much lower than in wild-type plants. The grains also contain more water at maturity than in wild-type plants, possibly because of the osmotic potential exerted by sucrose.

The major storage proteins of cereals are globulins and prolamins (including the glutelins). The latter form a large and complex group, but the various types are thought to have arisen in the course of evolution by gene duplication and subsequent divergence. The nomenclature of this group is confusing, because some authorities still distinguish between prolamins and glutelins (whereas the latter are actually a sub-class of the former) and because equivalent proteins in different species have been given different names. Thus *gliadin* in wheat is the homologous protein to *hordein* in barley. Further, there is in general use the term **gluten**, used to refer to the prolamins of wheat, barley and rye in relation to their baking properties.

The specific biochemical properties of prolamins in these three cereals, and especially in wheat and barley, make them very elastic, ideal for making a dough that rises (in the presence of yeast) before baking and helping to maintain the shape of the baked material after baking. Unfortunately, however, some people are allergic to gluten, a condition known as celiac disease, and thus must avoid ingesting the prolamins of wheat and barley. Rice and oats have a lower prolamin to globulin ratio, and in general their prolamins are less antigenic than those of wheat and barley. Thus, many celiac disease sufferers can ingest oats and, especially, rice proteins without any problem – although in really severe cases, even the latter cause an allergic response.

As in eudicots, protein is laid down in protein bodies, which, along with the amyloplasts, occupy a large proportion of the cell. The protein bodies of cereals and other grasses are of two types (Figure 4.10). The more 'conventional' protein bodies are those in which the 11S globulins are laid down. These are the direct homologues of the eudicot 11S globulins and are the major storage proteins in, for example, rice and oats. Protein-containing vesicles are transported from the ER to the vacuole via the Golgi apparatus. As the vacuole fills, it fragments to form the individual protein bodies. The prolamins, by contrast, are stored in balloon-like distensions of the ER which form as the ER lumen fills with the newly synthesized proteins.

Lipid content of cereal endosperm is low, with oat showing a maximum of 8 per cent of dry weight. The tri-acyl glycerols are richer in unsaturated acids, especially 16:0 (palmitic acid), than those of most eudicots. As might be expected, the mode of synthesis of the tri-acyl glycerols is the same as in the eudicots. However, the amounts of oleosins produced are much smaller, in relation to the amount of stored lipid, than in eudicots. This means that much of the lipid is not contained within well-defined oil bodies but instead seeps amongst the protein bodies and amyloplasts in the large storage cells of the endosperm. The reason for this low oleosin content is not clear, but it may reflect less need for protection of the lipids (see above) in cells that will die during desiccation (see below) than cells that remain alive. This idea receives some support from the observation that fully-formed oil

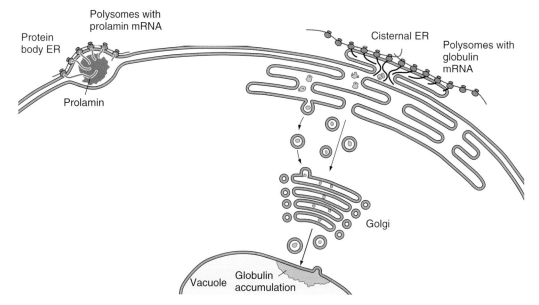

Figure 4.10 Formation of protein bodies in cereals. Prolamin-type and globulin-type (i.e. glutelins) exhibit different modes of synthesis and storage. Prolamins are laid down into the ER lumen, which balloons out and pinches off to make protein bodies. Globulins (glutelins) are transported from the ER via the Golgi to be deposited in the vacuole.
From Buchanan B. *et al.* (2002) Biochemistry and Molecular Biology of Plants, ASPB, Rockville, MD, p 1035

bodies do occur in the aleurone layer and in the scutellum (cotyledon), both of which remain alive in the dry seed.

At a relatively late stage of grain maturation in cereals, programmed cell death (PCD) occurs in the endosperm cells. Indeed, this process starts while deposition of storage reserves is being completed. It has been observed that cells undergoing the nuclear phases of PCD (chromatin condensation, DNA fragmentation, loss of nuclear integrity – see Chapter 3, section 3.5.2) may still be laying down storage compounds. PCD is a highly regulated process involving the activity of a specific set of genes – including, among others, those coding for PCD-associated proteases and deoxyribonucleases. It is regulated by the gaseous plant hormone ethylene, which works within the developmental framework that is controlled by ABA. Thus both hormones are essential for the occurrence of PCD in this system.

Immediately prior to the onset of PCD, the genes encoding the enzymes of the ethylene synthesis pathway, ACC synthase and ACC oxidase (Chapter 3) are up-regulated, as are genes that encode components of the ethylene signal transduction pathway (Chapter 3, section 3.6.6). However, an intriguing question remains: how do the cells of the aleurone layer resist the onset of PCD? It is likely that the aleurone and the main endosperm differ in their sensitivity to ethylene, possibly because the aleurone cells lack all or part of the ethylene signal transduction pathway. This is, therefore, another example of the importance of position within a tissue or organ, with cells that are adjacent to each other undergoing very different fates.

So, as in non-endospermous eudicots, seed maturation in a group of endospermous monocots, the cereals (and, indeed, all grasses) is a highly regulated process, with each phase giving way to the next in an ordered manner. Again, in similarity to the eudicots, master regulator genes are involved in this, the most important of which is *Viviparous-1* (VP1), which encodes a transcriptional activator involved in induction by ABA of maturation-specific genes late in seed development. Interestingly, the VP1 protein also acts as a transcriptional repressor of several genes involved in germination, and thus has a role in prevention of precocious germination.

4.6 Seed coat

In the previous sections, the term 'seed' has been used to cover the mature embryo (in non-endospermous eudicots) or the mature embryo and endosperm (in cereals). However, the seed also contains a component derived directly from maternal tissue, namely the seed coat or testa, formed from the integuments of the ovule. This also undergoes desiccation. In some species, final water content may be slightly higher than in the embryo, while in other species, the testa is a very dry, very hard outer layer. There are also plants, including the cereals, in which the testa is very rudimentary. Here, the outer layer of the seed is actually the pericarp, i.e. the fruit coat which is derived from the wall of the ovary.

4.7 'Recalcitrant' seeds

The developmental pattern discussed above may be regarded as typical, and seeds exhibiting this pattern (and in particular the desiccation phase) have been termed 'orthodox' seeds. Nevertheless, there are also many species in which mature seeds still have a high water content and are metabolically active; indeed, they are damaged by dehydration. They are termed *recalcitrant* (which in normal English usage means 'obstinately disobedient' or 'objecting to restraint') because, from the point of view of the grower, breeder or seed merchant, such seeds are difficult to deal with, especially as they cannot be stored for long periods.

There are many examples of plants with recalcitrant seeds in the moist tropics and among plants of moist habitats. However, recalcitrant seeds are also produced by some tree species in temperate zones, including oak (*Quercus robur*) and white willow (*Salix alba*).[xi]

4.8 Apomixis

In the late 1830s, a single female plant of *Alchornea ilicifolia*, an Australian member of the family *Euphorbiaceae*, was brought back to the UK and planted at Kew Gardens, where it caused a botanical surprise by producing fertile seeds. The phenomenon was fully documented over several years; it was the first properly described occurrence of **apomixis**, the asexual formation of seed from the maternal tissue of the ovule in the absence of meiosis and fertilization.

[xi] Also known as cricket-bat willow.

Figure 4.11 An apomictic species, Orange Hawkweed (*Hieraceum auranticum*). Like many apomictic plants, *Hieraceum* species also reproduce via stolons.
Photograph by Nancy Zack.

Although its formal recognition was based on a single plant from Australia, the phenomenon is actually widespread amongst perennial angiosperms and, indeed, occurs in several species with which UK-based Kew botanists would have been familiar. To date, apomixis has been found in over 40 flowering plant families, although 75 per cent of apomictic species are in just three families, Asteraceae, Rosaceae and Poaceae. In many of these species, individual plants may produce seeds both sexually and apomictically. Figure 4.11 shows Orange Hawkweed, a member of the large and complex *Hieracium* genus, in which many species are frequently or exclusively apomictic. Like many other apomictic plants, several *Hieracium* species also reproduce clonally by producing stolons.

The genetic control of apomixis is of great interest, especially because introducing apomixis into crop species would be a method of ensuring true 'breeding' and maintenance of elite genetic stocks.

Comparison within wild species of apomictic and non-apomictic lines has so far failed to reveal a clear genetic difference. Interestingly, those genes which, in sexual reproduction, ensure that seed/endosperm development only occurs after fertilization, the *FIS* genes, are also required for apomixis. Currently it is thought that the proteins encoded by different members of the *FIS* gene family interact in different ways in sexual reproduction and apomixis.

4.9 Seeds and fruit

4.9.1 Introduction

The focus of this chapter is on seeds, but we should not forget that seeds are formed within fruit. A strict definition of a fruit is a ripened ovary that contains a seed or seeds. Thus, the cereal grain is actually a one-seeded fruit; its outer wall, as already noted, is the pericarp, the wall of the fruit. The 'seeds' of strawberries are also one-seeded fruit; they are borne on the surface of the swollen and ripened receptacle which, although regarded as a fruit in everyday usage, is not one. Neither is the fig, the fleshy part of which is derived not from the ovary but from the peduncle (flower stalk). Technically, such structures are 'false fruit', but their function is similar that of true fruit.

For most species, fruit development (including development of false fruit) depends on successful fertilization. As the embryo develops, it exports auxin to the tissues that will become fruit and this triggers the production of GA (see below). These hormones then participate together in regulating fruit growth, probably along with cytokinins. The latter are thought to be involved in regulating cell division in the ovary. Fruit development thus proceeds in step with the development of the enclosed seeds. Across the angiosperms, fruit exhibit a huge range of shapes and sizes, such that a comprehensive discussion of all types is not possible. Figure 4.12 and sections 4.9.2 to 4.9.5 give some indication of the range.

4.9.2 Grains of cereals

In cereals, and indeed in all grasses, individual flowers are borne on an inflorescence (the spike or ear) and, provided fertilization has taken place, each develops into a fruit, the 'grain', and it is the whole fruit that undergoes desiccation. In wild grasses, the mature fruit are shed from the inflorescence when ripe. However, in cultivated cereals, the grains remain on the ear and are thus readily harvested from the plant rather than being gathered from the ground. It is likely that early use of cereals in human nutrition involved discovery of natural mutants in which the grains were not shed from the ear.

4.9.3 Pods

Many angiosperms produce several seeds in elongated, dehiscent fruit, generally named pods (as with peas and beans), although the pod-like fruit of the Cruciferae, including *Arabidopsis thaliana*, are called siliques.

(a)

(b)

Figure 4.12 (a) Dispersal on the coats of animals is achieved by seeds or fruit armed with hooks, as seen in this picture of a 'seed' head of burdock, *Arctium lappa*. The 'seeds' are actually one-seeded fruit known as achenes. Photo: Mike Vallender. (b) Several different seed or fruit morphologies are adapted for wind dispersal, including the 'parachute' seeds of dandelion (*Taraxacum*), the small seeds of poppy (*Papaver spp*) in their 'pepperpot' fruit and the winged fruit of *Acer* species, illustrated in this picture of sycamore, *Acer pseudoplatanus*. Photograph taken in Haddenham, Buckinghamshire, UK by Margot Hodson.

Dehiscence of the dried pod generally occurs along lines or zones of weakness and, in many species, is violent enough to disperse the seeds some distance from the parent plant. Indeed, many of our British readers will be familiar with the frequent explosive cracking sound on heathland in summer as the pods of gorse (*Ulex europaeus*) and broom (*Cytisus scoparius*) suddenly split, leading to a rapid twisting of the two halves of the pod and the scattering of the seeds. Pod or silique shatter is similar in the Cruciferae, but in this family the two halves of the fruit, instead of twisting, fold back suddenly, detaching the fruit from the central rachis.

One important goal for the breeder of crops such as oil-seed rape is therefore to obtain varieties in which pod shatter does not occur (or is delayed long enough to allow mechanical harvesting of the mature crop).

An interesting variant of the pod is seen in the genus *Impatiens*, which includes species such as Himalayan balsam (*I. glandulifera*; widely naturalized or even invasive in the UK) and 'Busy Lizzie' (*I. walleriana*; widely grown as a colourful ornamental). The five-valved pod does not dry out as the seeds mature, but develops turgor pressure that is unevenly distributed along the pod's axis.

The pod thus becomes increasingly convex (Figure 4.13) and develops a tension along its length such that, when touched or brushed against, it splits explosively, usually separating completely from the flower stalks, and scatters the seeds for distances up to several metres. The old name for plants in the genus, 'touch-me-not', is thus very appropriate. Indeed, the Latin name of one species, the yellow balsam, is *Impatiens noli-tangere*.

4.9.4 Fleshy fruit: drupes

In drupes, typified by peach (*Prunus persica*) and plum (*Prunus domestica*) the three layers of the pericarp (the ovary wall) are readily distinguishable. The outer layer or exocarp forms the skin of the fruit, the mesocarp forms the flesh of the fruit and the endocarp develops into a woody layer (the 'stone' or 'pit') that encloses the seed. The fruit are attractive as food to animals, many of which eat the drupe, including the seed (which passes through the gut of the animal, protected by the endocarp), thus facilitating seed dispersal. Other animals do not eat the stone but instead discard it, but this may also aid seed dispersal.

Palm oil trees (*Elaeis guineensis*) also produce drupes, and these are discussed in Chapter 12, section 12.3.5 (see

Figure 4.13 The explosive seed pods of *Impatiens* (in this picture, *I. glandulifera*). When ripe, the pods explode on touch, scattering the seeds over a range of several metres and often, as in the picture, detaching completely the pod from the flower stalk.
Photograph by R. V. Albitsky, reproduced by permission under the Creative Commons Attribution Share Alike Licence.

also Figure 12.12c). In strict botanical terms, the coconut (the fruit of *Cocos nucifera*) is also a drupe. The mesocarp is not fleshy but fibrous, and it enables the fruit to float on the sea, again aiding seed dispersal.

4.9.5 Fleshy fruit: berries

In berries, including tomato (*Solanum lycopersicum*), blackcurrant (*Ribes nigrum*) and grape (*Vitis vinifera*), all of the pericarp is edible; the endocarp does not form a hard, woody layer. Most berries are multi-seeded and the seeds are again dispersed because animals eat the fruit. Some of the seeds are accidentally discarded during consumption of the fruit, but most pass through the animals' digestive systems. Indeed, this may aid subsequent germination of the seed. For example, tomato seeds are coated in mucilage, which is removed by passage through a mammalian gut, a feature that is clearly illustrated by the number of tomato plants that appear at sewage treatment works!

However, we also note that the widely propagated idea that passage through the gut of the Dodo was essential for germination of seeds of the Madagascan tree *Calvaria major* has been shown to be untrue[xii]. The guts

of several other fruit eaters are equally effective, and thus germination has continued to occur since the extinction of the Dodo.

The term *berry* also applies to fruit that, at least in everyday usage, would not immediately be recognized as such. Thus the fruit of the Cucurbitaceae, including courgette or zucchini (*Cucurbita pepo*), cucumber (*Cucumis sativus*) and watermelon (*Citrullus lanatus*), are in botanical terms berries (although, because of their harder exocarp, they are sometimes placed in a separate sub-class called pepos). The watermelon is a spectacular example of a feature of all fleshy fruit, namely the massive cell expansion that follows the completion of cell division. Individual cells may be up to 700 μm in length, while the volume of a mature watermelon may be as much as 30 litres.

4.10 Fruit development and ripening

Space does not permit discussion of the physiology of fruit development across the range of fruit types. We thus focus on one species, the tomato, which has been become an excellent model for understanding development and ripening in berries.

The tomato, in common with many fleshy fruit and fruit-like structures (such as fig and apple), undergoes a phenomenon known as the **climacteric**. Initially this was identified as a huge increase and then a decline in respiration rate, but it is now known to include many aspects of metabolism. The gaseous hormone ethylene is a key regulator of the climacteric (see Chapter 3, section 3.6.6 for discussion of ethylene signalling pathways) and of the range of physiological changes that occur during ripening.

In a typical commercial tomato variety, successful pollination is followed by a cell division phase lasting about two weeks, then a cell expansion phase (including DNA endoreduplication) which lasts three to four weeks. As described above, auxin, cytokinin and GA are involved in the regulation of these processes. This leads to the formation of the mature but unripe fruit. Ripening takes a further one to two weeks, although the ripening phase may take longer if environmental conditions are unfavourable (e.g. in outdoor-grown tomatoes in unseasonably cool weather).

Ripening has been described by Lucille Alexander and Donald Grierson of Nottingham University as '*a complex, genetically programmed process that culminates*

[xii]One of us (JAB) confesses to repeating this erroneous story in an earlier publication (Bryant, J., 1985, Seed Physiology, Edward Arnold, London).

Table 4.4 The major changes that occur during ripening of tomato fruit.

Very large increase in the rate of ethylene synthesis
Wide-ranging changes in gene expression
Large increase in respiration rate (the 'climacteric')
Loss of thylakoids and photosynthetic pigments
Degradation of chlorophyll
Synthesis of lycopene and other pigments
Changes in organic acid metabolism
Increases in activities of polysaccharide-degrading enzymes, especially polygalacturonase
Depolymerization of cell wall polysaccharides, especially the pectins and hemicelluloses
Fruit softening

---

in dramatic changes in colour, texture, flavour and aroma of the fruit flesh'.

Some of the changes that occur during ripening are listed in Table 4.4. Mutations in ethylene synthesis or sensitivity lead to a failure to ripen, as does treatment of fruit with inhibitors of ethylene synthesis or action. Study of the specific effects of the mutations, of the down-stream effects of the inhibitors and the use of 'antisense' and RNAi techniques (Chapter 3, text box 3.1) has led to our current understanding of the role of ethylene in climacteric fruit.

Ethylene production is autocatalytic – the synthesis of ethylene leads to the synthesis of more ethylene[xiii]. In the mature ripe fruit, there is a low but detectable level of ethylene production, leading to a massive burst in ethylene synthesis which, in turn, triggers the climacteric. Two key enzymes – ACC synthase (which makes the immediate precursor of ethylene, 1-aminocyclopropane-1-carboxylic-acid or ACC) and ACC oxidase (the final enzyme in the pathway, formerly known as the ethylene-forming enzyme) – increase hugely in activity via a marked increase in the transcription of the corresponding genes (*ACS* and *ACO*). Both exist as small

multi-gene families; specific family members are involved in different situations in which ethylene is synthesized.

In ripening, specific transcription factors up-regulate *ACS2* and *ACO1*. Mutations in either gene or their down-regulation by use of antisense RNA lead to a very marked reduction in ethylene production and a failure to ripen. Mutation of the relevant transcription factors also leads to ripening failure. Thus the non-ripening *RIN* genotype is actually a mutation in the gene encoding the MADS-box protein[xiv] that activates *ACS2*.

It is very clear, then, that the massive burst in ethylene production leads to the whole range of processes that contribute to ripening, and it is now equally clear that this involves extensive changes in the pattern of gene expression. As described in Chapter 3 (section 3.6.6), the EIN3 protein in the ethylene signalling pathway is a transcription factor which regulates the synthesis of other transcription factors, thus setting in place a regulatory cascade. Overall, it is estimated that ethylene has an effect on the expression of about 300 genes during fruit ripening. At least three different types of response have been discovered.

First, there are genes which are rapidly up-regulated in response to ethylene. These include genes encoding enzymes involved in the control of the amount of ethylene and which have been shown to possess an ethylene-response element in their promoters. They are up-regulated by a transcription factor (or factors) which binds to the response element (Figure 4.14).

Second, there are genes which are up-regulated more slowly, after a delay of several hours. An example of this is the gene encoding polygalacturonase, an enzyme which participates in cell wall softening (Figure 4.15). This gene does not possess a specific ethylene-response element in its

TAAGAGCCGCC

Figure 4.14 Ethylene response element (ERE). Genes that are responsive to ethylene possess in their promoters two (usually) copies of an 11bp sequence that contains two adjacent GCC motifs. The 5bp sequence immediately upstream of the GCC motifs shows some variation. Transcription factors that bind to EREs all possess a 59 amino acid domain that binds to the GCC motifs.

[xiii]There is reference to this in the Old Testament of the Bible. The prophet Amos describes himself as a 'dresser' of sycamore figs. This refers to nicking or puncturing the unripe figs (as illustrated in carvings from that era). The fruit were harvested four to five days after wounding. The physiological reason for this practice was obviously unknown at the time but we now know that wounding caused the synthesis of ethylene, which in turn catalysed the production of more ethylene, thus hastening fruit ripening.

[xiv]MADS box proteins form a family of transcription factors; the name comes from the initials of the first four members of the family to be discovered. The MADS box itself is a conserved 55–60 amino acid domain that exhibits DNA-binding activity.

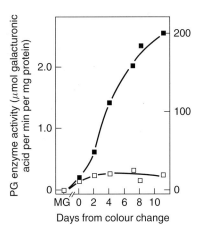

Figure 4.15 Development of polygalacturonase (a wall-softening enzyme) activity in tomato fruit (black squares) and suppression of its activity in GM plants carrying at antisense version of the polygalacturonase gene (white squares). MG = mid-green stage of development.
From Lea, P and Leegood, R (1999) Plant Biochemistry and Molecular Biology, Wiley, Chichester, p 318.

promoter, yet its up-regulation is dependent on ethylene. Genes such as this are developmentally regulated at a particular stage of ripening, but ethylene is required for the fruit to reach that stage of ripening. The transcription factors that regulate this and other late-responding genes have not yet been unequivocally identified, although there has been some recent progress.

Finally, there are some genes which are regulated post-transcriptionally in response to ethylene, for example by stabilization of the mRNA.

It is thus very clear that ethylene is the master regulator of the complex array of processes that contribute to ripening in climacteric fruit. However, there are many fleshy fruit, including cherry, orange and grape, that do not undergo a climacteric. Is ethylene involved in their ripening? Until recently the answer was not clear, but it is now known from research on ripening mutants and on gene expression that ethylene does indeed regulate ripening in these fruit.

4.11 Dormancy and quiescence

Dormancy may be defined as an inactive phase, during which growth and developmental processes are deferred. It is often a strategy for avoiding growth in

an unfavourable season. The environment in which a plant grows shows cyclical changes, in which seasons favourable for growth are usually separated by periods when growth must be very slow or even suspended. The success of a plant will therefore depend on resisting climatic changes and synchronizing its life cycle with the changing seasons.

A number of transformations occur between growth phases and dormant phases. Growing plants, as they approach an unfavourable season (in temperate climates the winter, but in other climates the summer), must make preparations for that season some time before it actually arrives. We therefore need to consider the factors which initiate dormancy and bring about the dormant state. A tissue in a true dormant state will only have its dormancy relieved by some specific environmental factors, which may be entirely different from those factors responsible for growth. In Chapter 6 (section 6.15) we will consider the other major type of dormancy in plants – that of buds – but here we continue with a discussion of seed dormancy.

For seeds, there are further functions of dormancy. During maturation, the seeds of many species become dormant and are unable to germinate. This prevents precocious germination, i.e. germination before the seed is shed from the parent plant (the role of the *VP1* gene in this has already been mentioned). However, this state of dormancy may not last very long, such that, soon after dispersal from the parent plant, dormancy is lost and the seed is now merely *quiescent*. If it is placed in suitable conditions for resumption of growth, it will germinate, as we discuss in the next section. This happens in many weed and ruderal species which are capable of rapid colonization of habitats (Chapter 11, section 11.1.3). It is also true of many of our crop and ornamental plants, in which breeders have selected against longer-term dormancy.

However, the seeds of many wild species remain dormant for a considerable period after being shed and will not germinate even when placed in ideal situations with regard to temperature, moisture and oxygen. Seed dormancy is thus an intrinsic block to germination and has been described by Ada Linkies and her colleagues as a *'characteristic of seeds that allows for long "distance" time travel'*.

Dormancy is a feature that allows long-term survival of seeds; it may prevent germination during, or immediately before, unfavourable seasons (as discussed above) and it

may also allow phased germination within a population of seeds (where the extent or depth of dormancy varies between individual seeds within the population). Five different types of dormancy have been described, as follows:
• *Morphological dormancy* is a feature of seeds in which the embryo, although differentiated, is very small compared to the size of the whole seed. On being exposed to conditions suitable for germination, the embryo needs to grow, perhaps for up to a month, before the radicle emerges from the seed coat. This type of dormancy is widespread in water lilies and other basal angiosperms, and is regarded as the ancestral type of angiosperm dormancy.
• *Physiological dormancy* is the most widespread form of seed dormancy. The hormone ABA, which has a key role in seed maturation, is a major positive regulator of physiological dormancy, acting via both the embryo and the seed coat. Thus, in many seeds, the dormancy may be broken by applying GA (which acts as an ABA antagonist) or by removing the seed coat (as seen in Table 4.5). Based on its distribution amongst angiosperm and gymnosperm families, physiological dormancy is thought to have been gained and lost several times during seed plant evolution. It is regarded as an adaptation to seasonal weather changes, ensuring that germination does not occur in conditions that will become unfavourable to seedling growth. Dormancy may be broken by 'stratification' of seeds under particular temperature regimes. Thus, in southern Europe, seeds of spring-flowering plants germinate in the autumn, after they have experienced the warmth of the Mediterranean summer. This ensures that seedlings are not exposed to summer drought. By contrast, seeds of northern European species, shed in late summer or autumn, require a period of cold weather before germinating, thereby avoiding the cold of winter (again as seen in Table 4.5). The changes that occur during the stratification period include an increase in the ratio of GA to ABA and a decrease in the expression of ABA-related genes such as *ABI3* (which has already been mentioned in connection with seed maturation). There is also an increase in the expression of GA-responsive genes, including some involved in energy metabolism and in the cell division cycle.
• *Morphophysiological dormancy* is a form of dormancy that combines a small embryo with physiological factors. It occurs in some of the less 'primitive' basal eudicots.
• *Physical dormancy* is a less widespread form of dormancy, in which the barrier to germination is the

Table 4.5 Effects on germination of moist storage of Norway Maple (*Acer platanoides*)[+] fruits at 4° C or 17° C.

| | Time to 50% germination, days | | | |
| | Intact seeds | | Isolated embryos | |
Days after harvest	4°C	17°C	4°C	17°C
0	NG	NG	>20*	>20*
20	NG	NG	14	14
60	NG	NG	13	13
80	NG	NG	11	17
120	9	NG	3	23

Fruits were gathered at maturity and stored in moist sand at 4°C or 17°C. At the times specified, whole seeds or isolated embryos were dissected from the fruit and set to germinate at 17°C. NG = No germination.
Note that intact seeds from fruit stored at 17°C never germinated and it took between 80–120 days of cold stratification to break the dormancy of intact seeds. Isolated embryos from fruit stored at either temperature were able to germinate, although there is evidence of deterioration in embryos from fruit stored long-term at 17°C (based on results in Slater & Bryant, 1982).
*A small percentage of embryos isolated from freshly harvested fruit germinated. However, the figure of 50 per cent germination was never reached.
[+]In the UK, *Acer platanoides* is widely grown in parks, where it is valued as a large, shapely tree whose foliage exhibits beautiful autumn colours. However, seedling establishment is very rare because UK winters are, in general, not consistently cold enough to break the dormancy of intact seeds.

impermeability of the seed coat (or fruit coat) to water. Breakage of dormancy involves slow deterioration of, or damage to, the impermeable covering. There are several examples of plants with physically dormant seeds in the legume family, including wild pea (*Pisum elatius*), lucerne (*Medicago sativa*) and even soybean (*Glycine max*).
• *Combinational dormancy* results from a combination of physical and physiological dormancy and may require a complex interplay of factors for dormancy breakage.

4.12 Germination

4.12.1 General features

The seed is a remarkable structure, containing an individual plant whose growth has, in the majority of species, been suspended in late embryogenesis by intense desiccation. In an evolutionary context, this may be seen as the formation of a structure both for surviving unfavourable environmental conditions while metabolism and growth are switched off, and for dispersal of the plant while it is protected from damage by enclosure in the seed integuments (and in many plants, within the fruit).

However, rehydration of the seed changes everything. If the seed is not dormant, the uptake of water re-initiates growth in the embryo. This leads to emergence of the growing root and shoot axes from the seed integuments, and thence to the establishment of a seedling. The overall process is termed *germination*, although when germination rates are being assayed, for example in a study of dormancy breakage, it is conveniently defined as emergence of the radicle (seed root) from the seed coat. It is useful to divide germination into three phases, although boundaries between the phases are somewhat fuzzy.

Phase one is marked by rapid uptake of water, leading to immediate resumption of metabolic activity, including protein synthesis (which makes use of mRNA molecules synthesized during seed development). Repair of DNA, organelles and membrane systems damaged by the desiccation and subsequent rehydration is also initiated in this phase.

In phase two, transcription of genes is renewed and newly-synthesized mRNAs gradually take over from the stored mRNAs used in phase one. A very wide range of proteins is synthesized, typified by those that participate in the resumption of DNA replication and cell division. Repair processes continue in phase two, but DNA repair is finished by the end of this phase.

The transition from phase two to phase three is marked by the emergence of the radicle; if we are defining germination as emergence of the radicle, then phase three is, strictly speaking, a post-germination phase. In most species, radicle emergence results from cell expansion and elongation, although DNA replication and then cell division re-start at about the time of the phase two to phase three transition.

In phase three, radicle growth by division and expansion continues and there is extensive mobilization of reserves that were laid down during seed development. Finally, well after radicle emergence, and therefore a truly post-germinative event, the shoot emerges. The embryonic leaves expand and embryonic plastids become chloroplasts, leading to greening of the shoot and the establishment of the seedling as a photosynthetic organism.

As with seed dormancy, a balance between ABA and GA is involved in germination. Thus, addition

of ABA to a seed in phase one inhibits germination (via transcriptional regulation of the *AB13/VP1* gene which is normally involved in preventing precocious germination – see above). Inhibition of GA synthesis also inhibits germination, while the inhibitory effects of adding ABA can be neutralized by adding GA at the same time. In natural germination, ABA synthesis is down-regulated and ABA signalling is inhibited. GA synthesis increases and the GA signalling pathway is activated by the binding of GA to its receptor. The hormone-receptor complex then targets the inhibitory DELLA proteins (transcriptional regulators that inhibit growth) for breakdown (as described in more detail in Chapter 3, section 3.6.4), and thus GA becomes the main hormonal regulator of germination.

4.12.2 Mobilization of reserves

The seed development phase is, for most flowering plant species, a period of deposition of reserves. Germination, by marked contrast, is a period of active hydrolysis and mobilization of reserves. The varieties of reserves and locations for deposition have already been noted and there is corresponding variety in the mechanisms and regulation of mobilization. Lack of space precludes a totally comprehensive treatment of this topic and so, in this and section 4.11.3, attention is focused on two general types, namely the mobilization of reserves in the cotyledons of eudicots and the mobilization of reserves in the endosperm of the monocot family Poaceae (Gramineae), including cereals. However, before that discussion, a brief overview of biochemical mechanisms is needed.

Proteins

In Chapter 3 (section 3.4.5), we discussed the targeted breakdown of proteins that occurs in the regulation of gene activity and in hormone signalling pathways. The specific protein cleavages involved in activating some precursor proteins was also noted. Hydrolysis of protein reserves is, in contrast, a much less subtle process. Proteases and peptidases are secreted into the protein bodies and break down the stored proteins into their constituent amino acids for transport to the growing regions of plantlet.

Polysaccharides

As indicated earlier, starch is the most widespread storage polysaccharide, and in this section it is the exclusive

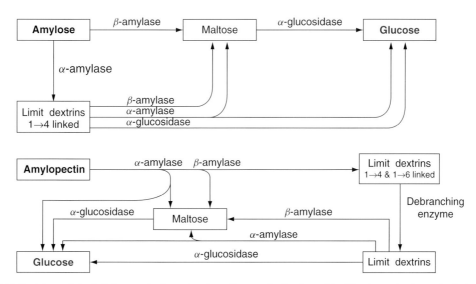

Figure 4.16 Breakdown of starch during germination and early seedling growth. Starch consists of two types of polymer of α-D-glucose. Amylase is formed of straight chains of glucose units joined by 1→4 linkages. Amylopectin is a branched polymer; the branches are formed by 1→6 linkages and all the other linkages are 1→4. Total breakdown of these molecules to yield glucose requires the cooperation of several enzymes.

focus of the discussion. There are four essential enzymes needed for the complete breakdown of starch to glucose (Figure 4.16).

Initial hydrolysis is carried out by α- and β-amylases. The former attacks amylase and amylopectin molecules at random, while the latter works its way along the starch chains, breaking off maltose units (these consist of two α-1→4 linked glucose units). In amylopectin, this stepwise processes is halted by the α-1→6 linkages that form the branch points. Thus, the end products of amylase action are maltose and short chains of glucose residues known as dextrins, some of which are branched. Hydrolysis is completed by a 'de-branching enzyme' and an α-glucosidase. For transport, the glucose is phosphorylated by hexokinase to give glucose-6-phosphate, which is then converted to sucrose.

Lipids

While the mobilization of polysaccharides and proteins presents no problems for the transport of the hydrolysis products, it is a different story for those plants that lay down triacylglycerols (TAGs, also known as triglycerides). Although these lipids represent a more efficient way of storing carbon than do polysaccharides, the major products of triacylglycerol breakdown are not readily transportable to the growing parts of the plant.

The initial phase of breakdown takes place in the oil bodies, in which lipases hydrolyze the ester linkages between glycerol and the fatty acids. The fatty acids are then subject to the normal oxidative pathway, β-oxidation, which in plants takes place in the microbodies or peroxisomes. This contrasts with the situation in animals, where β-oxidation occurs in mitochondria. This organellar location proves to be significant for the subsequent metabolism of the product of β-oxidation, namely acetyl-CoA (Chapter 2, section 2.14.6).

When fatty acids are oxidized as an energy substrate, the acetate group enters the Krebs cycle. In animals, this takes place in the same organelle as β-oxidation, but in plants the acetyl-CoA must be transferred from peroxisome to mitochondrion. However, there is a metabolic problem here when triacylglycerols are being metabolized to provide transportable carbon. When acetate enters the Krebs cycle, it does so by combining with the 4-carbon acid, oxaloacetate, to make the 6-carbon acid, citrate. In one turn of the Krebs cycle, two carbons are released as CO_2, and thus acetate entering the Krebs cycle cannot provide transportable carbon.

The answer to this puzzle came when it was realized that the same problem was encountered by bacteria that were able to utilize acetate as sole carbon source. These must be able to use the acetate both as a metabolizable

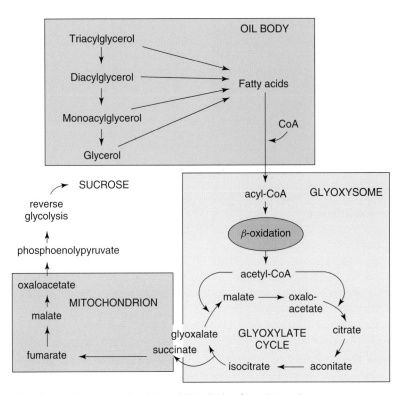

Figure 4.17 The glyoxylate cycle and its role in mobilizing lipids in fat-storing seeds.
From Lea, P and Leegood, R. (1999) Plant Biochemistry and Molecular Biology, Wiley, Chichester, page 131.

substrate for energy production and as a source of carbon for synthesis. They can achieve the latter via operation of a separate metabolic pathway, the glyoxylate cycle (Figure 4.17). Classic labelling experiments, using radioactive acetate, combined with enzyme assays, showed that plants that lay down lipids as a seed reserve do, indeed, operate this cycle. This is true both of those that use cotyledons as the site of deposition of reserves (such as soybean, squash and sunflower) and those that use endosperm (such as castor bean).

The glyoxylate cycle relies on the activity of some enzymes that also operate in the Krebs cycle, but it is especially characterized by the presence of two enzymes that operate specifically in the cycle, isocitrate lyase and malate synthase (Figure 4.17). Subcellular fractionation experiments have shown clearly that these enzymes are located in peroxisome-like organelles that were given the name glyoxysomes. However, these are not 'new' organelles – they are peroxisomes (which are the site of β-oxidation) that temporarily gain the glyoxylate-cycle-specific enzymes during germination. It is also clear that the two enzymes are synthesized *de novo* following up-regulation of transcription of the corresponding genes, and that these enzyme activities are lost when lipid mobilization is complete. The glyoxysome phase of peroxisome existence is thus transient.

4.12.3 Regulation of the mobilization of reserves

Hydrolysis of the stored reserves is obviously a very different pattern of metabolism from the deposition of the reserves during seed development. The regulation of the deposition phase has already been discussed, but there is also regulation of the hydrolysis phase to be considered. Is it simply a question of desiccation/rehydration acting as a sequential set of cues to bring about appropriate changes in the pattern of gene expression? And, in the plants that are one of the main subjects of discussion here, those eudicots that lay down reserves in cotyledons, what role does the growing axis (i.e. the root and shoot) of the embryo play in regulating cotyledonary metabolism?

After all, the embryo axis is the major sink for the hydrolysis products from the cotyledons.

Certainly, the changes in enzyme complement that are necessary for hydrolysis occur after the resumption of the growth of the axis, but this does not mean that the axis regulates cotyledonary metabolism. Indeed, experiments in which the embryonic axis is removed indicate that the axis has little effect on the development of hydrolytic activity in the cotyledons, and there has certainly been no demonstration of signalling molecules passing between the cotyledons and the growing axis. Once again, it seems very probable that, within the cotyledons themselves, as well as in root and shoot, it is the ABA to GA balance that is important, with ABA dominating during the deposition of reserves and GA dominating during their mobilization.

Among monocots, the role of plant hormones in the mobilization of reserves in cereals has been known for many years, even if knowledge of the underlying molecular mechanisms is more recent. The importance of cereal grains as dietary staples throughout the world is a good enough reason to focus on these species, but it is also true that cereal grains have proved very rewarding subjects of experimentation. As noted earlier in this chapter, the storage cells in cereal endosperm are dead by the time the grain is ripe, which poses a problem for hydrolysis of the reserves. However, the endosperm is surrounded by the aleurone, a layer of living cells, and it was long ago realized that the hydrolytic enzymes for breakdown of reserves nearly all come from these cells.

Further, the embryo itself clearly has a role to play in this. If cereal grains are cut in half so that both halves contain significant portions of the aleurone and endosperm, but only one half has the embryo, then when the half-seeds are hydrated, only in the half containing the embryo are the reserves hydrolyzed. Further, the embryo can be replaced by GA; indeed, if GA is added to the embryo-less half seed, it behaves as if an embryo were present. Further, the effect of GA is reversed by ABA, providing a classic example of GA and ABA acting in opposition on the same process.

So what is happening here? The simplest hypothesis is that GA secreted by the embryo after hydration up-regulates genes encoding hydrolytic enzymes and that these, in turn, are transported or secreted into the endosperm (Figure 4.18). Study of gene expression certainly gives support to that idea. Addition of GA to isolated aleurones leads to the presence of

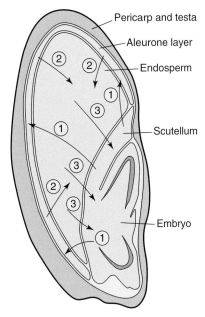

Figure 4.18 Diagram to illustrate control of the mobilization of reserves in germinating barley seeds. (1) The embryo secretes gibberellic acid (GA) to the aleurone layer. (2) In response to GA, the aleurone cells synthesize hydrolytic enzymes and secrete them into the endosperm. (3) The hydrolysis products are taken up by the embryo. The diagram is a very simplified version of the real thing. In particular, the reader should note that the embryo is growing actively during this time and that germination (defined as emergence of the radicle) occurs well before the hydrolysis of the reserves. From Bryant, J. (1985), Seed Physiology, Edward Arnold, London.

several additional mRNA species, including those encoding hydrolytic enzymes such as proteases and α-amylase. The modes of action of these enzymes have already been mentioned briefly. Here, attention is focused on regulation of transcription of the genes and especially on those encoding α-amylases.

There are three families of α-amylase genes in cereals – α-amy-1, α-amy-2, and α-amy-3. Families 1 and 2 are expressed in the aleurone, although some members of family 2 are also expressed in developing endosperm. The promoters of the aleurone-specific genes possess firstly a region immediately upstream from the gene required for localization of expression to the aleurone, and secondly a group of motifs (the GA-responsive complex or GARC) that together confer GA-inducibility and ABA-repressibility on the genes.

Within the complex, there is a short motif which forms the GA-response element (GARE). This is shown

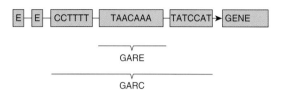

GARE

GARC

Figure 4.19 Most GA response elements (GAREs) are set within a broader domain – the GA response complex (GARC). There are some 'stand-alone' GAREs which are longer than those set within GARCs Key: E = enhancer. Note: the diagram is not to scale. From Lea, P and Leagood, R. (1999) Plant Biochemistry and Molecular Biology, Wiley, Chichester, p 323.

in Figure 4.19. As with other GA-signalling events, destruction of DELLA proteins allows the positive effects of GA. Here we note the de-repression of the gene encoding the GA-MYB transcription factor, which recognizes and binds to the GARE, thus up-regulating α-amylase gene transcription and the expression of other GA-responsive genes. Thus, according to Stephen Ritchie and Simon Gilroy '*MYB transcription factors appear to regulate a multitude of GA-induced genes in cereal aleurone.*'

β-amylase is also active in the endosperm but, unlike the α-amylases, this enzyme is not synthesized *de novo*. Indeed, rather remarkably, it persists in an inactive form in the dead endosperm cells and is activated by proteolytic cleavage, the activating protease originating from the aleurone.

Finally, calcium also participates in the effects of GA in the aleurone. Calcium is required to activate α-amylase and becomes bound to the protein in the ER. However, the concentration of Ca^{2+} in the aleurone is initially low. GA causes an increase in the activity of Ca^{2+} transporters in the plasma membrane and the ER, and in the amount of the calcium-binding protein, calmodulin (see Chapter 3, section 3.6.12). Intra-cellular and intra-ER concentrations of Ca^{2+} increase, and α-amylase is activated. Further, the secretion of α-amylase from the aleurone to the endosperm is also calcium-dependent.

4.12.4 Storage tissue – dead or alive?

In endospermous seeds, the emptying of the endosperm of its stored nutrients marks the end of its role. In plants in which the endosperm is alive at the start of germination, programmed cell death (PCD – see Chapter 3, section 3.5.2) is initiated and all the usable contents of the cell, including proteins and nucleic acids, are

hydrolyzed so that their components may be transported to the growing seedling. In cereals, where most of the endosperm is already dead, PCD occurs only in the living cell layer, the aleurone.

In plants that lay down the bulk of their reserves in cotyledons there are two main patterns (Figure 4.20). In many species, including lettuce (*Lactuca sativa*), Arabidopsis thaliana and courgette or zucchini (*Cucurbita pepo*), the cotyledons appear above ground and undergo a change in role from storage organs to photosynthetic organs. This is known as **epigeal** germination. In other species, such as pea (*Pisum sativum*) the cotyledons do not emerge above ground (**hypogeal** germination); like endosperm, they undergo programmed cell death. We also note that in some species that exhibit epigeal germination, for example French bean (*Phaseolus vulgaris*), the cotyledons do not become photosynthetic but undergo programmed cell death as in hypogeal germination.

4.13 Establishment

As we have already noted, if germination is defined strictly as radicle emergence, then the mobilization of reserves is a post-germination event. This is certainly the case when we consider the final exhaustion of the reserves. By this time, the seedling is well established (unless it has succumbed to pathogens at this vulnerable stage of its life), with functioning root, stem and leaves (as described in the next three chapters). At this point, we focus on the growth of the stem that brings the shoot tip above the ground and the transformation of the leaves into photosynthetic organs. It is a fascinating phase, involving an interplay between the phytochromes (and other light-sensitive pigments) and plant hormones.

The initial stages of this interplay are well illustrated by allowing pea seeds to germinate and the resultant seedlings to grow in the dark. The dark-grown seedlings are much taller and spindlier than those grown in the light. They are described as *etiolated*. The leaves have not unfolded and are pale yellow. In light-grown seedlings, the leaves have unfolded and expanded and have become photosynthetic. The extreme growth in the dark is driven by GA but, in the light, the elongation growth is inhibited following perception of the light by phytochromes. The phytochrome signalling pathway and the GA signalling pathway thus oppose each other in this situation. In

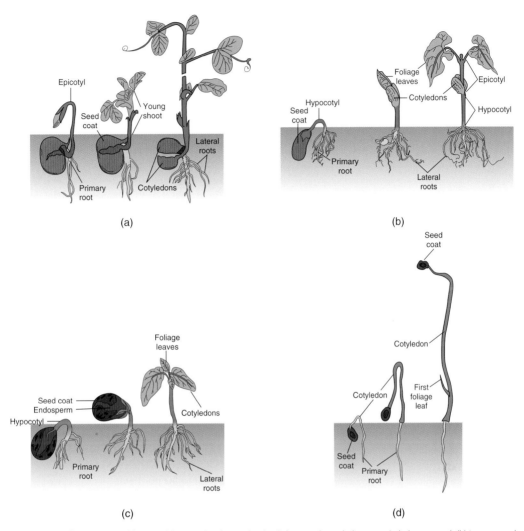

Figure 4.20 Styles of germination: (a) In pea (*Pisum sativum*), germination is hypogeal; cotyledons remain below ground. (b) In common bean (*Phaseolus vulgaris*), germination is epigeal; cotyledons appear above ground and, after mobilization of stored reserves, they wither and die. (c) In castor bean (*Ricinus communis*), germination is also epigeal but the cotyledons become photosynthetic and live on for some time. (d) Onion (*Allium cepa*) is a monocot that exhibits epigeal germination.
Reproduced with permission from Raven PH *et al.* (2005), Biology of Plants, 7th edition, Freeman, NY, pp 506–507

the dark, GA dominates; in the light the phytochromes dominate.[xv]

The interaction of the two signalling pathways centres on a transcription factor, PIF4 (PHYTOCHROME-

INTERACTING FACTOR4). This positively controls genes that promote cell elongation. PIF4 is destabilized by direct interaction with phytochrome B, thus repressing its growth-promoting function. It is also inhibited by binding with the DELLA proteins which, as we have already seen, are negative regulators in GA signalling. So, in the dark, in the *absence* of GA, PIF4 is not active; but in the *presence* of GA, still in the dark, it is active (because the DELLA proteins have been degraded – see

_____

[xv] In many plant species that colonize open ground as weeds or ruderals, germination itself is often light-dependent; this also involves phytochrome. Indeed, much of the early research on phytochrome was carried out on a variety of lettuce, *Grand Rapids*, in which germination is light-dependent.

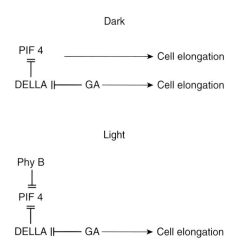

Figure 4.21 Interaction of GA and phytochrome in the early growth of seedlings. In the dark, the phytochrome-interacting factor, PIF4, promotes elongation. However, this is inhibited by the DELLA proteins. In the presence of GA, DELLA proteins are degraded (see Chapter 3.6, section 3.6.4). This allows the growth-promoting effects of both PIF4 and GA to occur, and the seedling becomes etiolated. In the light, PIF4 is bound by phytochrome-B, thus inhibiting PIF4's promotion of cell elongation. GA still causes degradation of DELLA proteins, thus allowing the 'normal' GA-mediated cell elongation. The seedlings are not etiolated.

section 4.11.1 and Chapter 3, section 3.6.4). Extensive elongation growth therefore occurs. However, in the light, even with GA present, the phytochrome-mediated de-activation of PIF4 very much reduces the extent of GA-induced elongation. This is illustrated in Figure 4.21

Another obvious role of phytochrome in these early stages of a plant's life is in the greening or de-etiolation process. Phytochrome A may be involved in the early stages as the shoot approaches the soil surface and begins to perceive light, but the main phytochrome here is phytochrome B (see Chapter 3, section 3.7.2). This acts, along with the blue-light receptor, cryptochrome (Chapter 3, section 3.74), in full daylight.

The mode of action of phytochrome in greening is fairly complex and is not yet fully understood. First it involves the COP1 protein. In the dark, this protein is located in the nucleus and represses photomorphogenesis by targeting a range of light-specific transcription factors for degradation (via the ubiquitin-proteasome pathway – see Chapter 3, section 3.4.5). In the light, COP1 migrates out of the nucleus; it thus has no access to those transcription factors and may indeed target itself for degradation.

Second, in the earliest responses to light, phytochrome A (and possibly B) interacts with PIF3 to regulate the synthesis of nuclear-encoded chloroplast components.

Third, phytochromes interact with PIF1. The effects of this interaction are undoubtedly complex. In general, PIF1 is a negative regulator of chlorophyll synthesis in the dark and its interaction with phytochrome relieves the negative regulation to promote greening. However, in darkness, PIF1 also positively regulates proteins that are involved in the control of chlorophyll biosynthesis, thus preventing the build-up of potentially toxic intermediates. Removal of this regulation in the light allows chlorophyll synthesis to proceed.

Finally, in a very interesting 'twist in the tale', ethylene has been shown to enhance greening. It acts via its 'normal' signal transduction pathway that ends with the EIN3 transcription factor (Chapter 3, section 3.6.6) which, in this system, cooperates with PIF1.

Three more things need to be said. First, as already mentioned, the blue-light receptor cryptochrome is also involved in greening. The way in which it does this is not yet clear. Second, for the greening process to proceed to completion, the balance between chloroplast gene expression and nuclear gene expression must be maintained (Chapter 3, section 3.3.2). Thus, in mutants of chloroplast development, the expression of nuclear-located photosynthesis-associated genes is markedly impaired, indicating that the chloroplast is able to signal to the nucleus (Chapter 3, section 3.3.2).

This leads to the third point. The balanced interaction between chloroplast and nucleus is a microcosm of what happens at the level of the whole leaf (see Chapter 7). The greening process takes place within leaves that respond 'morpho-physiologically' to light by unfolding. Zones of cell division, of cell expansion and of plastid division are established and leaf growth continues. These processes are coordinated in response to environmental cues (e.g. light) and developmental cues (e.g. position and age) by the actions and interactions of hormones, including GA, auxin, cytokinins and brassinosteroids and the light receptors. These signalling molecules thus participate in the integration of growth, development and activity of individual organs and in the whole plant.

This is the state that is reached when the seedling becomes fully established, which leads us to look, in the next four chapters, at the major organs of the functioning plant.

Selected references and suggestions for further reading

Alexander, L. & Grierson, D. (2002) Ethylene biosynthesis and action in tomato: a model for climacteric fruit ripening. *Journal of Experimental Botany* **53**, 2039–2055.

Angelovici, R., Galili, G., Fernie, A.R. & Fait, A. (2010) Seed desiccation: a bridge between maturation and germination. *Trends in Plant Science* **15**, 211–218.

Bill & Melinda Gates Foundation (2011): www.gatesfoundation.org/agriculturaldevelopment

Bowen, I.D. & Bryant, J.A. (1978) The fine structural localization of p-Nitrophenyl phosphatase activity in the storage cells of pea (*Pisum sativum* L.) cotyledons. *Protoplasma* **97**, 241–250.

De Smet, I., Lau, S., Mayer, U. & Jürgens, G. (2010) Embryogenesis: the humble beginnings of plant life. *Plant Journal* **61**, 959–970.

Linkies, A., Graeber, K., Knight, C. & Leubner-Metzger, G. (2010) The evolution of seeds. *New Phytologist* **186**, 817–831.

Richards, A.J. (2003) Apomixis in flowering plants: an overview. *Philosophical Transactions of the Royal Society, London, Series B* **358**, 1085–1093.

Ritchie, S. & Gilroy, S. (1998) Gibberellins: regulating genes and germination. *New Phytologist* **140**, 363–383.

Santos-Mendoza, M., Dubreucq, B., Baud, S., Parcy, F., Caboche, M. & Lepiniec, L. (2008) Deciphering gene regulatory networks that control seed development and maturation in Arabidopsis. *Plant Journal* **54**, 608–620.

Schlereth, A., Moller, B., Liu, W., Kientz, M., Flipse, J., Rademacher, E.H., Schmid, M., Juergens, G. & Weijers, D. (2010) MONOPTEROS controls embryonic root initiation by regulating a mobile transcription factor. *Nature* **464**, 913–916.

Slater, R.J. & Bryant, J.A. (1982) RNA metabolism during breakage of seed dormancy by low temperature treatment of fruits of *Acer plataanoides* (Norway Maple). *Annals of Botany* **50**, 141–149.

Sullivan, J.A. & Deng, X.W. (2003) From seed to seed: the role of photoreceptors in *Arabidopsis* development. *Developmental Biology* **260**, 289–297.

In the previous chapter, we outlined the early development of the embryonic root, the radicle. Now the details of root structure, growth and physiology will be considered, but first the main functions of the root will be summarized. Long distance transport of water, ions and sugars will mostly be considered in Chapter 6, but some aspects will be covered in the present chapter.

As one might imagine, knowledge of root structure and function has lagged some way behind that of the shoot. This is particularly the case for root development. The reason is quite obvious, in that it is more difficult to observe roots than shoots. Roots are the main sites for the uptake of water and minerals by most higher plants. There are exceptions, and some plants absorb substantial amounts of water and mineral ions through their leaves, but undoubtedly roots are most important in this respect. Once roots have absorbed water and mineral ions, they are also involved in transporting these up to the shoot. In return, roots receive sugars, the products of photosynthesis, transported from the shoot. These are used in respiration and for building more root tissue.

Some roots can become specialized storage organs (e.g. carrots), but all will function like this to some extent. Finally, roots are very important in providing a firm anchorage for the plant in the soil. Again there are exceptions, such as the epiphytes that grow on trees, but the vast majority of plants are rooted in the soil.

5.1 External morphology of roots

It is certainly the case that shoot morphology varies more than that of the root. However, there is still considerable variation in both the size and form of roots, both between plant species and as a result of environmental conditions. The largest root system is probably that of Pando, a clonal colony of aspen (*Populus tremuloides*) from Utah, USA. All of the trees are genetically identical and are connected by a single underground root system that covers $0.43\,km^2$. The longest roots are probably from desert plants, which can be up to 25 m in length.

Some plants have a taproot – one dominant root from which all other roots arise. Many trees have taproot systems, as have many herbaceous eudicots. On the other hand, in fibrous root systems there is no dominant single root, and often these roots all arise from the lower part of the stem. Fibrous root systems are often, but not only, found in grasses and cereals. Figure 5.1b shows the fibrous roots of date palm (*Phoenix dactylifera*) in the lower half of the specialized root laboratory.

Taproots can be seen as adaptations to obtain water from deep in the soil, and are found in many desert plants (see Chapter 10, section 10.4.2). They are also often adapted as storage organs, as in the parsnip (*Pastinaca sativa*) shown in Figure 5.2. Fibrous root systems tend to be mostly present in the surface parts of the soil, where most plant mineral nutrients are located. However, some plants have aerial roots that arise from the stems well above the surface of the soil, and in others aerial roots arise from beneath the soil (for instance the **pneumatophores** of mangroves – see Chapter 10, section 10.3.2).

5.2 Root anatomy

A transverse section of a root some distance behind the apex (Figure 5.3) will reveal the following tissues, going from the outside of the root and working steadily inwards:

1 The **epidermis** is the outermost layer of cells in the root. Within the epidermal layer, root hairs are often present, and these are important in increasing the surface area of the root for water and ion absorption.

Functional Biology of Plants, First Edition. Martin J. Hodson and John A. Bryant.
© 2012 John Wiley & Sons, Ltd. Published 2012 by John Wiley & Sons, Ltd.

Box 5.1 An unusual root laboratory

Some years ago, one of us (MJH) gave a seminar at Tel Aviv University. After the seminar Professor Yoav Waisel took me to see his rather special laboratory. It was a two-storey building in the botanical gardens called the 'Laboratoire central Sarah Racine pour la Recherche sur les Racines'. Yoav explained that a friend of his, Sarah Racine, had agreed to donate some money for research at the Botanic Gardens of Tel Aviv University. It was suggested that her donation could go to help construct a root laboratory to study *in situ* root growth (readers who have studied a little French will understand the connection; racine means root – so in English it would be 'The Sarah Racine (Root) Central Laboratory for Research on Roots').

However, the laboratory was not only remarkable for its name, as became clear when Yoav showed me around. First he took me into the upper storey. To be honest, this did not look much different to many other greenhouses I have visited. Quite a variety of shoots of plants were growing (Figure 5.1a), but nothing special. It was only when we visited the lower room that the true purpose of the laboratory became obvious. There we could walk among roots growing in a darkened room in a mist culture (Figure 5.1b). It was a wonderful environment for studying root growth, but had taken a lot of setting up. Just making sure that the mist culture never ran out of water or power involved all sorts of backup systems. Obviously, any failures would rapidly result in the death of plants that had been growing for many months. The roots grew well in this environment and were over two metres long. I was impressed!

(a)   (b)

Figure 5.1 (a) The Sarah Racine Root Research Laboratory at the Botanical Gardens of Tel-Aviv University, Israel. Date palm (*Phoenix dactylifera*). A shoot in the upper half of the laboratory. (b) Fibrous roots in the lower half of the laboratory.
Photos: Prof. Yoav Waisel.

Plant roots tend to be ignored by many plant scientists. They are more difficult to work on than shoots, and not as pretty! Yoav Waisel and his colleagues wrote an excellent book, *Plant Roots: The Hidden Half* (2001), in which they endeavoured to inspire more people to research root systems. So in this chapter we will take up the challenge and introduce you to some of the many fascinating aspects of root biology.

Figure 5.2 Taproot of parsnip (*Pastinaca sativa*) adapted as a storage organ.
Photo: MJH.

2 The layer of cells beneath the epidermis is the **hypodermis**, which has fairly recently been recognized as an important tissue for ion absorption in many plant species (see section 5.7.6).

3 The **cortex** is formed from several layers of apparently undifferentiated cells, often with large intercellular spaces between them. A transverse section of meadow buttercup root (*Ranunculus acris*) illustrating the cortex and the tissues internal to it is shown in Figure 5.4.

4 At the boundary of the cortex and the **stele** (where the vascular tissues are located) is a specialized layer of cells, the **endodermis**, which is a key control point for ion transport across the root (see section 5.7.5).

5 Inside the endodermis is a layer of cells called the **pericycle**. It is in this layer that meristems leading to the production of lateral roots are formed (section 5.3.4). Thus root branches do not arise on the surface as in shoots, but deep in the root. The root branches then have to grow through the cortical and dermal tissues before emerging into the soil.

6 Within the stele, the **xylem** and **phloem** can be found in association with various ground and fibrous cells. The xylem is responsible for long distance water and ion transport up to the shoot, while the phloem transports both sugars and ions. These transport systems will be discussed in detail in Chapter 6. The precise arrangement of the xylem and the phloem within the root varies between plant species; in woody species, considerable secondary development of the root tissues occurs.

Longitudinal sections through roots reveal four main parts (Figure 5.5), going from the apex backwards:

1 Root cap. This serves as a protection for the delicate tissues behind it as the root pushes its way through the soil. Its cells are continually sloughed off the surface, and are replaced by the meristematic cells behind. The root cap is also the main site of gravity perception in the root, a topic that will be covered in section 5.11.

2 Zone of cell division. This includes the various parts of the meristem in the root tip (the root apical meristem or RAM), and is responsible for the production of new cells (Figure 5.6).

3 Zone of cell elongation. As the name suggests, this is the area where growth is achieved by cells elongating, not dividing.

4 Zone of cell differentiation. Here the cells become functionally mature, particularly the vascular tissues. Lateral roots are often produced in this zone.

It should be pointed out that the above description of root anatomy is rather simplistic, and anatomy will vary with plant species, age, and environmental conditions.

5.3 Root growth

So far in this chapter, roots have mostly been considered as static entities – but they are not, and they grow. Primary growth is achieved by cell division in meristems and by cell elongation further back in the root. Even later, vascular tissues are formed and lateral roots begin to develop. In woody plants, roots may persist over several years and secondary growth and development occurs.

5.3.1 The root apical meristem

Meristems within the primary root are of two types: apical at the tip of the growing root (the RAM); and lateral emerging out of the root in the zone of cell differentiation. In the primary meristem at the root apex, the protoderm

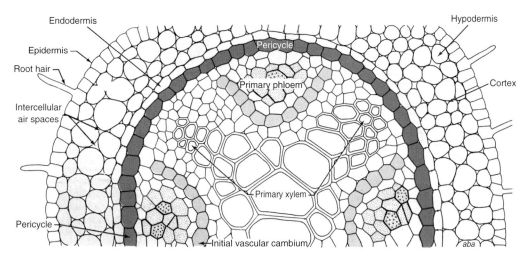

Figure 5.3 Cross section of a primary root.
From Weier *et al.*, 1974. (Reprinted with permission of John Wiley & Sons, Inc.)

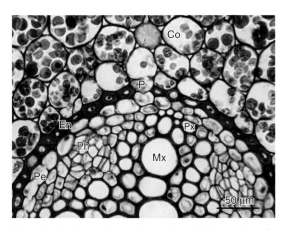

Figure 5.4 Meadow buttercup (*Ranunculus acris*). Cross section of triarch root, light micrograph. Key: Cortex (Co), Passage cell (P), Endodermis (En), Pericycle (Pe), Phloem (Ph), Protoxylem (Px), Metaxylem (Mx).
Photo: Prof. Thomas Rost.

produces more epidermal cells, the ground meristem produces cortical cells and the procambium produces stelar (vascular) cells (Figure 5.5). These meristems all produce cells that add to the cell lineages behind the apex. The **calyptrogen**, on the other hand, is the meristem responsible for the addition of cells to the root cap in grasses, and thus it works in the reverse direction from the other meristems in the root apex.

In eudicot roots, the epidermis and peripheral root cap are derived from the same tier of initials. One difficulty is

in deciding what exactly is the meristem. It is obvious that files of cells can be traced to an area just behind the root cap. The area which has been interpreted as the initiating centre of the root is relatively inactive during much of root development.

At the centre of the apical meristem, many roots, including most angiosperms, have a **quiescent centre** or zone. This consists of a variable number of cells (only four in *Arabidopsis thaliana* – see Figure 5.6), which divide infrequently compared to the cells around them. In this quiescent centre, little or no DNA synthesis occurs and very few mitoses. Ultrastructural features of this zone also indicate that the cells have a low metabolic activity, and the Golgi apparatus is poorly developed – for example, in *Zea mays*, the cell cycle time for cells in the quiescent centre is 170 hours, as opposed to 12 hours for the most rapidly dividing cells the root cap initials.

It is now recognized that the quiescent centre is not always 'quiet', and that at some stages in plant development cell divisions do occur here. However, in most mature roots, only very infrequent divisions are seen. So is it correct to include the quiescent centre in the meristem? It has been suggested that the root initials are located just outside the quiescent centre, but it has also been pointed out that even very occasional divisions in the zone would make the cells of the centre the initials of the root. Indeed, as mentioned in Chapter 4, some scientists regard these as the core 'stem cells' within the meristem.

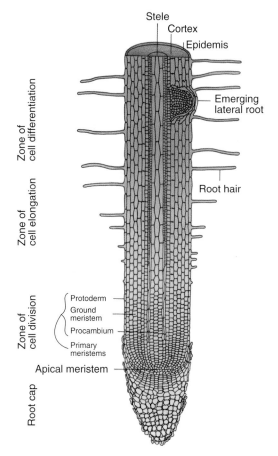

Zone of cell differentiation

Zone of cell elongation

Zone of cell division

Root cap

Stele
Cortex
Epidemis

Emerging
lateral root

Root hair

Protoderm
Ground
meristem
Procambium
Primary
meristems

Apical meristem

Figure 5.5 Primary growth of a root illustrating the four main parts, going from the apex backwards: root cap; zone of cell division; zone of cell elongation; and zone of cell differentiation.
From Campbell, Neil A., BIOLOGY, 2nd Edition © 1990. Reprinted by permission of Pearson Education, Inc., Upper Saddle River, NJ.

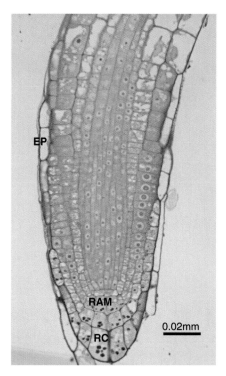

Figure 5.6 *Arabidopsis thaliana*. Longitudinal section of root tip, light micrograph. Key: Root apical meristem (RAM), Root cap (RC), Epidermis (EP).
Photo: Prof. Thomas Rost.

The description of a RAM presented here is an oversimplification, and meristems can be divided into two main types: 'closed', in which there are distinct boundaries between apical regions, and 'open', in which such boundaries appear to be absent. In a survey of 427 plant species, Heimsch & Seago (2008) found that within the main classification, there were many sub-types. Closed apical meristems were the only type present in 45 per cent of angiosperm families and occur in 63 per cent of all families. However, 36 per cent of angiosperm families have only open apical meristems, and these occur in 55 per cent of all angiosperm families.

RAMs can be either indeterminate or determinate. If a meristem is indeterminate, then it has the potential for the plant axis to grow indefinitely in length. The meristem will be active for variable periods of time and growth often only seems to cease due to local environmental factors. Some workers are, however, of the opinion that no roots and no meristems are indeterminate.

Determinate root growth implies that growth slows and stops after a pre-programmed time, and that this is under metabolic control. The RAM appears to become exhausted in determinate root growth and, as a result, all cells in the root tip differentiate. There are two types: *constitutive* if it occurs under any environmental conditions; and *inducible* if it can be induced by exposure to some conditions (e.g. phosphate starvation). It appears to function as an adaptation to water deficit in desert cactus species in the family Cactaceae. Cactus roots with determinate growth all show relatively short duration of primary root growth and early meristem exhaustion. Lateral roots of two species of *Opuntia* form determinate lateral roots a few centimetres in length. These roots rapidly cease growth as their apical meristem is exhausted, then new second-order lateral roots are formed close behind the root tip. Finally, 1 mm long 'root spurs' develop on these roots. These do not have a root cap, and root hairs eventually cover the tips. The increased root surface area provided by the spur roots almost certainly increase water uptake in the arid desert environment.

Other examples of determinate growth include:
• the **proteoid roots** in various species that are an adaptation to low soil mineral status;
• the roots of some hemiparasitic plants which respond to low concentrations of organic substances; and
• the roots of adhesive pads in climbing figs, which have a role in anchoring the plant to its substrate.

There are almost certainly many more instances of determinate root growth, but this phenomenon has only recently been recognized and it is a field of active research.

Control of primary root growth is not well understood, but the hormone auxin is involved at several stages, beginning with the early stages of embryogenesis (see Chapter 4, section 4.2). In those early stages, PIN1 and PIN 7 are involved in establishing polarity and patterning. Now PIN4, which encodes another auxin efflux carrier protein, is necessary for the RAM patterning and is expressed in the quiescent centre and surrounding cells. In PIN4 mutant embryos, the quiescent centre is replaced by irregularly dividing cells.

It is already clear that many genes are involved in the regulation and maintenance of the RAM. There are three major areas of involvement:
• pattern formation and maintenance of the quiescent centre;
• the cell cycle and differentiation;
• pathways concerned with plant hormones such as auxin.

5.3.2 Root elongation

Cell division in the RAM does add to root length, but nowhere near to the same extent as the elongation processes that occur further back in the zone of elongation. The primary roots of most cereals grown under favourable laboratory conditions can grow at about 20 mm per day, but this figure can increase to 60 mm per day in species such as maize. Expansion can be very large, and some xylem elements can increase in volume by greater than 30,000 times from their original size.

Root cells, like almost all plant cells, are encased in a carbohydrate cell wall. Plant cells cannot migrate within the plant because the cell walls are glued together, so cell division and expansion (and to a limited extent, cell death) are the only ways plant morphogenesis can occur. For a very long time, wall-loosening enzymes have been proposed to account for expansion of cells, but it is only relatively recently that the molecular basis of this process has begun to be understood (Chapter 2, section 2.2.3).

5.3.3 Formation of vascular tissue in roots

The procambium in the RAM produces the stelar (vascular) cells, and these gradually differentiate as the cells move backwards away from the meristem. In the centre of the stele is the primary xylem, which is made up of the protoxylem cells. This differentiates while young and close to the root apex. The metaxylem cells differentiate further back in the root and tend to be much larger.

The xylem cells often lose their cytoplasmic contents relatively close to the root apex and for the most part are empty, hollow, tubes. However, some of the root metaxylem elements in maize (called late metaxylem) retain living protoplasts and intact end walls until 20 cm or more after the root tip. The root phloem is found in bundles outside the xylem.

We will consider water, sugar and ion transport in the xylem and phloem in Chapter 6.

5.3.4 Lateral root growth

In the shoot (Chapter 6, section 6.3), branching patterns are, at least to some extent, initiated and controlled at the shoot apex. In the vast majority of roots, however, this is not the case, and the RAM has little direct role. The exceptions to this are some of the lower vascular plants such as the lycopods (Chapter 1, section 1.6) where branching is terminal, and the roots show determinate growth as described in section 5.3.1 above.

One of the main determinants of root system architecture is the production of lateral roots, which are also very important in adaption of the root system to particular environmental conditions. The observation that lateral roots are produced some distance from the tip of a growing root suggests either that a certain level of tissue maturity is required before initiation can occur, or that the parent root exerts some regulatory influence on the process of initiation. In the zone of differentiation, lateral meristems are formed in the pericycle, and these eventually grow through the root cortex and emerge as lateral roots. In eudicots, lateral roots are formed only from xylem pole pericycle cells – those cells lying over the developing xylem, while in many cereals, lateral roots form from the phloem pole pericycle.

The process of lateral root production is under hormonal control and it is clear that auxin plays a major role in this at all stages. In the basal meristem near the apex, pulses of auxin prime specific pericycle cells to become lateral root initial cells. These cells gradually move away from the apex as new cells are added, and they do not begin to form lateral roots until they reach a point in the zone of differentiation where this is possible. At this point, auxin triggers the degradation of the IAA14 protein, a repressor (see Chapter 3, section 3.6.3). Once the repressor is degraded, Auxin Response Factor (ARF) transcriptional activators, ARF7 and ARF19, activate Lateral organ Boundaries-Domain/Asymmetric Leaves-like (LBD/ASL) genes. LBD/ASL proteins then activate cell cycle and cell pattern genes, which lead to the formation of a new lateral root primordium (Figure 5.7).

Hormones other than auxin are also involved in lateral root initiation, and cytokinin acts as a negative regulator in many species. Plants with decreased cytokinin content have increased numbers of lateral roots. Lateral root formation is also affected by ethylene and brassinosteroids through interactions with the auxin-dependent pathway (see Chapter 3).

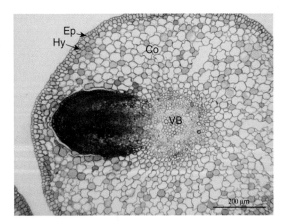

Figure 5.7 Cross section of a willow (*Salix* sp.) root, showing a lateral root emerging. Key: Epidermis (Ep), Hypodermis (Hy), Cortex (Co), Lateral Root (LR), Vascular Bundle (VB). Photo: Prof. Thomas Rost.

5.3.5 Secondary growth of roots

In Chapter 6 we will consider secondary growth and the production of wood in plant stems in some detail, but it is worth noting here that, in woody plants, secondary growth also occurs in roots. In the gymnosperms and the eudicots that have secondary thickening, the cambium – a layer of meristematic cells – is first seen on the inside of the phloem. It then produces secondary phloem to the outside and secondary xylem on the inside. At first, the cambium has an irregular shape, weaving between the poles of phloem and xylem, but it soon becomes circular in cross-section. Root secondary tissues are generally different from the stem, having fewer fibres but larger vessels and less distinct growth rings.

5.4 Soil chemistry and water relations

'*Soil is a natural body composed of minerals, organic compounds, living organisms, air and water in interactive combinations produced by physical, chemical and biological processes*'

Gerrard (2000).

So far we have considered the morphology, anatomy and development of the root in isolation from the medium in which roots normally grow – the soil. Soil science is a huge topic, and here we will just give a basic introduction to provide essential background.

Plants obtain most of their water and minerals from the soil, but soils also provide aeration for roots and support. A good supply of oxygen is essential for most roots. In waterlogged conditions, most plants have their growth reduced due to lack of oxygen, which can lead to plant death (see Chapter 9, section 9.3.2). In plants adapted to waterlogging, **aerenchyma** often develops (see Chapter 10, section 10.3.1). Finally, without soil, most plants would have great difficulty in maintaining an upright position.

5.4.1 Soil components

There are two major types of soil – organic and mineral. The organic soils are those that derive from marshes, bogs and swamps. A typical organic soil is peat. Organic soils include all soils with more than 20 per cent by weight organic matter, although this figure can be as high as 80 per cent. In terms of land mass covered, organic soils occupy a small fraction compared with mineral soils, but they are important in nature conservation and as stores of carbon. Mineral soils are those with organic matter content of 20 per cent or less, but usually below 10 per cent.

The major components of a mineral soil are the mineral matter, the organic matter, water and air. The mineral matter is usually, but not always, derived from the underlying rock material, while the organic matter represents the accumulation of decayed animal and plant material. In mineral soils, the mineral matter will occupy 45–60 per cent of soil volume and the organic matter from a trace to 10 per cent.

Water and air together can take up about 50 per cent of soil volume. The proportions of these two are subject to great natural variations and have a considerable effect on plant growth. If the soil is too dry, the plant may desiccate; if it is too wet, the plant may become waterlogged.

5.4.2 Soil water potential

Some plants absorb water through the leaves, while some grow in environments where there is little soil, but most obtain nearly all of their water from soil. In Chapter 2 (section 2.3.4), we considered the water potential of plant cells, which indicates the direction of water movement across cell membranes. It is also desirable to be able to quantify the forces acting in the soil, and soil water potential is thus defined as: '*the amount of work that must be done per unit quantity of pure water in order to transport reversibly and isothermally an infinitesimal quantity of water from a pool of pure water at specified elevation and at atmospheric pressure to the soil water (at a specified point)*' (SSSA, 1996).

Total soil water potential (Ψ_{soil}) is a composite of several sub-potentials:

1 **Matric potential**. The adsorption, adhesion and surface tension forces within the soil together make up the matric potential. Pure water will move into and up dry soil, a phenomenon known as *capillarity*. By definition, pure water has zero water potential so, if it moves into soil, this must have a negative water potential. The soil matric potential is always negative and is usually the major factor influencing movement of water into plant roots.

2 **Gravity potential**. This potential is due to the position of water in a gravitational field. It is a minor component in dry soil, but it is very important in water-saturated soil and it accounts for movement of water from higher to lower elevations. Usually, the soil water is considered as higher than a reference pool of pure water, so water moves downwards towards pure water and the gravity potential is positive.

3 **Pressure potential**. This is directly proportional to the hydrostatic pressure exerted by the soil column and is similar to that on a submarine when it is submerged. Pressure potential can be important at depth, but only in saturated soil. This potential is positive.

4 **Osmotic potential**. This arises from the salts dissolved in soil moisture. In very moist normal soils, the osmotic potential may be the predominant component of the total, although it will be rather small. As soil dries, the matric potential becomes dominant. Osmotic potential is a very important factor in saline soils, where the salt concentration may be responsible for considerable plant growth and yield reductions (see section 9.5). Osmotic potential is negative.

The total water potential of the soil can therefore be represented as follows:

$$\Psi_{soil} = \Psi_m + \Psi_\pi + \Psi_p + \Psi_g$$

where: m = matric; π = osmotic; p = pressure; and g = gravitational

Matric and osmotic potentials are negative, while pressure and gravitational potentials are positive. In almost all cases, the first two factors outweigh the last two and total potential is negative. Soil water potential is a key factor in determining water flow through the plant, and this is a topic that we shall return to in Chapter 6. Low soil water potential is an important factor in plant water stress, which will be considered in Chapters 9 and 10.

5.4.3 Cation exchange capacity

Most of the mineral elements are taken up by plants in soluble form and originate from the soil. One important place where cations are stored in the soil is on cation exchange sites. Cation exchange reactions in soils occur mainly near the surfaces of colloidal-sized clay and humus particles called micelles. Each micelle can have thousands of negative charges that are neutralized by adsorbed cations.

Of the soil mineral matter, only the clay fraction makes a significant contribution to cation exchange. Larger mineral particles (e.g. sand and silt) cannot form colloids and have only a fraction of the area needed for cation exchange. Mineral soils usually have some organic matter, and organic soils have a considerable amount. The other source of cation exchange capacity is the organic colloids. Humus is not a single compound and does not have a single structural makeup, but about 85 per cent or more of its negative charge is due to the dissociation of carboxyl or phenolic groups.

Thus, cations such as potassium and calcium are held on cation exchange sites in the soil, while anions (e.g. nitrate) can be lost from the soil fairly easily. The cations on the exchange sites can be exchanged for other cations, and these cations are also available for uptake by roots.

5.5 Plant mineral nutrition

Some plants absorb some nutrients through the leaves, but the vast majority of plants obtain most of their nutrients from the soil, and particularly from the **rhizosphere** (the region that is directly influenced by root secretions and associated soil microorganisms). The exception is carbon dioxide (CO_2), which is fixed in the leaves through photosynthesis (see Chapter 7, section 7.4).

Plant nutrition studies have a long history. Humans first took an interest in the topic in Neolithic times (8–9,000 years BP), when plants were first cultivated in the Middle East. Homer mentioned the use of manure on the land in his *Odyssey* 3,000 years ago. The Bible has many references to dung as a fertilizer (e.g. Luke 13: 6–9), and the Romans and Greeks used quite sophisticated crop rotations. At that time, no-one knew why crop rotation or using dung as a fertilizer 'worked', increasing plant growth. Knowledge was acquired by trial and error.

It was only with the flowering of scientific investigation in the 17th century that we began to understand the reasons for improved plant production as fertilizers were applied. The great chemist Justus von Liebig (1803–73) was the first to realize that plants absorbed nutrients as inorganic compounds from the soil. He detected nitrogen, sulphur, phosphorus, potassium, calcium, magnesium, silicon, sodium and iron in plant material and considered that all were essential for healthy plant growth.

By the early 1900s, however, it was recognized that the presence or concentration of an element in a plant do not necessarily indicate essentiality. Plants will absorb almost any mineral element that they are presented with, sometimes taking them up in minute amounts. To determine whether an element is essential, scientists grow plants in hydroponic cultures, where elemental concentrations are easier to control than in soil. Culture solutions lacking one element are prepared, then growth is compared with plants grown in full culture solution. If the plants in the culture solutions lacking the element do not grow well, show deficiency symptoms and eventually die, then the element is deemed essential for plant growth.

In 1939, Arnon and Stout formalized the definition of an essential mineral element as follows:
• A given plant must be unable to complete its life cycle in the absence of the mineral element.
• The function of the element must not be replaceable by another mineral element.
• The element must be directly involved in plant metabolism – for example, as a component of an essential plant constituent such as an enzyme – or it must be required for a distinct metabolic step such as an enzyme reaction.

This definition is still largely accepted today, and a list of essential elements has gradually been constructed (See Table 5.1).

From Table 5.1 it can be seen that the essential elements have been broken down into macronutrients (may exceed one per cent of the dry weight of the plant, but are often nearer 0.1 per cent), micronutrients (less than one,

Table 5.1 Essentiality of mineral elements for higher and lower plants.

Classification	Element	Higher plants	Lower plants
Macronutrient	N, P, S, K, Mg, Ca	+	+ (except Ca for fungi)
Micronutrient	Fe, Mn, Zn, Cu, B, Mo, Cl, Ni	+	+ (except B for fungi)
Micronutrient and 'beneficial' elements	Na, Si, Co	+/−	+/−
	I, V	−	+/−

to several hundred, parts per million in most plants) and beneficial elements. The macronutrients and micronutrients are all essential for plant growth, and absence of any one of these will mean that a plant cannot complete its life cycle. The last five elements in the table, Na, Si, Co, I and V, are in a 'grey zone'. In some plants, these elements are essential, but in others they appear not to be. With such elements, it is often very difficult to determine if they are essential as they may be required in very small amounts. It is problematic to ensure that these elements are completely absent from a culture solution and are not present at low levels as a contaminant.

We will not dwell on the roles of the macronutrients here as they will be covered in other parts of this text, but their main roles are summarized in Table 5.2. Many micronutrients are involved as enzyme activators, or as co-factors at the active sites of enzymes (e.g. copper in phenol oxidases). The beneficial elements include silicon, which is often used as a strengthening material (particularly in grasses and cereals) and has roles in decreasing the effects of toxic elements and preventing pathogen infections.

Most of the elements listed in Table 5.1 are essential for plant growth over a certain range of concentrations in the nutrient medium, but above a certain concentration will become toxic. This is particularly the case with micronutrients like copper and zinc, where the difference between deficiency and toxicity levels may not be very great. Finally, many elements are neither macronutrients, micronutrients or beneficial, but are only known for their toxic effects. Examples include lead, cadmium, arsenic and aluminium. In some natural soils, and in some areas that have been affected by human activities, metal toxicity can affect plant growth, and we will return to this topic in Chapter 9 (section 9.6.1) and Chapter 10 (section 10.6).

Table 5.2 Main roles of macronutrients.

Element	Some roles
N	Component of proteins, nucleic acids, etc.
K	Enzyme activator, osmoticum
Ca	Cell wall strength, secondary messenger
P	Present in ATP, nucleic acids, phospholipids, etc.
Mg	Enzyme activator, component of chlorophyll
S	Coenzymes. Present in cystine, cysteine and methionine

5.6 Movement of nutrients to the root surface

Having considered root structure, soil chemistry and plant mineral relations, the next topic to cover is the movement of nutrients to the root surface. There are three components in this:

1 **Interception**. This arises as the roots grow into new soil. As they do so, they directly encounter some of the ions present in the soil solution. Because roots generally only occupy 1 per cent of the soil volume, only about 1–2 per cent of nutrients are available for uptake at the root surface using this mechanism.

2 **Mass flow**. In section 6.8, we will consider long distance transport of water in the plant. Water is lost from the leaves of the plant. This is replaced from the xylem transpiration stream, and the water for this stream is supplied by the roots. This pathway can be extended out into the soil. Soil water is moved by mass flow to the root surface and, of course, ions are brought along in the water. Most nitrogen comes to the root surface in this way.

3 **Diffusion**. Ions in soil are in constant motion and will tend to diffuse into areas where the concentration is low. This might be near the root surface after interception uptake. At field capacity, the rate of diffusion per day through the soil is 0.13 cm for potassium and only 0.004 cm for phosphate. However, diffusion is very important for the uptake of ions such as these, which are only present at low concentrations.

5.7 Absorption of water and nutrients

Now we turn to the topic of ion transport in the root. First we will consider these processes at the cellular level, and then look at long-distance transport.

5.7.1 The apoplast and free space

As we have seen, ions are brought to the root surface by a variety of mechanisms. However, ion uptake into root cells is not always restricted to the external layer of cells. The **apoplast** of the root constitutes all compartments beyond the plasma membrane, including the interfibrillar and intermicellar space of the cell walls. The other compartment is the **symplast**, which is composed of the cytoplasm and the plasmodesmata connecting the cells together to form a 3-D network. Un-lignified cellulose cell walls make a system of microcapillaries which offer little resistance to ion and solute movement. This space is freely penetrated and is directly accessible to the external solution. Uptake into the apoplast is reversible, non-selective, independent of metabolism and essentially passive. In young roots, the free space (volume available for passive solute movement) can represent about 10 per cent of the total root volume.

The primary cell walls consist of cellulose, pectins, hemi-cellulose and glycoprotein (see Chapter 2, section 2.2). These compounds form a network which contains pores. The pores are quite large (up to 5.0 nm in diameter) and should not restrict the movement of ions in these spaces (hydrated K^+ = 0.66 nm diameter, Ca^{2+} = 0.82 nm). However, high molecular weight compounds will definitely be restricted by the network. The pectins in the root cell walls have many carboxylic acid groups ($R.COO^-$). These fixed anions act as cation exchange sites, in a similar way to those in the soil; cations are attracted to these carboxylic acid groups and anions are repelled.

Thus, ions do not simply pass through the apoplast, but many are adsorbed en route to the plasma membrane. If a root that is loaded with ions (e.g. K^+) is put into water, a large amount of ions will be washed out – the so-called water extractable fraction. A further fraction can be washed out if the root is then exposed to a salt solution with which the adsorbed ions can exchange the exchangeable fraction. In essence, the root apoplast can be divided into two sections:

- Water-free space (mobile ions in the aqueous phase).
- What is known as the Donnan free space (the exchangeable fraction adsorbed to immobile cation exchange sites).

The ions in the apoplast are then available for uptake through the plasma membrane into the cell protoplasts. If there are no barriers within the root apoplast, then essentially ion uptake can occur into the protoplasts of cells anywhere within the cortex, so the area for uptake is vastly increased in comparison with that provided by the root surface.

5.7.2 Active and passive transport

We have considered ion movement up to the membranes of the cells in the root, and now we need to discuss ion transport across membranes. In Chapter 2 (section 2.3) we covered membrane structure and saw that membranes have a number of important properties. The most significant in the present context is that they are selective in what they allow into the protoplast.

Ion transport across membranes is not always an active process. Passive transport involves solutes being more concentrated on one side of a membrane than another and diffusion from a higher to a lower chemical potential. This is a 'downhill transport' and it can be maintained by lowering ion activity in the cytoplasm. For example, ions could be adsorbed onto fixed charged groups or could be incorporated into organic structures (e.g. phosphate in nucleic acids), or they could be sequestered in specific cell components. In all three situations, the effective concentration of the ion in the protoplast would be reduced and a gradient favouring downhill transport would be maintained. Obviously, 'uphill transport' against a gradient of potential energy will require an energy-consuming mechanism – a 'pump' – in the membrane (see section 5.7.3 below). However, to determine whether active transport has to take place, we need to know:

1 the concentration of the ion on both sides of membrane (the chemical potential gradient); and

2 the electrical potential gradient (difference in millivolts (mV)) across the membrane.

The electrical potential gradient can be measured between the cell sap and the external solution using a microelectrode, and these potentials have been found to be strongly negative values. The concentration at which ions are at electrochemical equilibrium in the external solution with those in the vacuole can be calculated using the Nernst equation:

$$E(mV) = -59 \, Log_{10} \frac{\text{Conc. inside (vacuole)}}{\text{Conc. outside (external solution)}}$$

This equation can be used to calculate the internal concentrations at equilibrium if the electrical potential difference between the nutrient solution and the cell sap and the concentration of ions in the external solution is known. Those same concentrations can be measured experimentally by extracting some cell sap with a micropipette for analysis. The difference between the calculated and the experimental values will allow us to determine whether active transport is required, and in which direction.

What does this mean for the concentrations of ions in cells? Even a relatively low potential difference between the external solution and the cell vacuole allows a tenfold difference in concentrations between the inside and outside, and the ions are still at electrochemical equilibrium.

Table 5.3 Experimentally determined and calculated ion concentrations due to the electrical potential differences in oat roots.

Ion	Experimental concentration (mM)	Calculated concentration (mM)
Potassium	66	27
Sodium	3	27
Calcium	3	1400
Chloride	3	0.038
Nitrate	56	0.076

The external nutrient solution composition was 1.0 mM KCl, 1.0 mM $Ca(NO_3)_2$ and 1.0 mM NaH_2PO_4. The electrical potential difference between the cell vacuoles and the external solution was determined as −84 mV. Data from Higinbotham *et al.*, 1967.

One could think of it like this: the cell has many negative charges that attract cations but repel anions. Of course, anions, which are negatively charged, will nearly always require an active uptake process.

Table 5.3 shows an example taken from an experiment where oat seedlings were grown in a nutrient solution. If the electrical potential difference between the external nutrient solution and the cell sap is determined, and the concentration of ions in the solution is known, then the Nernst equation can be used to calculate the internal concentrations at electrochemical equilibrium. Ion concentrations were also determined experimentally in extracted sap, and these can be compared with the calculated concentrations. From this, it can be inferred that potassium requires active uptake for the experimental concentration to be reached. The calculated equilibrium values for sodium and calcium are much higher than the values measured experimentally. This suggests either that the plasma membrane restricts permeation or that the ions are pumped back to solution. For calcium, an active extrusion pump would require enormous amounts of energy, so it seems likely that physicochemical factors such as size and charge partly restrict movement of calcium across the plasma membrane. As might be expected, the data suggest that both anions would require active uptake against an electrochemical gradient.

5.7.3 Ion channels, carriers and pumps

We can distinguish several means by which solutes may traverse the plasma membrane. First, small uncharged molecules, including water, can diffuse freely through the lipid bi-layer and will thus enter the cell if the

concentration gradient is in their favour (or will leave the cell if the gradient runs in the opposite direction). Diffusion of some substances may be aided by carrier proteins or by channel proteins in a process known as *facilitated diffusion*. As their name implies, channel proteins provide an aqueous channel through which solutes such as positively charged ions may diffuse, each type being selective for a particular solute. These channels operate as gates that are open for brief periods of time and then closed again, thus controlling the extent to which small polar solutes can actually travel down their concentration gradient into (or out of) the cell. Carrier proteins, on the other hand, do not make channels but bind specific solutes and undergo a conformational change in order to deliver the solute across the plasma membrane.

It has been suggested that protons (H^+ ions) should be regarded as one of the 'energy currencies' of living cells, along with ATP, NADH and NADPH. The involvement of proton gradients in ATP synthesis in both chloroplasts (Chapter 2, section 2.5.3 and Chapter 7, section 7.4.4) and mitochondria (Chapter 2, section 2.14.2) certainly supports that suggestion, as does the use of ATP to pump protons out of plant cells.

The formation and maintenance of electropotentials is partly due to fixed negative charges in the cytoplasm. However, electrogenic pumps, which transport an ion in one direction without an accompanying ion of the opposite charge, are the driving force behind the main transport mechanism in plants. These are mainly proton pumps, membrane-bound ATPases that pump H^+ out of the cytoplasm.

In Chapter 2 (sections 2.5 and 2.6) we noted that ATP synthase occurs in the membranes of mitochondria and chloroplasts and that, as protons move down their electrochemical gradient, the energy is used to drive ATP synthesis. The proton pumps work in the reverse manner. Mediated by a membrane-spanning ATPase and the energy from ATP hydrolysis, protons are pushed out through the protein complex. This pumping produces an electrical potential across the plasma membrane which, under most conditions, is between −80 and −100 mV. As we saw in section 5.7.2 above, electrochemical gradients feature strongly in the regulation of trans-membrane transport. The export of calcium ions is also mediated by linkage to ATP hydrolysis but, in this instance, the ATPase does not span the plasma membrane.

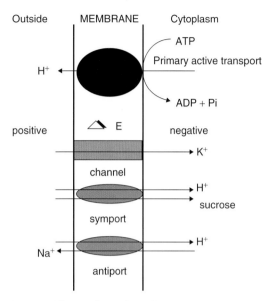

Primary and secondary
active transport

Figure 5.8 Primary and secondary active transport across membranes. This diagram shows the primary active transport of protons across the membrane. ATP energy is used and the transport generates a membrane potential, ΔE. The proton gradient is used to drive other transport systems. Here uptake is through a uniporter (K^+) or a symporter (sucrose), and efflux is through an antiporter (Na^+).
From Flowers & Yeo, 1992. (Reprinted with kind permission from Springer Science+Business Media B.V.)

The export of H^+ ions and of calcium ions are examples of *primary active transport* (see Figure 5.8). It is the pumps that create the electrochemical gradient across the membrane, and the inside of the cell becomes more negative than the outside. This gradient can then be responsible for *secondary active transport*, which can occur in two ways. First, ions can move down the electrochemical gradient through selective channels. A large amount of potassium uptake can be accounted for by this mechanism. The second way is due to the proton pump generating an electrochemical gradient of protons. The energy available when the protons return across the membrane down their free energy gradient can be coupled to the movement of another solute against its gradient in free energy. The solute (anion, cation or non-electrolyte) may move in the same direction as the protons when the carrier is a **symporter**, or in the opposite direction when the carrier is an **antiporter**. Two examples of this are the linkage of

sucrose uptake with the influx of H^+ ions in the proton-sucrose symporter and the linkage of sodium ion efflux with H^+ ion influx in the proton-sodium antiporter.

Evidence for proton pumps producing an electrogenic component of membrane transport in plants is now overwhelming. Interestingly, they are more similar to bacteria and fungi in their transport systems than to animal cells, which seem to use mostly ion pumps (e.g. the sodium-potassium pump) driven by ATPases, and primary active transport.

5.7.4 Ion movement across the root

We have now covered how ions are absorbed by individual cells. In this section, we will consider ion movement across the root. As we have already noted, this can be divided into two types: apoplastic in the free space; and symplastic through the cytosol and plasmodesmata. In the symplasm, it is thought that ions move across the cortex through the endodermis to the stele, while photosynthetic products (sugars) move in the opposite direction to provide respiratory substrates for the cortical cells. The mechanism of movement is thought to be diffusional in the cytoplasm, with cytoplasmic streaming or cyclosis aiding the movement. Inhibition of this streaming inhibits transport of ions across the root.

Until relatively recently, it was believed that the soil solution could penetrate freely into the surface layers of the cortex and thus reach the endodermis without crossing any membrane or high resistance barrier en route. This would mean that each cortical cell is bathed in a solution that is very similar to that in the soil solution.

The concentration of the ions in the xylem sap (within the stele), can be very different from that of the external solution. It is also known that there are electrical potential differences between the external solution and the xylem sap. There must be a barrier or barriers that are the inner limit to the free space, or such gradients could not exist. The most often cited barrier is the endodermis.

5.7.5 The endodermis

The endodermis can be defined as: '*a sheath of cells surrounding the vascular tissues of all roots and of some stems and leaves. The endodermal cells are characterized by having Casparian strips that prevent the diffusion of material between the cortex (or mesophyll) and the vascular tissues* (Mauseth, 1988).

The **Casparian strips** are lignified bands in the radial walls of the endodermal cells (Figure 5.4). These are considered to constitute a barrier to water and ion movement in the apoplast, and there are two lines of evidence for this – direct and indirect. Several direct approaches (autoradiography, lanthanum uptake, x-ray microanalysis and fluorescent dyes) have suggested that the endodermis is a barrier. In all cases, substances have been observed to move through the apoplast up to the Casparian strips in the endodermal walls and then stop. The plasma membrane is fused to the wall at the Casparian strip at this point, preventing further movement (Figure 5.9).

Indirect free space measurements in some roots have shown that the free space in excised roots is higher than that in intact roots. This suggests that a radial barrier has been broken in the excised root, making it directly accessible to external solution. The free space value for the cortex alone is the same as for the intact root, implying that this radial barrier is located at the endodermis.

However, this simple picture of an endodermal barrier is, in fact, more complicated, as the endodermis undergoes anatomical development as the root tissues mature, which affects radial ion movement. There is a developmental sequence in which, after the endodermis forms, the Casparian bands are laid down in the endodermal walls, and later suberization (deposition of waxy substances) of all the walls occurs.

What difference do these changes make to ion transport across the root? Back in the 1970s, David Clarkson and his co-workers developed systems to feed radioactive

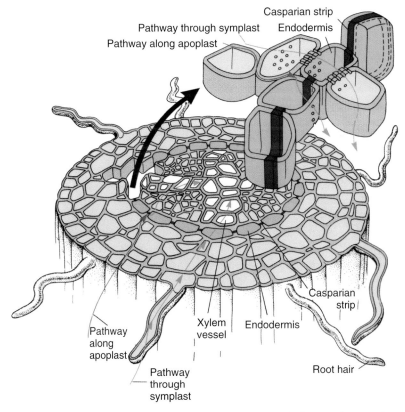

Figure 5.9 Pathway of mineral transport in roots. This diagram shows that there are two paths for mineral transport in roots: symplastic, through the cell protoplasts; and apoplastic, through the cell walls and intercellular spaces. The apoplastic pathway is blocked at the Casparian strips in the endodermis, and at this point ions must cross the plasmalemma if they are to continue into the stele.
From Campbell, Neil A., BIOLOGY, 2nd Edition © 1990. Reprinted by permission of Pearson Education, Inc., Upper Saddle River, NJ.

ions to different parts of the marrow (squash, *Cucurbita pepo*) root. They then analyzed what was translocated to the shoot. Radioactivity detected in the shoot came principally from the segment exposed to labelled culture solution. Potassium translocation was similar over the whole length of the root, but calcium was only rapidly and extensively translocated from the first 140 mm of root behind the tip.

Why the difference? Clarkson investigated the endodermis at 140 mm behind the root tip and found that this was where the endodermis became suberized. The K^+ ions, which are rapidly absorbed by cortical root cells, can travel in the symplastic route very easily, and changes to the endodermal walls do not make much difference. However, Ca^{2+} ions, being divalent, are only sparsely taken up into cortical cells and are mostly transported in the apoplast. It seems that when the endodermal plasma membrane becomes inaccessible to the apoplast, Ca^{2+} transfer to the stele is effectively stopped because of its negligible movement in the symplast.

Thus the idea that the endodermis was the major barrier to apoplastic transport in the root became established, and this model is certainly still the major one to be found in most text books. However, that has now all changed with discoveries concerning the exodermis.

5.7.6 The exodermis

Beginning in the 1980s, Carole Peterson led a group at the University of Waterloo, Canada and began work which ultimately undermined the simple classical model outlined above. Peterson's early work involved the development of new stains, which have revealed features of cell structure in a layer of cells beneath the epidermis known as the hypodermis. Peterson showed that the roots of many species had a Casparian strip in the hypodermal layer and, when this was the case, the layer was known as the **exodermis** (Figure 5.10).

In fact, Peterson reported that only 6 per cent of species had no hypodermis (e.g. legumes), while 3 per cent had a hypodermis with no wall modifications. The majority of species (88 per cent) had a hypodermis with a Casparian band. It might be expected that these Casparian strips would cause a blockage to apoplastic flow at this point. Initially this idea caused some controversy, but it now appears that the exodermis represents a barrier of variable resistance to the radial flow of both water and solutes.

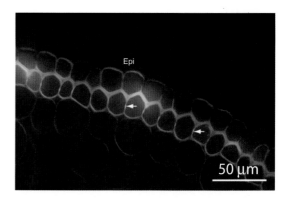

Figure 5.10 Casparian bands (arrows) in *Glyceria maxima* exodermis stained with berberine-toluidine blue in basal parts of a root. Key: Epi, epidermis.
Micrograph from Soukup *et al.*, 2007. (Reprinted with the permission of Blackwell Publishing)

Like the endodermis, changes in structure and anatomy of the exodermis occur in developing roots. In the exodermis, however, the extent and rate at which the Casparian bands and suberin lamellae are laid down in exodermal walls depends on the environmental conditions in which the plant is growing. These apoplastic barriers do not work in an all-or-none manner. By changing the amount and chemical nature of the barriers, regulation of water and solute transport is achieved (See Hose *et al.*, 2001).

5.7.7 Ion release into the xylem

So far, we have followed ion movement from the soil solution across the cortex and through the endodermis, and now we must consider the topic of ion release into the xylem. The xylem vessels lose their contents to become dead water-conducting tubes within a relatively short distance of the root tip. The mechanism of long-distance water and ion transport will be considered in Chapter 6, and here we will only consider the loading of ions into the xylem in the root.

The xylem vessels may be dead and empty, but the living xylem parenchyma cells and contact cells surrounding them have direct access to the vessels. They play an important role in the loading of the xylem. The ultrastructure of these cells (dense cytoplasm, well-developed endoplasmic reticulum, numerous mitochondria and wall invaginations) suggests that they are metabolically active.

The precise mechanism of transport into the xylem is still under debate. As long ago as 1938, Crafts and

Broyer suggested uphill transport in the symplasm to the endodermis, and in the stele a 'leakage' into the xylem. They considered that the lower oxygen tension in the stele might prevent active transport of ions into the xylem, but this idea has since proved to be incorrect. ATPase activity has been observed in the membranes of the xylem parenchyma cells using cytochemical techniques. Cation movement from the symplasm of the cortex to the xylem appears to occur down an electrochemical gradient and primary H^+-ATPases are thought to be responsible for generating this gradient. The plant hormone ABA is known to inhibit K^+ loading into the xylem.

Loading of K^+ into the xylem is regulated in a different way from K^+ uptake from the external solution. The two types of K^+ channel in the root are AKT1 for K^+ uptake from the medium, and SKOR for K^+ loading into the xylem. SKOR channels are selective for K^+, also transport Ca^{2+}, but are impermeable to Na^+. The gene encoding the K^+ channel that causes loading into the xylem, *SKOR*, is in the Shaker superfamily. Plants that are treated with ABA show a fast decline in the amount of *SKOR* mRNA, which explains the inhibition of xylem K^+ loading by ABA. Water flow into the xylem elements is aided by the presence of aquaporins (see Chapter 2, section 2.3.4) in the surrounding parenchyma cells acting as water channels.

5.8 Mycorrhizae

So far we have considered roots as entities that are made up of one organism – the plant – but in fact many roots are symbiotic associations between the plant and a fungus. These are known as mycorrhizae. The majority of seed plants have mycorrhizal associations, including all gymnosperms and most angiosperms (see Chapter 1, section 1.6). In ectotrophic mycorrhizae, fungal hyphae form a thick sheath over the whole root surface, and sometimes they penetrate between the cortical cells (Figure 5.11). These are the most common type in tree species. Vesicular-arbuscular mycorrhizae do not form such a dense mass over the root surface, but do penetrate the root cortical cells to form vesicles and arbuscules. When the hyphae penetrate the cortical cells, they do not rupture the plasma membrane or the tonoplast, but these membranes surround the hyphae to form arbuscules.

Mycorrhizae hugely increase the surface area available for water and mineral uptake by the plant root. They are particularly important in increasing the uptake of phosphate, which is generally present in the soil at low concentrations and/or availability. There is also some evidence that mycorrhizae act to protect roots from toxic metals. The fungal symbionts obtain carbohydrate from the plant tissues, so there is a cost to the plant.

5.9 Root nodules and nitrogen fixation

Most plants absorb nitrate and ammonium ions from the soil, and these are then converted into organic compounds by a series of enzyme reactions (see Chapter 2, section 2.14.8). The Earth's atmosphere is 78 per cent nitrogen, but plants and animals cannot directly use it; the very strong triple bond between the two nitrogen atoms makes the gas quite inert. As far as is known, no eukaryotic organisms are able to fix nitrogen directly from the atmosphere. Rather, it is fixed as ammonium during thunderstorms and by certain groups of bacteria, some of which are symbionts with higher plants in specialized root nodules. These symbiotic associations are the topic of this section. We will consider the legumes in detail, but species in other genera also form symbiotic nodules.

About 80 per cent of the species in the legume family have been found to have nitrogen-fixing nodules containing the symbiotic bacterium, *Rhizobium*. Most of the species in the two sub-families Papilionoideae and Mimosoideae form nodules, but only a third of the more primitive and largely tropical Caesalpinoideae do so. The legumes include peas and beans, but there are also many trees and shrubs.

Most legumes, and indeed most nitrogen fixers, are colonizers of nutrient deficient habitats, coming in early in succession. Having symbionts does have a cost to the host, and these plants often transport large amounts of carbon to their roots. Therefore, legumes have a real advantage in nitrogen-poor soils, but are often out-competed later in succession, when nitrogen is abundant in the soil and the costs of keeping a symbiont are too much. Legumes are not only sown as crops, but often with grass to increase yields in pasture. The nodules and roots release some nitrogenous compounds into the soil even when alive, and even more is released on death and decay.

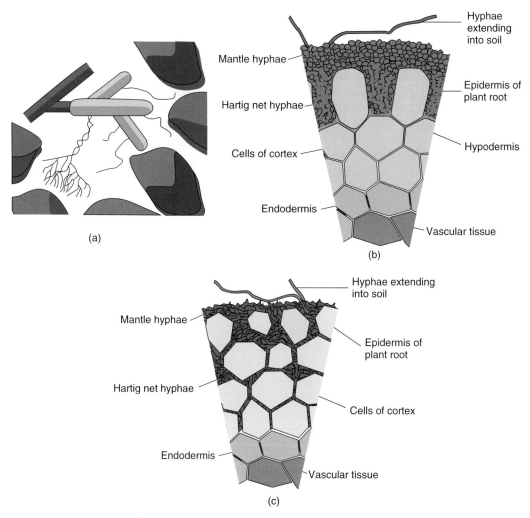

Figure 5.11 A typical root infected with an ectomycorrhizal fungus (a). In some cases the sheath encircles the root without penetrating the root cortex (b), while in other associations, the cortex is penetrated up to the endodermis (c).
From Scott, 2008. (With kind permission from John Wiley & Sons Ltd.)

5.9.1 Nodule formation in *Rhizobium*

We must now consider how the nodules are formed. A particular *Rhizobium* species is often only effective with one type of legume. The free-living bacteria are aerobic and live saprophytically in the soil until they infect either a root hair or a damaged epidermal cell. Legume roots produce flavonoid molecules, which attract the bacteria in the soil to the root surface. The flavonoids also signal the rhizobia to produce sugar-containing compounds

called Nod (nodulation) factors. These factors then bind to a specific receptor on the root hair cells, which initiates a signalling pathway that causes the root hair to curl near the tip around the rhizobia.

Infection threads formed from Golgi vesicles grow in the root hair and the bacteria multiply extensively. The threads extend and penetrate the cortical cells, and the rhizobia then take control of the cell division in the cortex. During this time, the rhizobia change shape and enlarge

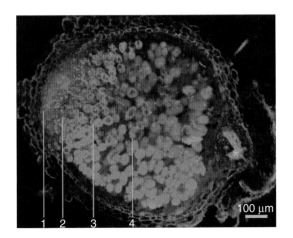

Figure 5.12 Longitudinal section of a mature *Medicago truncatula* root nodule. Regions of the nodule visible using confocal microscopy: 1, nodule meristem; 2, prefixation zone; 3, interzone; 4, nitrogen-fixation zone. *, Stele of root.
Micrograph from Haynes *et al.*, 2004. (Reprinted with the permission of Blackwell Publishing)

to become **bacteroids**. In the inner cortex, the bacteria are released into the cytoplasm and stimulate some cells to divide. This leads to the formation of a mature root nodule (Figure 5.12).

5.9.2 The mature nodule

The mature nodule consists mainly of tetraploid cells containing bacteria and some diploid cells without bacteria. Typically, each nodule will contain several thousand bacteroids. Four to six bacteroids are grouped together and are surrounded by a peribacteroid membrane. Between the membrane and the bacteroids is the peribacteroid space. Nitrogen fixation in the root nodules takes place in the bacteroids. The host plant provides carbohydrate, largely as sucrose, via the phloem, and the bacteroids provide the plant with nitrogen.

A very important protein, **leghaemoglobin**, is located in the plant cytoplasm. The molecule is red because of the attached haem group that is similar in structure to that in blood haemoglobin. This haem is attached to the colourless globin protein. Leghaemoglobin is present at high concentrations ($700\,\mu$M in soybean nodules) and gives the nodules a pink colour. Globin is produced by the host plant in response to infection, and the haem is made by the bacteria.

The rate of nitrogen fixation by the nodules is correlated with leghaemoglobin concentration. Leghaemoglobin has a very high affinity for oxygen – around ten times greater than human haemoglobin. It is thought that leghaemoglobin acts as a buffer system in the nodule, only allowing oxygen into the nodules at a very controlled rate. As we will see below, nitrogenase, the enzyme responsible for nitrogen fixation, is highly sensitive to oxygen.

5.9.3 Nitrogenase

The overall reaction for the reduction of molecular nitrogen to ammonia in root nodules is essentially the same as the Haber process which is used to produce nitrogen fertilizers. However, the Haber process requires high temperature, high pressure and metal catalysts, while the bacteroids carry it out at room temperature and pressure, in water and with oxygen present. The enzyme responsible for fixation in the bacteroids – nitrogenase – is thus a remarkable enzyme with unusual properties.

The nitrogenase complex can be separated into two components: a molybdenum-iron protein and an iron protein. The iron protein is the smaller and has two identical subunits (30–72,000 Da), containing one 4Fe-4S cluster. The iron protein is very sensitive to oxygen, which irreversibly inactivates the protein with a half-life of 30–40 seconds.

The molybdenum-iron protein has four subunits totalling 180–235,000 Da. It has two molybdenum atoms per molecule in two Mo-Fe-S clusters, and some Fe-S clusters. It is also sensitive to oxygen, with a half-life of 10 minutes. In the reaction catalyzed by nitrogenase, flavodoxin or ferredoxin reduces the iron protein, which in turn reduces the molybdenum-iron protein (Figure 5.13). The molybdenum-iron protein then reduces nitrogen to ammonia. The total cost of reducing one N_2 molecule is eight electrons transferred and 16 ATPs hydrolyzed.

As we have mentioned, nitrogen fixation is very sensitive to oxygen. The protein leghaemoglobin is partly responsible for the protection of the bacteroids, but it is almost certainly the structure of the nodules that is most important in preventing excess oxygen reaching the nitrogenase enzyme.

The ammonia produced through fixation is translocated out of the bacteroids and is further metabolized

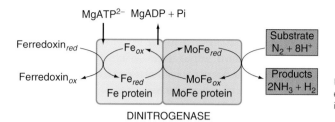

DINITROGENASE

Figure 5.13 The nitrogenase reaction in bacteroids, where electron flow is from left to right. The main electron donor is ferredoxin, which is reduced via respiration.

before it can be used by the host. In the cytoplasm of the bacteroid containing cells in many plant species, the ammonia is converted into organic compounds (see Chapter 2, section 2.14.8). The major compound in temperate legumes (e.g. peas, clover, lupins), is asparagine, while ureides dominate in legumes of tropical origin (e.g. soybean, cowpea). These compounds are transported in the xylem to the leaves, where they are broken down to ammonia and incorporated into amino acids.

5.9.4 Genetic factors in nitrogen fixation

The genetics of nitrogen fixation is extremely complex, and what follows is a very simplified overview. The bacterial genes involved in infection or nodulation are designated NOD. Mutations in these produce bacteria that cannot infect plants or cannot form nodules. *NOD* genes are plant genes encoding for nodulin proteins, and the genes encoding the early plant products are *ENOD* genes. These *NOD* and *ENOD* genes are expressed at various developmental stages of the nodule, leading to the production of leghaemoglobin and the enzymes involved in ureide metabolism, for example. The *nif* genes are responsible for the fixation process, including those for nitrogenase and several regulatory genes, and they only become active when the nodule is well formed. Mutations in these will prevent nitrogen fixation.

There is considerable interest in introducing *nif* genes into plant species that do not normally fix nitrogen, in an effort to increase plant yields. One idea is to introduce the genes into the chloroplast, and this appears to have a number of advantages:

• Chloroplasts are sites of ATP synthesis.
• The chloroplast is a major site for ammonia assimilation.
• The chloroplast has genes that are similar to those in prokaryotes (Chapter 1, section 1.3 and Chapter 3, section 3.3.2).

However, even if this manipulation can be achieved, there will remain problems for nitrogenase to operate in the chloroplast environment (e.g. oxygen sensitivity). Recent work has shown that some free-living bacteria are able to fix nitrogen using vanadium-containing enzymes, and that some species have enzymes that are not sensitive to oxygen. This has suggested that it is possible that some eukaryotic organisms already have nitrogenase. If this were the case, then genetic modification of higher plants for nitrogen fixation would be made much easier.

5.10 Tropisms

The final part of this chapter concerns the mechanism of the downwards curvature of roots in a gravitational field. This is one of a category known as growth movements or tropisms. We have already had cause to mention tropisms previously in Chapter 3, when we considered auxins (section 3.6.2) and phototropins (section 3.7.5). In this section, we will give some general background to tropisms before turning to the specific case of gravitropism in roots in section 5.11.

Earlier in this chapter, we considered the growth and development of roots. Here, growth resulted in an increase in mass and morphogenesis, but growth can also be involved in the response to external stimuli. Plants can alter the orientation of their parts in space in relation to external stimuli, either through growth movements or localized turgor changes. Growth movements, which occur in response to unidirectional stimuli and result in the positioning of the plant part in a direction related to the direction of the stimulus, are termed tropisms.

There are several possible responses to a stimulus. If the growth response is directly towards the stimulus, the reaction is positively orthotropic, but if it is in the opposite direction from the stimulus, it is negatively orthotropic. If it is at an angle to the stimulus, it is plagiotropic, and if it is at a right angle, it is a diatropic response.

In general, a tropic response can be divided into a number of stages separable in space or time:

1 Perception of the external stimulus.

2 Transduction of the stimulus in the sensitive region, which involves the induction of a metabolic change.

3 Transmission of an internal stimulus to a reaction site.

4 A reaction which is generally expressed through differential growth on two sides of an organ.

This scheme is applicable to any tropism, and we will see further examples in Chapter 6, but now we will concentrate on gravitropism in roots.

5.11 Gravitropism in roots

It has been known for a long time that if a root is placed horizontally, it tends to bend down towards gravity. Cells on the top side of the root some way back from the tip elongate faster than those on the bottom side, causing the curvature. We now need to consider what is known of the four stages outlined above in the particular instance of gravitropism in roots.

5.11.1 Perception

The most likely way for a plant organ to detect its orientation with respect to gravity is by movement of one or more of its component parts. In the human ear, we have otoliths, granules of calcium carbonate, which are involved in human gravity perception. When a root is turned from a vertical to a horizontal position, the starch-containing amyloplasts (see Chapter 2, section 2.5.4) in the root cap cells move to take up positions on the bottom edge of the cell. This observation led Haberlandt to propose the **statolith** hypothesis in 1900, with the idea that amyloplasts are involved in gravity perception. Over the years, there have been alternative hypotheses suggested, but this remains the most likely idea. However, the precise mechanism has still to be determined.

It was recognized that gravity was perceived near the root tip, but it was not until the experiments of Barry Juniper and his colleagues in the early 1960s that it was proven that gravity was detected in the root cap (Figures 5.5 and 5.6). Juniper surgically removed the root cap from maize roots. This had no effect on the rate of root elongation, but it completely prevented perception of

gravity, and the decapped roots did not bend down. The root cap regenerated after about 36 hours, whereupon ability to perceive gravity returned. The root cap cells were found to be not highly vacuolate, variable in shape, and about twice as long as broad. The columella cells in the centre of the root cap have been shown to make the greatest contribution to root gravitropism. They have numerous mitochondria and Golgi bodies and the nuclei are often amoeboid in shape. The organelles of greatest interest within these cells are, however, the amyloplasts. The number of amyloplasts varies between species, but there can be 50 or more in a cell. Quite obviously, plants without amyloplasts in the columella cells of the root cap would be expected not to respond to gravity. Most interest in plants without amyloplasts has centred on so-called starchless mutants. It appears that in most cases, the gravitropic response is greatly reduced in the mutants, but that it is not totally eliminated, suggesting that there is another mechanism operating at a lower level. After years of uncertainty, it appears that the statolith hypothesis involving the amyloplasts of the root cap columella cells is still the most likely to be correct in respect of the major gravitropic response.

5.11.2 Transduction and transmission

The transduction (induction of a metabolic change in the sensitive region) and transmission phases are also unclear. It is interesting that another old idea, the Cholodny-Went hypothesis is still dominant here (see Chapter 3, section 3.6.2). It proposed that lateral transport of auxin is induced by directional stimuli – in this case, gravity – and leads to asymmetrical auxin distribution within plant organs. In a root growing vertically, the concentration of auxin reaching the growing cells is equal on all sides. When a root is moved to a horizontal position, more auxin is found on the lower side than the upper. In roots, the auxin concentration is inhibitory on the lower side and optimal on the upper side, and hence downward curvature occurs. In shoots, the *higher* concentration of auxin on the darkened side stimulates growth. Thus, mutants that are defective in auxin transporters show an impaired response. Experiments with labelled auxin also suggest that in horizontally oriented root caps, the dominant movement of auxin is in a downwards direction.

Selected references and suggestions for further reading

Brady, N.C. & Weil, R.R. (2007) *The Nature and Properties of Soils*. 14th ed. Prentice Hall, NJ.

Campbell, N.A., Reece, J.B. & Mitchell, L.G. (1999) *Biology*. Addison Wesley Longman Inc. Menlo Park, CA.

Cheng, Q. (2008) Perspectives in biological nitrogen fixation research. *Journal of Integrative Plant Biology* **50**, 786–798.

Clarkson, D.T. (1974) *Ion transport and cell structure in plants*. McGraw-Hill, London.

Cutler, D.F., Botha, C.E.J. & Stevenson, D.W. (2008) *Plant anatomy. An applied approach*. Blackwell Publishing, Oxford.

De Boer, A.H. & Volkov, V. (2003) Logistics of water and salt transport through the plant: structure and functioning of the xylem. *Plant, Cell and Environment* **26**, 87–101.

Epstein, E. & Bloom, A.J. (2004) *Mineral Nutrition of Plants: Principles and Perspectives*. 2nd Ed. Sinauer Associates Inc., USA.

Flowers, T.J. & Yeo, A.R. (1992) *Solute Transport in Plants*. Chapman and Hall, London.

Gerrard, J. (2000) *Fundamentals of Soils*. Routledge, London.

Haynes, J.G., Czymmek, K.J., Carlson, C.A., Veereshlingam, H., Dickstein, R. & Sherrier, D.J. (2004) Rapid analysis of legume root nodule development using confocal microscopy. *New Phytologist* **163**, 661–668.

Heimsch, C. & Seago, J.L. Jr. (2008) Organization of the root apical meristem in angiosperms. *American Journal of Botany* **95**, 1–21.

Higinbotham, N., Etherton, B. & Foster, R.J. (1967) Mineral ion contents and cell transmembrane electropotentials of pea and oat seedling tissue. *Plant Physiology* **76**, 249–253.

Hopkins, W.G. (1999) *Introduction to Plant Physiology*, 2nd ed. John Wiley & Sons, New York.

Hose, E., Clarkson, D.T., Steudle, E., Schreiber, L. & Hartung, W. (2001) The exodermis: a variable apoplastic barrier. *Journal of Experimental Botany* **52**, 2245–2264.

Marschner, P. (ed) (2012) *Marschner's Mineral Nutrition of Higher Plants*. 3rd ed. Academic Press, London.

Mauseth, J.D. (1988) *Plant Anatomy*. Benjamin Cummings, Menlo Park, CA.

Morita, M.T. & Tasaka, M. (2004) Gravity sensing and signaling. *Current Opinion in Plant Biology* **7**, 712–718.

Nibau, C., Gibbs, D.J. & Coates, J.C. (2008) Branching out in new directions: the control of root architecture by lateral root formation. *New Phytologist* **179**, 595–614.

Scott, P. (2008) *Physiology and Behaviour of Plants*. John Wiley & Sons, Chichester, UK.

Shishkova, S., Rost, T.L. & Dubrovsky, J.G. (2008) Determinate root growth and meristem maintenance in angiosperms. *Annals of Botany* **101**, 319–340.

Soil Science Society of America (SSSA) (1996) *Glossary of Soil Science Terms*. SSSA, Madison, WI.

Soukup, A., Armstrong, W., Schreiber, L., Franke, R. & Votrubová, O. (2007) Apoplastic barriers to radial oxygen loss and solute penetration: a chemical and functional comparison of the exodermis of two wetland species, *Phragmites australis* and *Glyceria maxima*. *New Phytologist* **173**, 264–278.

Waisel, Y., Eshel, A. & Kafkafi, U. (2001) *Plant Roots: The Hidden Half*. 3rd ed. Marcel Dekker, New York.

Weier, T.E., Stocking, C.R. & Barbour, M.G. (1974) *Botany: An Introduction to Plant Biology*. 5th ed. Wiley International, New York.

CHAPTER 6

Stems

We have seen that the roots anchor the plant in the soil and are the main means of absorbing water and mineral ions. The main functions of most stems are to support the leaves and flowers of the plant (e.g. holding the leaves in positions where they can obtain maximum light) and to provide transport systems connecting the roots with those leaves and flowers.

6.1 Structure of the stem

Stems consist of an alternating system of nodes, the points where leaves are attached and internodes, the stem segments between the nodes. In the angle between each leaf and the stem is an axillary bud, which has the potential to form a branch shoot. Most of the axillary buds of a young shoot are dormant, and most of the growth occurs at the terminal bud. We will first consider the structure of the young stem, and so-called primary growth.

6.2 The young stem

A plant anatomist interested in the internal structure of a particular stem will invariably start by taking a transverse section. Here, even more so than in the root, the anatomy depends on the phylogenetic group investigated. A transverse section of a typical young eudicot stem (alfalfa) at the primary growth stage is shown in Figure 6.1.

Working from the outside, stems have the following structures:

1 The epidermis, which may be single layered or multi-layered. The outer surface of the young stem is covered by a waxy cuticle and may be ornamented. Stomata also occur on the surface of some stems, as can trichomes and hairs.

2 The hypodermis is the layer of cells beneath the epidermis. Sometimes, but not always, it has a structure distinct from the cortical cells inside it.

3 The cortex consists of all the cell layers beneath the hypodermis and outside the vascular bundles. The exact inner boundary of the cortex is often indistinct. The outer cortical cells may contain chloroplasts.

4 The endodermis is indistinct in most stems and is sometimes called an endodermoid layer. Even if present, the cells may lack Casparian strips. **Hydrophytes** often have a well-developed endodermis (See Chapter 10, section 10.3.2).

5 The vascular bundles are often present in one or more rings in dicots. At the outer edge of each bundle there is sometimes a thick layer of phloem fibres. The conducting tissues of the phloem are interior to the fibres, while in the xylem, several large vessels are obvious. The vascular cambium is a meristematic tissue that produces xylem on one side and phloem on the other. This leads on to secondary growth (see section 6.5). When the xylem is first produced, the tissues are live, but the contents soon die, producing the long, hollow empty tubes which are the main water-conducting elements in the plant. The phloem, on the other hand, remains alive, and it is the main sugar-conducting system in the plant. In monocots, such as maize, there is no obvious ring of vascular bundles, and these are scattered in distribution.

6 In the centre of the eudicot stem is a relatively undifferentiated tissue, the pith.

A transverse section of a young stem is useful for gaining an impression of the main internal tissue types but, to investigate how the shoot grows and produces leaves and other structures, a longitudinal section through the shoot apex is needed. The plant is a three-dimensional structure and one has to meld the cross and longitudinal section images together to visualize the actual structure.

Functional Biology of Plants, First Edition. Martin J. Hodson and John A. Bryant.
© 2012 John Wiley & Sons, Ltd. Published 2012 by John Wiley & Sons, Ltd.

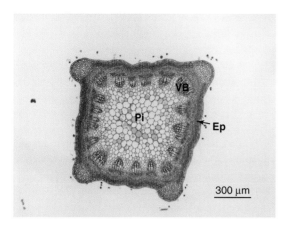

Figure 6.1 Alfalfa (*Medicago sativa*) stem in cross section. Key: Epidermis (Ep), Vascular Bundle (VB), Pith (Pi). Photo: Prof. Thomas Rost.

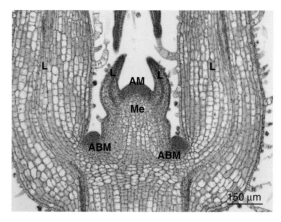

Figure 6.2 Coleus (*Solenostemon scutellarioides*) shoot tip in longitudinal section, showing the shoot apical meristem (SAM). Key: Leaf (L), Apical meristem (AM), Meristematic zone (Me), Axillary bud meristem (ABM). Photo: Prof. Thomas Rost.

6.3 The shoot apical meristem

After fertilization, embryonic differentiation gives rise to shoot and root apical meristems at the opposite poles of the embryonic body (Chapter 4, section 4.2). Higher plants continue growth throughout their lives as a result of these persistent localized growth centres. We considered the root apex in Chapter 5, section 5.3.1. There are also meristems at the shoot apex, and in monocots there are additional intercalary meristems at the internode and leaf base.

The general structure of a eudicot shoot apical meristem (SAM) is shown in longitudinal section of Coleus in Figure 6.2. The terminal meristem is situated at the distal end of the shoot axis and can have a range of geometric forms, from conical or dome-shaped to flat or even slightly depressed. The diameter of the initiating region also differs greatly between species, being as large as 3,500 μm in cycads and smaller than 50 μm in some flowering plants. Most species of angiosperms have SAM diameters in the range 100–250 μm. There is also considerable variation within the same species, depending on the age of the plant, the season and various environmental factors.

The outer layers of the apex, including the epidermis, are more regular in shape than those internally. The first layers are the *Tunica*, where the plane of cell division is almost always anticlinal (normal to the plane of the tissue). The number of cell layers in the *Tunica* is one

to five. In the remainder of the tissue, the *Corpus*, cell division planes are apparently random. The youngest leaf primordia appear as small bumps at the edge of the apical dome, followed by successively older leaves. It is important to note that leaf and bud primordia are initiated in the *Tunica* layers, and not in deep-seated tissues as in the roots (see Chapter 5, section 5.3.4).

The period of time that elapses between the formation of one primordium and the next is called a **plastochron**. In the garden pea (*Pisum sativum*), this length of time is about 48 hours, but it can be a week or longer.

The leaf and bud primordia commonly arise singly on the apical dome of the stem, and the initiation point follows a circular pattern around the apex. The leaves are initiated in a spiral sequence, and then the stem elongates between the formation of two successive primordia. The spiral pattern formed has often been shown to follow a very precise mathematical relationship. This positioning of leaves is called phyllotaxis, and Figure 6.3 shows the arrangement of the leaves on a potato (*Solanum tuberosum*) shoot in transverse section. The lines are called **parastichies**, which are sets of spiral patterns. In this example, two main sets of spirals can be seen running in opposite directions around the apex. In one set, successive primordia differ in serial number by 2 (e.g. 2, 4, 6, 8), while in the other the difference is 3 (e.g. 3, 6, 9). This type of phyllotactic arrangement is known as a 2 + 3 arrangement. It has been found that there are several such arrangements, many of which can be described by

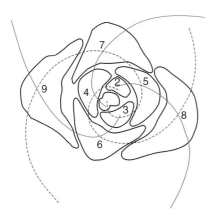

Figure 6.3 Phyllotaxis. This figure shows the arrangement of the leaves on a potato (*Solanum tuberosum*) shoot in transverse section. The lines are parastichies, which are two sets of spiral patterns running in opposite directions around the apex.
From Steeves, 1989. Reproduced by permission of Cambridge University Press.

two successive terms of the **Fibonacci series** (1, 1, 2, 3, 5, 8, 13, 21, 34, 55 . . . . . .). Common examples are 2 + 3, 3 + 5, 5 + 8 and 8 + 13. One further observation is that the angle between each new leaf primordium is on average constant at the so-called 'ideal' or Fibonacci angle of 137.5°. Thus the angle remains the same but, because of different relative growth rates of the apex and leaf primordia, a whole range of patterns is possible.

Three possible reasons have been put forward for Fibonacci spirals:

1 Some mathematical models have suggested that the spirals are needed for optimal light capture by leaves.

2 As long ago as 1873, Hubert Airy had the idea that phyllotaxis was the solution to a space packing problem.

3 The most recent idea is that a molecular mechanism is responsible for phyllotaxis and that this, in some way, constrains the branching patterns that are available. We will consider this below.

The primordia we have been discussing can develop in several ways. They can have fairly limited growth, as in the cases of foliage leaves, scales, bracts, and floral structures, or they can have potentially unlimited growth if they become axillary buds.

6.3.1 Control of primordium development

It has been clear for some time that the patterns of primordium development outlined above must be under

hormonal and gene control, but it is only recently that we have begun to gain some understanding of these processes.

The apical meristem itself is characterized by the activity of KNOX genes that are members of the KNOTTED family of homeobox transcription factors. These are recognized to play a major role in the organization and preservation of meristematic cells. Primordium initiation at the edge of a SAM involves the inactivation of meristematic identity and activation of a number of different genes that typify the developing primordium. The downregulation of KNOX genes is an important stage that sets the scene for future development in the new primordia.

It has now been conclusively demonstrated that the presence of high local concentrations of auxin is involved in the outgrowth of primordia. In *Arabidopsis*, two gene families (LAX and PIN-FORMED (PIN)) encoding auxin carriers have been identified (see also Chapter 4, section 4.2 and Chapter 3, section 3.1). These auxin carriers seem to be involved in creating areas of auxin concentration, where the primordia are formed, and areas where primordium development is inhibited due to decreased auxin concentration.

6.3.2 Mechanism of apical dominance

Terminal buds often inhibit the growth of axillary buds, and this is known as apical dominance. Thus the branching pattern of a shoot could potentially reflect the phyllotaxis, but many of the initiated buds never grow out, being inhibited by the apical bud. When the apical meristem is cut off, laterals below it grow out, and this is used when pruning to encourage the growth of a shrub or tree.

Thimann and Skoog showed in 1933 that auxin originating from the apex inhibited axillary bud growth. Conventional decapitation studies and the characterization of loss of apical dominance in mutant plants have shown that auxin plays an important role in apical dominance (see also Chapter 3). A mutant of Arabidopsis, *bushy and dwarf1 (bud1)*, produces more branches than wild-type plants due to a loss of apical dominance, resulting from decreased polar auxin transport. The Arabidopsis MAP KINASE KINASE7 (MKK7) is encoded by the *BUD1* gene, and MKK7 is part of the MAP kinase cascade which regulates IAA transport and, thus, apical dominance.

Although auxin obviously has an important role in apical dominance, it may not be the only hormone involved, and it seems likely that cytokinins and carotenoid compounds such as strigolactones may also have a part (see Chapter 3).

6.4 Shoot organizational forms

Shoots do not always show their phyllotactic arrangement of leaves because of apical dominance which prevents lateral leaves and branches growing out. The major shoot organizational forms result very much from an interaction between the genotype and environment. In 1978, Hallé, Oldeman and Tomlinson put forward 23 architectural models (Hallé *et al.*, 1978), and these have subsequently been further developed. Here are three examples of very different shoot organizational forms:

1 Coconut palm (*Cocos nucifera*). This has a principal meristem which grows continuously (Figure 6.4). There are no lateral branches, and the reproductive structures are borne laterally.

2 Pedunculate oak (*Quercus robur*) has a principal meristem growing rhythmically (Figure 6.5). Lateral branches are formed rhythmically and repeat the pattern of the main axis. The reproductive structures are born on lateral branches in lateral positions.

3 The boojum tree or cirio (*Fouquieria columnaris*) is a member of the Fouquieriaceae that grows in the deserts of the Baja California Peninsula and the Sierra Bacha of Sonora, Mexico (Figure 6.6). It is a xerophyte (Chapter 10, section 10.4.2), and the tallest recorded plant was over 26 metres in height, with an estimated age of more than 700 years old. The central trunk has a tapering-columnar form that has great amounts of non-lignified xylem parenchyma and forms a water storage tissue. The central trunk bears a series of short lateral branches that have leaf-producing short shoots.

All of these branching patterns above are presumably under hormonal and genetic control, but as yet we have a very incomplete understanding of shoot development.

6.5 The mature stem

Primary growth contributes to the elongation of the axis, while secondary growth contributes to lateral growth.

Figure 6.4 Coconut palms (*Cocos nucifera*), growing in Sri Lanka. Photograph: Mr Ken Redshaw © University of Leeds. Image from Centre for Bioscience (Higher Education Academy) ImageBank.

Annual plants undergo primary growth for one season, reproduce and then die. Once the primary stem is formed, its diameter only increases due to cell expansion. Many plants, however, are perennials that are able to grow and reproduce for many years.

Now we must consider the process of secondary growth, where the diameter of the plant axis is increased by the activities of meristems known as vascular and cork cambia. Figure 6.7 shows how primary growth (section 6.2 above) can change to the secondary growth that is responsible for the production of wood (secondary xylem). One of the events that terminates primary growth is the initiation of the vascular cambium. Once this is active, the secondary xylem is formed to the inside and the secondary phloem to the outside. At the end of the season, this growth stops and there is a period of dormancy (usually

Figure 6.5 Pedunculate Oak (*Quercus robur*). This tree, known as the Carroll Oak, is growing at Birr Castle, Birr in County Offaly, Ireland. It at least 400 years old and has a girth of 6.5 m. Photo: MJH.

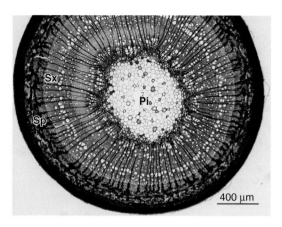

Figure 6.7 *Prunus* stem in cross-section, showing secondary growth. Annual rings are present. Key: Secondary xylem (Sx), Secondary phloem (Sp), Pith (Pi).
Photo: Prof. Thomas Rost.

Figure 6.6 The boojum tree (*Fouquieria columnaris*) grows in the deserts of the Baja California Peninsula and the Sierra Bacha of Sonora, Mexico.
Photo: MJH. Used with the permission of Boyce Thompson Arboretum, Superior, Arizona, USA.

over winter). Then, in the spring, the vascular cambium is reactivated and begins to produce another layer of xylem on the inside (secondary xylem) and another layer of phloem on the outside (secondary phloem). This process is then repeated annually.

6.5.1 The xylem

In general, the xylem elements produced at the beginning of the growth season are larger than those towards the end of the season, which leads to the phenomenon of tree rings. The conducting tissue of the xylem consists of **tracheids** (Figure 6.8) and **vessel elements** (Figure 6.9), which together make up the tracheary elements.

Both the tracheids and vessel elements are thickened, hollow and dead at maturity. The tracheids have no perforation plates in their primary walls, while the vessel elements have perforation plates at each end. Vessel elements (members) are connected end-to-end, with the terminal elements not having perforation plates and instead having bordered pits. In this way, a vessel has a finite length which is fairly specific to a species. Tracheids are more associated with primitive plant groups, whereas vessels are more common in the angiosperms. The conifers have only tracheids and never have vessels, but they are a highly successful group. Evolution of vessel elements is undoubtedly more recent, but they are not necessarily superior to tracheids in all environments.

Associated with the xylem vessels and tracheids are two cell types: the xylem parenchyma, which is involved

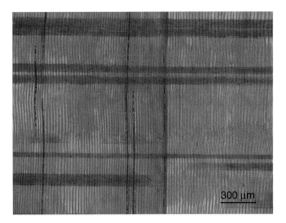

Figure 6.8 Juniper (*Juniperus* sp.) longitudinal section of wood showing tracheids in secondary xylem.
Photo: Prof. Thomas Rost.

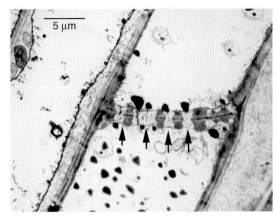

Figure 6.10 Transmission electron micrograph of pea (*Pisum sativum*) phloem sieve tube cells in longitudinal section showing sieve plate with large pores. The white callose material (arrowheads) blocking the pores is probably a fixation artefact.
Photograph: Dr Gordon Beakes © University of Newcastle upon Tyne. Image from Centre for Bioscience (Higher Education Academy) ImageBank.

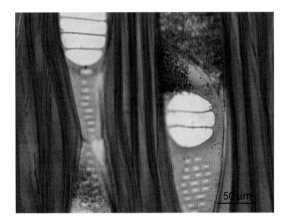

Figure 6.9 Tulip tree (*Liriodendron tulipifera*) longitudinal section of wood showing vessel elements in secondary xylem. Note the scalariform perforation plates.
Photo: Prof. Thomas Rost.

in storage and in loading and unloading the xylem; and fibres, which are responsible for strengthening the xylem.

6.5.2 The phloem

Phloem is a tissue that is made up of mostly living cells and tends only to be active for a few seasons. It is often crushed as the xylem expands, so it forms a relatively thin band outside the vascular cambium. The phloem is composed of a number of cell types: sieve elements, companion cells, phloem parenchyma, fibres and often sclereids.

Again, there are two main types of phloem sieve elements: sieve cells, which are found in nearly all non-angiosperms, can be 3.5 mm long and have pores over much of their surface; and sieve tube members, which are in nearly all angiosperms, are only up to 150 μm in length and have distinct sieve areas where sieve pores are concentrated in their end walls.

Sieve tube members join end-to-end to form sieve tubes. The anatomy of phloem sieve elements is far more difficult to study than that of xylem, as it is a delicate living tissue (Figure 6.10). Much uncertainty over its structure has arisen because of the production of artefacts during the preparation of tissue for the study, and this ambiguity has impacted on ideas about the mechanism of phloem transport (see section 6.9 below).

Early in differentiation, sieve elements contain all of the organelles that most plant cells have: nuclei, vacuoles, plastids, mitochondria. Endoplasmic reticulum is associated with the plasmodesmata which traverse the side and end walls of the sieve element. During development, sieve pores (usually around 1 μm in diameter, but they can reach 14 μm) develop in the end walls around the sites of the plasmodesmata as cell wall material is removed. Many of the organelles begin to degenerate, usually including the nucleus. Plastids and mitochondria become less numerous and may change in structure.

Box 6.1 The cork oak

Cork oak (*Quercus suber*) is an evergreen oak tree of moderate size (Figure 6.11) that is native to southwest Europe and north-west Africa. The tree has a cork layer of bark that develops a considerable thickness and is heavily suberized. This thick bark is very resistant to forest fires. After a fire, the branches of cork oak, protected by cork, quickly re-sprout (see Chapter 9, Figure 9.4), while many tree species rely on seeds or re-sprouting from the base of the tree (Chapter 10, section 10.2.4). This quick regeneration of a canopy seems to be an evolutionary advantage compared to other species.

Cork oak is the main source of cork for wine bottle stoppers. The trees live about 150–250 years and the first cork is usually cut from trees that are 25 years old. Trees can then be harvested to produce cork every 9–12 years and, in its lifetime, a tree can be cut around 12 times.

At the time of each harvest, the bark has grown to 3–4 centimetres thick. Circular cuts are made and the bark is peeled off. No machinery is involved, and skill is required for the labourers to harvest the bark without harming the tree. The outer bark is extracted, but the inner bark remains on the tree and a new layer of cork re-grows, making it a renewable resource.

Half of the world's cork harvest comes from Portugal. Since the mid-1990s, there has been a decline in the use of cork in favour of alternative wine closures such as synthetic plastic stoppers and metal screw caps. However, natural cork allows oxygen to interact with wine, and this is particularly important in red wines which are laid down with the intention of ageing. As one might imagine, there is a considerable debate amongst those in the wine industry as to which closures are best for which purposes!

Figure 6.11 Cork oak (*Quercus suber*) growing on the slopes of Picota, Monchique Mountains, Southern Portugal. The outer bark is cut off for commercial cork production, but the inner bark remains on the tree, and a new layer of cork re-grows. Photo: MJH.

Callose, a polymer consisting of glucose units linked in a $\beta 1 \rightarrow 3$ configuration, is deposited in the matrix of the primary wall around the borders of the plasmodesmata. These then begin to increase in diameter leading to the formation of the sieve pores. The cells also synthesise a large amount of P-protein. If the cells become damaged, this P-protein is rapidly moved to the sieve pores and blocks them, preventing leakage. The companion cells are closely associated with the sieve elements, have a dense cytoplasm and are involved in the loading and unloading of the phloem.

6.5.3 The cork cambium

In annual plants, the epidermis and hypodermis do not undergo further development, but in many plants these layers are replaced by the **periderm**. The latter is a mixture of cells from the epidermis, cortex and secondary phloem that turn into meristematic tissue known as the *phellogen* or cork cambium. This meristem then produces **phellem** or cork cells. These cells undergo modifications to become water- and enzyme-resistant, and then they die. The periderm, plus the remaining primary tissues and secondary phloem, then constitutes the bark. In some plants, bark can be several centimetres thick, and it provides considerable protection against fires and pathogenic organisms (see Box 6.1).

6.6 The tallest, largest and oldest plants

Secondary growth enables plants to continually increase in size every year, and it is among the trees that we find some of the tallest, largest and oldest plants.

The tallest trees in the world are the coast redwoods (*Sequoia sempervirens*), the tallest of which is about 115 m high, and can be found in Redwood National Park, California. Reports that an Australian mountain ash (*Eucalyptus regnans*) at Watts River, Victoria, reached 132.6 m in 1872 have now been discounted as a probable exaggeration, although living trees have been recorded at nearly 100 m in height. Almost certainly, the coast redwood does hold the record for height.

The giant sequoia (*Sequoiadendron giganteum*) which grows in the Sierra Nevada mountains of California is thought to be the largest (by volume) tree on earth (Figure 6.12). Mature individuals range in height from 50 to 85 m, with the largest giant sequoias reaching heights of 94.8 m. Trees up to 17 m in diameter are known. Giant sequoia is therefore commonly considered the most massive living organism (the largest recorded volume is 1,487 m³), although most of the tree consists of dead, not

Figure 6.12 Giant sequoia (*Sequoiadendron giganteum*). North Grove Trail, Calaveras Big Trees State Park, California, USA. Photo: MJH.

living tissues. Giant sequoia also has some of the oldest living organisms, with the oldest recorded being 3,266 years.

Until recently, the Great Basin bristlecone pine (*Pinus longaeva*), growing on the White-Inyo mountain range of California, was considered the oldest tree. In the *National Geographic* of March 1958, dendrochronologist Edmund Schulman announced his discovery of trees that were over 4,000 years old. He found the oldest trees at elevations of 3048–3354 m, often growing in apparently impossible locations. In 1957, one of these trees, dubbed 'Methuselah', was, using tree ring data, found to be 4,723 years old. The environment of the bristlecone pine, though difficult, does not have very large fluctuations. The rainfall of the mountains is low, at 300–330 mm per annum, so little soil erosion occurs, and the lack of ground cover means few fires.

Methuselah held the undisputed record for the world's oldest known living tree until 2004. Then Leif Kullman found a Norway spruce (*Picea abies*) in Dalarna Province, Sweden, that had been growing for 9,550 years on a mountain at an altitude of 910 m. The tree started to grow at the end of the last ice age, but the shoot is only 4 m tall and is not that old. However, the root system, when carbon-dated, suggested that this *was* the longest-lived tree in the world. The stems have a lifespan of around 600 years but, when one dies, another appears from the root stock. So the spruce is able to clone itself, and this explains its longevity.

Clonal plants are known for their long life spans. The individuals may die, but the clone may live on almost indefinitely. The oldest living clonal plant described so far is the Tasmanian shrub, *Lomatia tasmanica*, which has been dated to an age of at least 43,600 years. The shrub reproduces exclusively asexually, and it is a critically endangered species.

6.7 Ageing and senescence

Having considered some of the oldest known plants, it is appropriate to investigate the topic of ageing and senescence in plants. Senescence can be defined as: '*The period between maturity and death of a plant or plant part*' (Tootill (Ed.), *Penguin Dictionary of Botany*, 1984).

Andrew Watkinson gives a more detailed definition: "*Senescence is defined by evolutionary biologists as the decline in age-specific survival and fecundity that reflects*

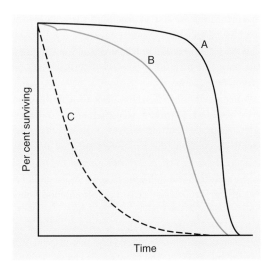

Figure 6.13 Survival curves showing when plants die over a period of time. A: the curve for annual species. B: the curve followed by most mammals. C: the curve followed by oak trees. From Woolhouse, 1972. (By permission of Oxford University Press.)

declines in the performance of many different physiological functions in individuals of sufficiently advanced age' (Watkinson, 1992).

Here we will cover whole plant aspects of this topic; cellular aspects were covered in Chapter 3 (section 3.5.2) and ageing, senescence and abscission of leaves will be addressed in Chapter 7 (section 7.11).

As we have seen, some plants can live for thousands of years. The higher plants with the shortest lifespan, going from seed to seed in a matter of a few weeks are the desert **ephemerals** (see Chapter 10), which often survive as seed for many years, then burst into life after rainfall. How do we explain these differences in plant age, and what are the controlling factors? Figure 6.13 shows three potential curves for plant mortality observations:

A. Plants in a field of maize or any other annual plant show this type of curve. Some individuals are lost when the plants are at the seedling phase. Then there is a period of reasonable constancy, followed by all of the plants dying when seed ripening occurs.

B. This is the type of curve for many animal populations, including our own. An early phase of relatively high mortality is followed by a more gradual decline, and then a phase when deaths occur more regularly. In the developed world, this phase now happens between the ages of 70 and 80 for humans.

C. This is a pattern based on a stand of oak trees. It shows a steady mortality, with the number of plants dying over a given time interval being a function of those alive at the beginning of that time. This same curve has been found to apply to glass tumblers in a cafeteria. It does not mean that senescence processes are not occurring, but rather that these processes may be concealed by the occurrence of catastrophic events in the environment which remove certain individuals by chance.

6.7.1 Plant senescence patterns

There are three categories of plants according to their senescence patterns:

Semelparous plants

These are **monocarpic plants** with a single reproductive event (see Chapter 8, section 8.3); when the plant flowers, it then dies. These include not only the annual plants, but also some quite long-lived plants. Some of the bamboos will show vegetative growth for years, flower once and then die. The survivorship curves of such plants all illustrate a conspicuous increase in mortality following flowering. In semelparous plants, all of the meristems of the plant die following flowering, and thus the whole plant dies. The mechanism of senescence in these plants is not known, but it is possibly related to the mobilization of plant resources to make developing seeds.

Iteroparous plants

These are **polycarpic** plants with multiple reproductive events. They include the trees and shrubs and they can be divided into two types:

1 Plants with a single meristem (e.g. palms) seem to be inhibited by the fact that leaf number remains constant as size increases. Thus, very old palms gradually taper and their yield declines. This is probably due to the same amount of photosynthetic area having to support larger and larger amounts of respiratory tissue.

2 The majority of trees and shrubs have multiple meristems. In most cases, fecundity seems to increase with age. Mortality declines with age until the tree reaches about 15–20 cm diameter at breast height, then it becomes fairly constant. The death rate curves seem to be U-shaped, with an increase in the rate for older plants. This is associated with a decrease in metabolism, a reduction in growth rate and a loss of apical dominance. Most of the problems seem to arise from greater size. This leads to an increasing

burden of respiratory material, an increased distance for movement of water and minerals, and greater susceptibility to herbivores and pathogens. The consequences are a decreased amount of photosynthesis relative to the size of the tree, an increase in dead tissue (necromass) and an increased likelihood of wind throw. However, in many species, the main body of the plant may fall over but it is always possible that it may rejuvenate. The remaining apices, particularly those at the base of the tree, often shoot again. We saw this phenomenon with the old spruce tree in Sweden (see Section 6.6). This is also the basis of coppicing, a practice that removes necromass.

Clonal plants

Here rejuvenation takes place continually during clonal growth, as older tissues decay and new tissues are produced. Plants such as the buttercup and strawberry escape the biomechanical constraints and accumulation of necromass found in the trees and shrubs. Some scientists have argued that there is no real evidence of senescence in clonal plants, but there are some cases where decrease in reproductive rate does happen over time. Bamboos are clonal, yet all die after reproducing, while smaller grasses often show an age-associated decrease in output.

There is no doubt, though, that many clonal plants can live for thousands of years (see section 6.6 above). They seem more likely to be killed by fire, competition or disease than any intrinsic factor. One could, therefore, suggest that some of the clonal plants and trees are almost immortal.

6.8 Long-distance xylem transport

We considered water movement in plants at the cellular level in Chapter 2 (section 2.3.4) and loading of ions into the xylem of roots in Chapter 5 (section 5.7.7). The xylem is the major long-distance water transport system, but it is also a major transport system for all of the essential macronutrients and micronutrients (see Chapter 5, section 5.5). Xylem sap often contains some sugars and amino acids, but these are not major components.

Transport in the xylem involves many of the same principles as transport at the cellular level. Water movement occurs from areas of high water potential to areas of lower water potential. Any hypothesis to explain water

movement must be able to account for transport to the top of the tallest of trees, and we have seen that the tallest redwood tree is over 115 m tall. A total of five hypotheses have been put forward to explain long distance water movement in plants:

1 *Water is pumped up the xylem elements.* This was one of the earliest ideas. As we have seen, mature xylem elements and tracheids are dead and there are no living cells to carry out any pumping. So this hypothesis has been discarded.

2 *Water moves up xylem elements via capillarity.* Capillarity results from the adhesion of water molecules to the sides of small tubes, and it is the phenomenon behind water movement up a wick. Tracheids range in diameter from $10-20\,\mu\text{m}$ in conifers, while xylem vessels are $100\,\mu\text{m}$ in diameter in woody eudicots, which is much too wide for any major rise by capillarity. Even the smallest-diameter tracheids of $10\,\mu\text{m}$ would only allow a height of rise of 3 m. This is nowhere near enough for tall plants.

3 *Water is pushed up xylem elements by atmospheric pressure.* Atmospheric pressure can support a column of water, which is easily demonstrated by filling a long closed tube with water and placing the tube open end down in a tub of water. Two opposing forces determine the height of rise of water in the system: gravity and air pressure. Equilibrium is reached when the water column is 10.4 m high. At heights above that, the column cavitates and a partial vacuum filled with water vapour forms at the top of the tube. It is difficult to conceive how this mechanism might be compatible with plant structure, but even if it reached 10.4 m, this could not account for tall trees.

4 *Water is pushed up by root pressure.* Herbaceous plants often show the phenomenon of **guttation** early in the morning, when their leaf edges have droplets of water around their edges. This is caused by root pressure. Minerals are actively absorbed by roots at night and are secreted into the root xylem. This influx of solutes decreases the water potential of the xylem sap, causing water to flow in behind – thus, the pressure in the xylem increases. As there is little transpiration at night, the pressure eventually forces the water out of **hydathodes** (water pores) on the leaves. In most plants, the amount of water lost through guttation is quite small and, although root pressure can push water several meters up a plant, again it cannot account for the movement of water to the top of tall trees.

5 *Water is pulled up plants by evaporation.* Although this has had critics in the past, it is now almost universally

agreed that the most important mechanism for long distance water transport in the xylem is the cohesion-tension theory first proposed in 1894 by Dixon and Joly. We now need to consider this in detail, particularly looking at the case of the tallest trees.

Water movement through a tree is always from areas where the water potential is highest to areas where it is lower. The water potential of the atmosphere is approximately −50 MPa at 70 per cent relative humidity and ≤100 MPa in hot arid conditions. Solar-powered transpiration from the leaf dries the air spaces, then the cell walls of the mesophyll. These cells then obtain water from the cells of the internal leaf tissues, which in turn obtain moisture from the xylem. This whole process continues down to the tips of the roots and out into the soil.

We saw in Chapter 5 (section 5.6) that water moves to the root surface by the process of mass flow. Soil water potential is approximately zero when the soil is saturated, and below −10 MPa in dry soil (even lower in saline soils). Under ideal conditions (so-called field capacity), soil water potential is about −0.3 MPa. Thus the flow of water through the tree is from the soil to the atmosphere, down a gradient of water potential. This appears to be a good scheme, but there are a number of questions that need to be answered before the cohesion-tension theory can be accepted:

Is there a gradient in water potential?

One ingenious way of proving that this is the case is to shoot twigs down from various heights with a rifle and assess their xylem tensions using a pressure bomb. The water potential is measured by applying pressure to the leafy end of a cut shoot that is just enough to force water out of the cut surface. This pressure can then be equated with shoot water potential. As might be expected, the xylem in the twigs from the higher tree branches has lower water potentials.

Is water under a negative tension in the xylem?

If this were the case, then we would expect the diameters of roots and tree trunks to shrink when transpiration is at its greatest. This *is* the case. Minimum diameter is seen in the afternoon, when water stress is greatest and maximum diameter in the early morning. This change is partly due to shrinkage of the xylem but, at higher water potentials, it can also be due to shrinkage of living cells in the phloem, cortex and pith. At lower water potentials,

the shrinkage is more due to the xylem, as all the possible shrinkage of the living cells has occurred.

Can columns of water in the xylem withstand the tensions required?

For some time, one of the worries about the cohesion-tension theory was the incredible tension that the columns of water must be under, and it was thought that **cavitation** would be inevitable. However, it has been calculated that the columns would only need to withstand −2 MPa for water to reach the top of a tall tree, and in reality the thin columns of water can endure at least −30 MPa due to the very strong cohesion of water molecules.

What about cavitation?

Despite the evidence above for the strength of the xylem water columns, there are times when cavitation does occur, and air bubbles or partial vacuums block the xylem elements, preventing water flow. When sensitive microphones are placed against a tree trunk, clicking noises can often be heard. These indicate cavitation events, which increase when the tree comes under mechanical stress while bending in the wind.

Another key time for cavitation is when the xylem sap freezes during the winter (see Chapter 9, section 9.2.2). Small plants can use root pressure to refill the bubbles; tall trees cannot do this, but it seems that water can detour around obstructions through pits in the xylem walls. That water transport can be diverted around obstructions is supported by experiments using overlapping cuts from opposite sides of a tree trunk. Provided that the cuts are further apart than a distance that is characteristic for each plant species, water can move around the cuts, and the water potential of the foliage is hardly affected. Also, many trees mostly use the new xylem that is added every spring for water conduction, while the older xylem is not functional in that way, but supports the tree.

6.9 Translocation in the phloem

As we saw in Section 6.5, the phloem is an entirely different system to the xylem, with the principal difference being that phloem is a living tissue. The other major difference is the makeup of the solutes transported. Phloem is the major transport system for sugars (mainly sucrose), but it is also important in the transport of some minerals.

Box 6.2 An upper limit to tree height?

It now appears almost certain that the cohesion-tension theory is correct and that water is able to reach the top of the tallest tree in this way. It seems, however, that there is an upper limit to the height of a tree, and George Koch and his colleagues carried out some fascinating work on coast redwoods (*Sequoia sempervirens*) to determine this (Koch, 2004: see Figure 6.14).

Although water columns in the xylem can supply trees over 100 m tall, this is not without its problems for the redwoods. As we noted above, the xylem in twigs from the higher tree branches had lower water potentials, as low as −1.84 MPa at the highest sampling point of 108 m. Reduced water potential led to a decline in leaf turgor in the highest leaves, analogous to that seen in water stressed plants (see Chapter 9, section 9.4). Turgor is important in leaf growth and, not surprisingly, the leaves on the highest branches were much smaller. Stomatal conductance is also reduced in the higher leaves, which limits photosynthetic potential. It was thus possible to estimate a theoretical maximum tree height by calculating the height at which the variables examined would reach a limiting value. For instance, cavitation was estimated to become a major problem when xylem pressure dropped to −1.9 MPa, which was suggested to be likely at 122 m. Likewise, maximum photosynthesis was predicted to drop to zero at 125 m. Taking all the parameters into account, the maximum height possible was estimated at 122–130 m – similar to the tallest trees recorded in the past.

Kempes and colleagues took a somewhat different approach (Kempes et al., 2011), designing a model that was applicable to all tree species across the different environments of the USA. The model combined scaling laws with energy budgets and the factors affecting them, including precipitation, temperature and solar radiation. They were also able to use the model to predict what would happen as climate changes (see Chapter 12, section

Figure 6.14 Coastal redwood trees (*Sequoia sempervirens*). Founders Grove, near Weott, northern California, USA. Photo: MJH.

12.2.2), finding that an increase in temperature of 2°C would result in a decrease in average maximum tree height of 11 per cent. This approach could be useful in determining the likely impacts of global warming on timber production.

Because phloem is a living tissue, this does put some constraints on the minerals transported, and while potassium is translocated in large quantities, both sodium and calcium are present at low concentrations.

More recently it has been recognized that phloem is also involved in the transmission of alarm signals such as volatile derivatives of jasmonic acid (Chapter 3, section 3.6.9). These signals are triggered when the plant is attacked by herbivores (Chapter 11, section 11.2.1) or pathogens (Chapter 11, section 11.3.9). RNA and protein molecules are also translocated in the phloem (e.g. 'florigen': Chapter 8, section 8.4.4).

There was a considerable debate 20–30 years ago about the mechanism of transport in the phloem. Now the debate is almost over and the mass flow hypothesis, first proposed by Münch in 1930, is thought to be correct. The theory can be explained by a simple physical model (Figure 6.15). Two cells, A and B, have membranes that are permeable only to water and are connected by a glass tube. Cell A is the 'source' and cell B is the 'sink'. Cell A contains a highly concentrated sucrose solution, while cell B contains a lower concentration. When the two cells are placed in a water vessel, water enters cell A down a water potential gradient as a result of osmosis. This produces a turgor pressure in cell A and causes solution to move along the tube. This will force water out of cell B. When the concentrations of sucrose in A and B are equal, flow stops. However, if sucrose could be removed from B

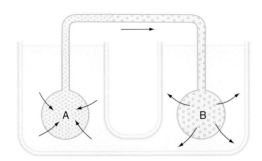

Figure 6.15 A simple physical model to explain the mass flow hypothesis proposed by Münch.

by respiration or growth, then flow in the system would be continuous.

The physical model above can be transferred to the plant. Carbohydrates (mainly sucrose) are produced in mature leaf as a result of photosynthesis (Chapter 7, section 7.4). Water from the xylem is absorbed by cells containing sucrose in the leaf as a result of osmotic forces. This brings about increased turgor pressure in these cells. Sucrose is used up in areas such as roots, developing seeds and young leaves, to produce growth for respiration and in storage. This leads to a decreased turgor pressure in the cells of these organs.

There is thus a gradient in turgor between the 'source' (mature leaf) and the 'sink'. Mass flow between the two occurs via the phloem. While the flow of water and minerals is always upwards in the xylem, direction of flow in the phloem can change during the season. So, for instance, flow can be predominantly downwards from the leaves to the roots early in the season, but switch to upwards when flowers and seeds are produced later on. The loading and unloading of phloem has received a fair amount of attention in recent years. Loading involves H^+/sucrose symporters that are fuelled by H^+-ATPases.

6.10 Biological clocks in plants

Humans keep the time by wearing watches, and we look at clocks regularly throughout the day. However, we also have biological clocks. Think what happens when we take a long-haul flight – our biological clocks go out of synch with reality in the well-known phenomenon of jetlag. Plants also have biological clocks, and the aims of this section are to introduce the various types of clocks that plants have, to assess the importance of these clocks in adapting plants to their environments and to discuss what is known about the mechanisms behind these phenomena.

Plants show several types of biological clock. These processes are different from others in biology as they not only consume time but they also measure it, dividing it up into units. Here we will briefly introduce high-frequency rhythmical processes, and then consider in more detail rhythms related to daily (**circadian**) cycles. Clocks are also involved in photoperiodism, which is adjustment to season by measurement of day length (see Chapter 3, section 3.7.2 and Chapter 8, section 8.4.1), and in interval timers that occur in dormancy of a fixed duration.

6.10.1 High-frequency rhythmical processes

High-frequency processes, including breathing and heart-beat in animals, do divide up time, but they are not true clocks as they are not reset by external time cues. The speed at which they run is regulated not by time but by temperature or physiology. Relatively little is known about this type of process in plants. However, small oscillatory movements in bean (*Phaseolus*) leaves of about 10–15 mm have been observed with a period of about 10 minutes. These are very distinct from the movements of leaves with a 24-hour cycle, where the whole leaf folds down. The physiological significance of these small movements is obscure.

6.10.2 Rhythms related to daily (circadian) cycles

Daily cycles have been most studied, and are therefore the ones we will concentrate on. These biological clocks in plants have three features:

Endogenous or innate

The plant is capable of generating a rhythm with a period length of approximately 24 hours. There are a considerable number of processes in higher plants which show daily or circadian (from the Latin '*circa*' – about; and '*dies*' – day) rhythms. These include: gene expression; many enzyme activities; respiration; photosynthesis; cell division; growth rate; root pressure; leaf position; flower movements; stomatal movements; and dark carbon dioxide fixation in CAM plants.

Light is known to interact very considerably with some of these processes. One of the best known examples is the regular daily movement of certain leaves and flowers. This was first described for leaves in the tamarind tree (*Tamarindus indica*) by Androsthenes (one of Alexander the Great's generals) in the 4th century BC on the island of Bahrain in the Persian Gulf. In 1729, De Mairan, the French astronomer, worked on leaf movements of the sensitive heliotrope (probably *Mimosa pudica*), as the plants folded their leaves down at night and raised them to the horizontal during the day. He was the first to show that the movements did not depend on a continual cycle of light and dark, and that the rhythm would continue for several days in constant darkness. Linnaeus came up with the ingenious idea of a 'flower clock' in 1751, so that people could wander the countryside and know the time by observing the different opening and closing times of various flowers.

Figure 6.16 shows the main features of a circadian rhythm. During daily cycles of light and dark, the period is entrained (reset daily) to 24 hours. However, the free-running form of the rhythm occurs under stable conditions (e.g. darkness), and the period is not then

24 hours. For instance, in 1832, de Candolle showed that the free-running period of *Mimosa pudica* was between 22 and 23 hours, which is typical of circadian rhythms. Free-running rhythms tend to range from 22 to 28 hours and will occur in both constant dark and constant light. Most rhythms stop after a few cycles in constant light. The fact that the underlying rhythms are not 24 hours is part of the evidence for their endogenous nature.

A phase is any point or part of the cycle – so, in a flower, we could have open and closed phases. In constant darkness, the amplitude and the range become damped, so we would expect less movement of leaves and less distinct opening and closing of flowers, but they would still happen.

Temperature compensation

The effect of temperature is that reactions tend to go faster in warmer environments (within limits for enzymatic reactions). Thus, chemical reaction rate doubles with each $10°C$ increase in temperature ($Q_{10} = 2$). It is therefore surprising that biological clocks, in general, seem to be very little affected by temperature. However, being sensitive to temperature would render a clock

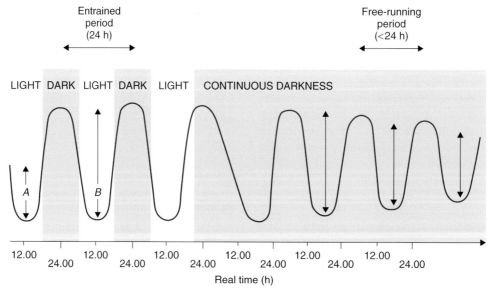

Figure 6.16 A simple circadian rhythm. On the left, in daily cycles of light and dark, the plant is entrained to a 24-hour period. When moved into conditions of continuous darkness, the plant switches to the free-running rhythm with a period that is not 24 hours. In the example shown, the phase is delayed as the third light period was extended. In continuous darkness, both the amplitude (A) and the range (B) are damped. From Hart, 1988. (Reproduced with kind permission from Springer Science+Business Media B.V.)

useless, as it would speed up with higher temperatures; modern watches all have mechanisms to compensate for temperature changes.

Birds and mammals would not be expected to have many problems in this regard, as they maintain relatively constant body temperatures (except during hibernation). Plants, however, almost all adopt a temperature near to that of the environment (See Chapter 9, section 9.2). They tend not to show complete temperature compensation as in animals. For example the free-running period of runner bean (*Phaseolus coccineus*) leaf movements is somewhat influenced by temperature. In 1931, Bünning showed that the period of leaf movement in runner beans exhibited a Q_{10} of 1.2.

Temperatures of 0–5°C, just above freezing, usually retard plants' biological clocks for the length of cold exposure. After prolonged cold, re-warming usually resets the clock.

Resetting

To be useful to an organism, clocks must be able to be reset, just as our watches are, in order to synchronize them with events in the environment. As we have seen, many organisms have free-running cycles that are somewhat different to 24 hours. If no resetting occurred, then the organism would gradually drift off solar time.

Daily resetting (or phase shifting) is known as *entrainment*. Generally, one particular feature of the environment is used as a time cue, and this is known as a *zeitgeber* or time-giver. By far the most common cue is sunset or sunrise or both. Day/night alternation is used as the primary zeitgeber, presumably because light is the most regular and reliable oscillating factor in the environment. So, under natural conditions, exposure to regular cycles of light and darkness entrains the free-running circadian rhythm to 24 hours.

Most circadian rhythms will entrain to light/dark cycles as short as 18 hours or as long as 30 hours, but attempts to force them further away from 24 hours usually result in the breakdown of the cycle.

6.10.3 The mechanism of the clock

It has been clear since the early experiments of Charles and Francis Darwin in 1880 that circadian rhythms were inherited rather than imprinted with a 24-hour period

during development by exposure to diurnal cycles. Later, Bünning's crossing experiments in the 1930s showed that free-running period length of bean hybrids was intermediate between those of the parents. It was clear that the clock was under genetic control and, in the 1970s, there was considerable progress in elucidating the mechanism in animal systems. At around the same time, evidence was obtained that some enzymes in plant cells exhibited circadian peaks in activity. However, it was not until 1985 that Kloppstech, working on pea, found that three transcripts encoding the light-harvesting chlorophyll a/b binding (LHCB) protein, the small subunit of Rubisco, and an early light-induced protein, all showed a circadian rhythm in abundance.

By the early 1990s, experiments on *Arabidopsis* showed that transcription rate and transcript accumulation of LHCB and a number of other genes were under circadian control. It is now known that many processes in *Arabidopsis* conform to circadian rhythms, including about 10 per cent of all genes regulated at the level of mRNA abundance.

However, all of these rhythmic processes represent the 'hands' of the clock, and it is only relatively recently that the mechanism behind the clock has begun to be elucidated. Essentially it is now thought that timekeeping depends on the synthesis of proteins that regulate their own production through feedback control. These proteins are transcription factors that inhibit transcription of the genes that code for the transcription factors themselves. They undergo cyclical changes in concentration and inhibit their own production. The genes responsible were first identified in fruit flies and bread moulds, and more recently in plants. It now seems likely that at least three interlocking feedback loops are involved in the circadian oscillator, but more work is still required to determine all of the components.

Thus there is now plenty of evidence for biological clocks operating in plants, and we are beginning to understand their mechanism. It seems the clocks allow plants to synchronize their activities. Biological clocks ensure that all flowers of a particular species are open at a particular time of day and are available for visits by pollinators. Biological clocks also allow plants to sense and anticipate changes in the seasons.

6.11 Phototropism – how do stems curve towards the light?

In Chapter 5 (section 5.10), we introduced tropisms in general and then considered gravitropism and the response of roots to gravity (section 5.11). Stems show a number of tropic responses. A phototropism is where a plant organ responds to directional light by undergoing a directional growth response. Generally, stems and other aerial parts of plants are positively phototropic (i.e. bend towards the light source), while roots and other underground organs are negatively phototropic (i.e. bend away from the light source) – but there are many exceptions. Phototropism has an interesting history, with early studies conducted by Charles Darwin, and it was also through studies of phototropism that auxins were discovered (see Chapter 3, section 3.6.2).

In the 19th century, it was observed (in France) that plants growing behind bottles of red wine did not bend towards the light. Later, Charles Darwin found that when light was transmitted through a red solution of potassium dichromate, again plants did not bend towards the light. However, it was Blaauw, in 1909, who showed that blue light was the most effective.

Blaauw also demonstrated that for both common oat (*Avena sativa*) coleoptiles and the **sporangiophores** of the fungus *Phycomyces*, the Bunsen-Roscoe reciprocity law holds over a wide range of light intensities and time. This law essentially means that the same quantity of light (irradiance level × duration of irradiance) is required to produce the same photochemical effect if only one photochemical receptor is operating. So the extent of response is related to the amount of light received. This law applies when exposing a coleoptile to a long period of light at low intensity or a short burst of light at high intensity; providing the quantity of light is the same, the same overall response is observed.

The fact that the reciprocity law is 'obeyed' suggested that only one photochemical receptor was involved, and so the search was on to find the receptor. As a much greater response was obtained with blue light than any other, this ruled out phytochrome (see Chapter 3, section 3.7.2) as the pigment involved in phototropism – so a pigment was needed that strongly absorbs light in the blue part of the spectrum. In the mid-1990s, phototropins were

recognized to be the pigments involved, and these were described in Chapter 3 (section 3.7.5).

As in the root responses to gravity (Chapter 5, section 5.11), the Cholodny-Went hypothesis is still accepted as the means by which a directional stimulus of light produces a growth response in stems (see Chapter 3, section 3.6.2). Lateral transport of auxin is induced by the directional light stimulus, and this leads to asymmetrical auxin distribution in the stem, with more auxin being present on the side away from the light. In shoots, this difference causes the stem to bend towards the light as the auxin stimulates the cells on the side away from the light to expand faster.

Despite rapid progress since the discovery of phototropins, the transmission phase is still unclear, and how phototropin activation by blue light leads to increased auxin on the shaded side of the stem is uncertain. A gradient of phot1 (a phototropin) autophosphorylation has been observed across oat coleoptiles when exposed to unilateral irradiation, with the greatest phosphorylation happening on the side nearest the light. How the phosphorylation gradient across the stem can bring about a lateral gradient in auxin is still a matter of active research.

6.12 Gravitropism in stems

In addition to phototropism, most stems also show negative gravitropism, tending to bend upwards away from the gravitational force. As we saw in Chapter 5 (section 5.11.1), the most widely accepted hypothesis for gravity perception in roots is the statolith hypothesis, involving movement of amyloplasts in the root cap. Work with Arabidopsis mutants strongly indicates that the site of gravity perception in the stem is the endodermis; mutants where the endodermis does not differentiate properly are unable to respond to gravity. The endodermal cells have been shown to have sedimenting amyloplasts that are analogous to those in the root cap.

Once more, the Cholodny-Went hypothesis is the favoured idea for producing curvature, with this time auxin accumulating on the lower side of the stem to cause bending. The events that link movement of the amyloplasts in the stem endodermis with the development of an auxin gradient are uncertain.

6.13 Thigmotropism

In some plants that lack strengthening tissues in their stems to maintain an upright position, thin appendages called tendrils grasp the objects they encounter, and thus the plant attains an upright position (e.g. beans and peas). Many plants respond to touch, and we will consider here **thigmotropism** in tendrils. In thigmotropism, the direction of movement is determined by the surface stimulated. Again, Charles Darwin was a pioneer in this area of research. He observed that contact coiling begins between 30 seconds and five minutes after tactile stimulation, but tendrils of most species straighten if the contact stimulus is removed prematurely. Coiling of pea tendrils can be stimulated by stroking with a glass rod, and the magnitude of the response increases with the number and frequency of the strokes. Stroking the upper surface of pea tendrils is ineffective but, if both sides are stroked less curvature is noted.

Much less work has been done on thigmotropism than on other tropisms, but there is some information about the mechanism. After being touched, the plasma membrane of the epidermal cells is distorted, causing an action potential to be conducted through the neighbouring cells. It is uncertain how this leads to the bending response, but again work on *Arabidopsis* has given some clues. Four specific genes (TCH1-4) are rapidly up-regulated when the plant is touched. Of these, TCH1 encodes for a calmodulin (a calcium-binding protein), TCH2 and TCH 3 for calmodulin-like proteins, and TCH4 encodes a xyloglucan endotransglucosylase/hydrolase that affects cell wall properties (see Chapter 2). Thus, three of the genes encode proteins that are implicated in calcium signalling, and it seems likely that this is involved in the response of plants to touch.

6.14 Nastic movements

Nastic movements differ from tropisms. They also involve unequal growth on opposite sides of an organ in response to environmental stimuli, but they are not elicited by directional stimuli. The direction of the response is determined by the organ. Characteristically, nastic reactions take place in organs with bilateral symmetry, whereas radially symmetrical organs execute tropic movements. If a plant organ shows higher growth rate on the upper surface, this is termed epinasty, while higher growth rate on the lower surface is known as *hyponasty*. Gravinastic, photonastic, thermonastic, thigmonastic and chemonastic movements are all known. Here we will only outline a few of the nastic growth responses. If knowledge of the mechanisms underlying the tropisms is incomplete, then this is even more so with the nastic responses.

6.14.1 Nyctinastic plants

Plants whose leaves or leaflets assume a vertical orientation in the dark are known as **nyctinastic** (Greek: night, close). These leaves open in the daylight. The opening and closure are on a circadian rhythm (see above), and these are so-called sleep movements. The legumes show these movements particularly well.

What is the explanation for these movements? It is quite obvious that there is adaptive value in orientating leaves in a horizontal daytime position to catch the most light, but the reasons for closure at night are less certain. In 1881, Darwin suggested that the sleep movements could help prevent excessive heat loss to the open sky at night, and there is some evidence in favour of this idea. Another suggestion is that these movements prevent moonlight from interfering with time measurement in photoperiod-sensitive plants. At a latitude of $50°$, moonlight can reach 0.3 lux, and up to 0.9 lux at the equator. Light intensities as low as 0.1 lux can influence time measurement in some plants. Sleep movements in legumes reduces moonlight incident on the upper leaf surface by up to 95 per cent.

Some data have accumulated concerning the mechanism of these movements. Unlike the tropic movements and some other nastic movements, these are totally reversible. They involve the **pulvinus**, a joint-like swelling situated at the petiole base, and leaf movement is caused by changes in the volume of motor cells. The flux of potassium ions across the plasma membrane of the motor cells means that as much as 60 per cent of the ions move from the flexor side to the extensor side of the pulvinus and back again during a complete circadian cycle. This is followed by a substantial water flux that results in swelling or shrinking of the motor cells. The opening and closing of the potassium channels involved in nyctinastic leaf movement is presumably linked in some way to the circadian oscillator described above.

6.14.2 Seismonastic responses

The nyctinastic movements are fairly slow responses occurring through a whole circadian period. The so-called **seismonastic** responses, however, are very rapid, being propagated 1–2 seconds after mechanical stimulation. We will consider two examples: the rapid movements of plants such as *Mimosa pudica*, and the seismonastic response of the Venus flytrap (*Dionaea muscipula*), a carnivorous plant.

A limited number of higher plants can exhibit very rapid seismonastic movements in addition to slow sleep movements. The most famous example is the sensitive plant, *Mimosa pudica*, which very rapidly moves within 1–2 seconds of the plant being shaken or touched (Figure 6.17). The bigger the stimulus, the greater the distance the signal travels. The whole leaf folds downwards if the stimulus is strong enough.

Seismonastic movements take place at the pulvinus. The leaflets rapidly fold together as a result, and the petioles drop. Following the stimulus, the petioles return to the non-stimulated position after only 8–15 minutes. If a plant is stimulated approximately every 30 minutes, it will return to its original shape. It seems fairly certain that, in seismonastic plants, predation is discouraged by making leaves less conspicuous and exposing thorns on the stems.

The mechanism of seismonastic movements cannot involve the redistribution of plant hormones, as this occurs too slowly. The investigation of seismonasty has led to the discovery that electrical signals can propagate through plants and are an important means of communication in the plant body. It is uncertain how important electrical signals are in other plant processes, and only the seismonastic responses are presently accepted to be transmitted in this way.

The stimulus is transduced into an action potential that propagates through the phloem and protoxylem at about $2–4\,cm\,sec^{-1}$. Movement begins 10–20 ms after the pulvinar action potential is initiated. The rapid pulvinar movements are brought about by potassium being rapidly lost from shrinking cells, causing a loss of turgor. Water corresponding to about 25 per cent of the cell volume is lost via osmosis within one second. Aquaporins (see Chapter 2, section 2.3.4) in the plasma membrane of the pulvinar cells are probably responsible for this rapid water movement. As these movements are reversible, recovery occurs when potassium ions are re-absorbed into the pulvinar cells. H^+ ATPase activity is high in pulvini, probably to sustain these ion fluxes. As the ions move in, water follows and turgor pressure is restored.

The final topic in this section on growth movements concerns the Venus flytrap (Figure 6.18) and the rapid movements that it demonstrates when trapping insects. Venus flytraps grow in nitrogen-poor environments such as bogs, and the trapping of flies and spiders and their subsequent digestion is seen as an adaptation to these conditions. Charles Darwin described the plant as 'one of the most wonderful in the world'.

Some parts of the closure mechanism are now clear. Again, hormones are not involved here as the response is far too quick (about 100 ms). There are six sensory hairs on the surface of the trap, which consists of two modified leaves. If one hair is touched twice, or two hairs are bent once each, this triggers the trap to close. When a hair is touched, a single action potential is evoked, which moves across the trap tissue at $20\,cm\,sec^{-1}$. Two such signals are needed to cause the trap closure.

The trap-snapping mechanism entails a multifaceted interaction between elasticity, turgor and growth. The lobes of the trap are convex and bent outwards in the open state, but after tripping they rapidly change to concave, forming a cavity in which the insect is digested. The reasons most often given for the fast response are a permanent wall loosening induced by release of acid, and a rapid decrease in turgor in motor cells in the midrib

Figure 6.17 Leaves of *Mimosa pudica*, which are touch-sensitive. (Photograph: Dr Gordon Beakes © University of Newcastle upon Tyne. Image from Centre for Bioscience (Higher Education Academy) ImageBank.

Figure 6.18 Venus flytrap (*Dionaea muscipula*). Traps are formed by modified leaves.
Photograph: Dr Gordon Beakes © University of Newcastle upon Tyne. Image from Centre for Bioscience (Higher Education Academy) ImageBank.

between leaves. However, these mechanisms alone are not thought to be able to explain the speed of closure, and it has recently been suggested by Forterre *et al.* (2005) that elastic deformations of the trap lobes may have a role. The rapid shutting of the trap is now considered to arise from a mechanical snap-buckling instability, the beginning of which is actively controlled by the plant.

6.15 Bud dormancy

Dormancy was defined in Chapter 4, section 4.10, and was discussed in relation to seeds. Here we focus on bud dormancy. As we saw in section 6.3, the growth of the stem is initiated in meristems at the tip of the shoot. These are centres of cell division activity and they have high rates of metabolism during active growth. They are also centres of differentiation, giving rise to leaf and flower primordia, and are known as buds. In perennial plants, the buds undergo successive cycles of activity and dormancy. Bud dormancy studies have focused on three questions: the environmental signals that stimulate the onset of dormancy; the metabolic changes that are responsible for the reduced activity; and the signals that start up of growth again at the correct time.

6.15.1 Bud formation

When the winter buds begin to form, stem cell elongation stops and the production of leaves may change to give rise to a short succession of scales, which are tightly appressed around the bud. These changes typically occur well before the onset of winter. The close-fitting shape and resistant texture of the bud scales gives protection to the over-wintering stem tip. After the scales have formed, the meristem then produces a succession of vegetative leaf or flower primordia, which remain within the bud as miniature organs, growing out only after the release of the bud from dormancy. In some species, all the leaves which will function during the next season are preformed in the winter bud.

The onset of bud dormancy occurs at the same time as leaf fall, decreased cambial activity and increased cold hardiness. Bud dormancy seems to be a short-day response (see Chapter 8, section 8.4.1), initiated by the short days at the end of the summer. In most cases phytochrome is involved, and it is the leaves that perceive the day length. However, at least in birch (*Betula*), the response is localized in the apical bud.

During the onset of dormancy, changes in the levels of plant hormones have often been observed, and these are involved in the process. The hormones ABA, GA and cytokinin play important but antagonistic functions in all dormancy types. The varying dormancy responses of different species have been categorized as: paradormancy (growth inhibition by distal organs (apical dominance)); endodormancy (growth inhibition by internal bud signals); and ecodormancy (growth inhibition by transitory adverse environmental circumstances). A bud may be controlled by one or all of the signals that regulate these characteristics of dormancy.

Little is known of the physiology of dormant buds, but they are characterized by low respiration and an inability to grow even if temperature, oxygen and water supply are adequate.

Figure 6.19 Beech (*Fagus sylvatica*) buds bursting in the spring to reveal new leaves and flowers, Oxford, England. Photo: MJH.

6.15.2 Bud break

The key factor in breaking bud dormancy in plants growing in temperate and cooler regions is a period of low temperature (Figure 6.19).

Temperatures at or near freezing seem to be most effective in breaking dormancy. The amount of chilling required varies with the species, cultivar and even the location of the buds on the trees. Thus, apple, pear, and cherry require seven to nine weeks of chilling, while the American plum needs 22 weeks, the apricot four to six weeks and persimmon only four days. These may be related to the original provenance of the trees. The persimmon can therefore be grown much further south than the other fruit trees.

In some species, there is ecotypic variation. For instance, in the sugar maple, 12 weeks of chilling is required to break bud dormancy in plants from Southern Canada, while only a few weeks are required in plants collected from near the southern edge of its range. Although temperature in the winter varies greatly in temperate areas, this does not seem to affect dormant tissues, as they seem to be able to sum the periods of cold.

Selected references and suggestions for further reading

Binkley, S. (1997) *Biological Clocks: Your Owner's Manual.* Harwood, Amsterdam.

Carraro, N., Peaucelle, A., Laufs, P. & Traas, J. (2006) Cell differentiation and organ initiation at the shoot apical meristem. *Plant Molecular Biology* **60**, 811–826.

Chehab, E. W., Eich, E. & Braam, J. (2009) Thigmomorphogenesis: a complex plant response to mechano-stimulation. *Journal of Experimental Botany* **60**, 43–56.

Forterre, Y., Skotheim, J. M., Dumais, J., & Mahadevan, L. (2005) How the Venus flytrap snaps. *Nature* **433**, 421–425.

Génard, M., Fishman, S., Vercambre, G., Huguet, J-D., Bussi, C., Besset, J. & Habib, R. (2001) A biophysical analysis of stem and root diameter variations in woody plants. *Plant Physiology* **126**, 188–202.

Graf, A., Schlereth, A., Stitt, M. & Smith, A. M. (2010) Circadian control of carbohydrate availability for growth in Arabidopsis plants at night. *Proceedings of the National Academy of Sciences, USA* **107**, 9458–9463.

Hallé, F., Oldeman, R. A. A. & Tomlinson, P. B. (1978) *Tropical trees and forests: an architectural analysis.* Springer-Verlag, Berlin.

Hart, J. W. (1988) *Light and Plant Growth.* Unwin Hyman, London.

Horvath, D. P., Anderson, J. V., Chao, W. S. & Foley, M. E. (2003) Knowing when to grow: signals regulating bud dormancy. *Trends in Plant Science* **8**, 534–540.

Kempes, C. P., West, G. B., Crowell, K. & Girvan, M. (2011) Predicting maximum tree heights and other traits from allometric scaling and resource limitations. *PLoS ONE* **6**(6), e20551. doi:10.1371/journal.pone.0020551

Kloppstech, K. (1985) Diurnal and circadian rhythmicity in the expression of light-induced plant nuclear messenger RNAs. *Planta* **165**, 502–506.

Koch, G. W., Sillett, S. C., Jennings, G. M. & Davis, S. D. (2004) The limits to tree height. *Nature* **428**, 851–854.

Kuhlemeier, C (2007) Phyllotaxis. *Trends in Plant Science* **12**, 143–150.

Mauseth, J. D. (1988) *Plant Anatomy.* Benjamin Cummings, Menlo Park, CA.

McClung, C. R. (2006) Plant circadian rhythms. *The Plant Cell* **18**, 792–803.

Monshausen, G. B. & Gilroy, S. (2009) Feeling green: mechanosensing in plants. *Trends in Cell Biology* **19**, 228–235.

Munné-Bosch, S. (2008) Do perennials really senesce? *Trends in Plant Science* **13**, 216–220.

Owen, J. (2008) Oldest living tree found in Sweden. *National Geographic News.* (April 14, 2008) http://news.nationalgeographic.com/news/2008/04/080414-oldest-tree.html

Prusinkiewicz, P. & Remphrey, W. (2000) Characterization of architectural tree models using L-systems and Petri nets. In: Labrecque, M. (Ed.): *L'arbre – The Tree 2000: Papers presented at the 4th International Symposium on the Tree*, pp. 177–186.

Steeves, T. A. & Sussex, I. M. (1989) *Patterns in Plant Development*. 2nd ed. Cambridge University Press, Cambridge, UK.

Steudle, E. (1995) Trees under tension. *Nature* **378**, 663–664.

Tasaka, M., Kato, T. & Fukaki, H. (1999) The endodermis and shoot gravitropism. *Trends in Plant Science* **4**, 103–107.

Thomas, P. (2000) *Trees: Their Natural History*. Cambridge University Press, Cambridge, UK.

Tootill, E. (Ed.) (1984) *Penguin Dictionary of Botany*. Penguin Books, London.

van Bel, J. E. (2003) The phloem, a miracle of ingenuity. *Plant, Cell and Environment* **26**, 125–149.

Wang, Y, Li, J. (2006) Genes controlling plant architecture. *Current Opinion in Biotechnology* **17**, 123–129.

Watkinson, A. (1992) Plant senescence. *Trends in Ecology and Evolution* **7**, 417–420.

Woolhouse, H. W. (1972) *Ageing Processes in Higher Plants*. Oxford University Press, Oxford.

Leaves are the most obvious structures of most plants for the majority of the growing season. They show a great variety in morphology and can be adapted to a wide variety of environments. They are the main photosynthetic organs of most plants, although stems are also usually photosynthetic.

7.1 External morphology of leaves

Leaves can vary in size from small scales to the large compound leaves of the palms, which can be 3 m long (see, for example, the coconut palms in Chapter 6, Figure 6.4). We considered the gymnosperm, Ginkgo (*Ginkgo biloba*), in Chapter 1, Box 1.2. This tree has unique fan-shaped leaves that have veins radiating out into the leaf blade, sometimes splitting but never forming a network (Figure 7.1).

The leaves of eudicots usually consist of a flattened blade and a stalk, the petiole, which joins the leaf to the node of the stem. The palmately lobed leaves of Japanese maple (*Acer palmatum*) are a typical example (Figure 7.2).

Grasses and many other monocots lack a petiole, and the leaf forms a sheath that envelops the stem (e.g. the maize leaves shown later in this chapter in Figure 7.11).

Leaf form can vary in many ways: the margin can be smooth or jagged; the texture can be soft and pliable or tough and durable; thickness can vary greatly, and some leaves are succulent (e.g. the ice plant shown in Figure 7.14); the presence of hairs and glands on the surface is also variable; and colours range through all shades of green, but also many other colours depending on the secondary pigments present.

Leaves can also be modified into a whole variety of other structures including: bud scales (Chapter 6, section 6.15.2); sepals and petals (Chapter 8, section 8.2); tendrils

for support (Chapter 6, section 6.13); spines and thorns for defence against predators (Chapter 11, section 11.2.1; Figure 6.18); and storage organs.

Simple leaves have undivided blades, while compound leaves are sub-divided into leaflets split up along a main or secondary vein. In compound leaves, each leaflet can appear to be a simple leaf; it is the position of the petiole that identifies a compound leaf. Some families of flowering plants, such as the legumes, are characterized by compound leaves (e.g. *Mimosa pudica*; see Chapter 6, Figure 6.17). These have the advantage of maximizing the surface area available for photosynthesis while minimizing the mechanical constraints that a large simple leaf might experience.

The earliest land plants were the bryophytes (Chapter 1), and these were almost certainly species without leaves. Leaves seem to have evolved at least five times since the invasion of the land. There are several groups of liverworts and mosses in which the gametophyte generation (see Chapters 1 and 8) has leafy shoots that must have evolved from their leafless ancestors. Among the vascular plants, leafy shoots are found in the sporophytes of the lycophytes, moniliformopses (horsetails and ferns) and angiosperms, and they seem to have evolved independently in all of these.

7.2 The anatomy of the leaf

We will now consider the main internal tissues of the leaf, working from the outside. Figure 7.3 shows the structure of a typical eudicot leaf. The leaf is surrounded by its epidermis, which covers the surface. The epidermis is the layer that is the first line of defence against herbivory (Chapter 11, section 11.2.1) and pathogens (Chapter 11,

Functional Biology of Plants, First Edition. Martin J. Hodson and John A. Bryant.
© 2012 John Wiley & Sons, Ltd. Published 2012 by John Wiley & Sons, Ltd.

Figure 7.1 Ginkgo (*Ginkgo biloba*) has fan-shaped leaves that are normally 5–10 cm long. Here the leaves were photographed in the autumn in Paris, France.
Photo: MJH.

Figure 7.2 Japanese maple (*Acer palmatum*) is a woody plant native to Asia. The leaves, here photographed in the autumn, are palmately lobed and are 4–12 cm long and wide.
Photo: Margot Hodson. Used with the permission of Westonbirt Arboretum, Tetbury, Gloucestershire, UK.

the case around the stomata of the lower epidermis, as it allows CO_2 to diffuse in to the photosynthetic sites.

Finally, we must briefly mention the vascular tissues in the veins of the leaves. The veins are often surrounded by specialized bundle sheath cells. The two most important tissues are the xylem and the phloem, which, as we saw in Chapter 6, are the long-distance transport systems in the plant.

section 11.3.7). It is also very important in preventing water loss and it secretes a waxy cuticle onto the surface.

Within the epidermis are the stomata (singular: stoma – the word *stoma* originates from the Greek word for mouth). Stomata are important both in allowing CO_2 to enter the leaf for photosynthesis and in controlling water loss. We will return to these topics in section 7.10.

Inside the leaf, the dominant tissue is the **mesophyll**, from the Greek *mesos* (middle) and *phyll* (leaf). This layer consists mainly of cells with many chloroplasts, and it is the area where most photosynthesis occurs in a plant (we will consider this in section 7.4). In many eudicots, there are two types of mesophyll cell: palisade and spongy. The palisade layer is beneath the upper epidermis and consists of cells that are columnar in shape. Beneath this is the spongy mesophyll, which has more loosely packed cells with greater air spaces between them. This is particularly

7.3 Control of leaf growth and development

In Chapter 6 (section 6.3), we noted that leaf primordia are formed at the flanks of the shoot apical meristem (SAM). We also saw that the SAM is characterized by the activity of KNOX genes that are members of the KNOTTED family of homeobox transcription factors that play a major role in the organization and preservation of meristematic cells (Chapter 4, section 4.2 and Chapter 6, section 6.3.1).

It also appears that KNOX genes are important in determining leaf morphology. In the tomato sub-genus, KNOX genes are expressed in species that have dissected leaves, but not in simple-leaved species. In transgenic tomatoes that have increased KNOX expression, this leads to super-dissected leaves. Later work has showed that while KNOX expression in leaf primordia demonstrates a relationship with dissected morphology at the early

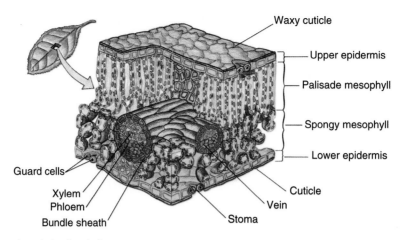

Waxy cuticle

Upper epidermis

Palisade mesophyll

Spongy mesophyll

Lower epidermis

Guard cells

Xylem

Phloem

Bundle sheath

Cuticle

Vein

Stoma

Figure 7.3 Anatomy of a typical eudicot leaf.
From: Campbell, Neil A., BIOLOGY, 2nd Edition © 1990. Reprinted by permission of Pearson Education, Inc., Upper Saddle River, NJ.

stages of leaf development, this is not always the case with final leaf morphology. Secondary morphogenesis occurs in some species, which leads to cell differentiation and growth at leaf margins, obscuring the existence of any leaflets.

Moreover, some plants, including pea (*Pisum sativum*), do not express KNOX genes in the leaf primordium. These use a second pathway involving the UNIFOLIATA (UNI) gene, which encodes a member of the LEAFY (LFY) family of transcription factors. Recent work has also emphasized that constant communication between the SAM and the developing leaf is important, and that treating them as separate entities is unwise.

The processes described above will lead to the development of a small but identifiable leaf. During this early stage, growth is mainly achieved by the manufacture of new cells that make organized tissues, generating a flattened leaf structure. In eudicot leaves, the blade and petiole are distinct by this stage but, in monocot leaves, the sheath is formed only after the blade has grown to a sizeable length.

Zones of cell division and cell expansion are much easier to identify in monocot leaves than eudicot leaves. In the latter, cell division and expansion zones often overlap to a great extent, both spatially and temporally. Cell expansion is far more important than cell division in producing the surface area of the mature leaf. When cell division stops, leaf cells carry on increasing in volume, reaching 20–50 times that of the meristematic cells from which they were derived. Differences in cell number

and cell volume both contribute to variation in leaf size between species.

7.4 Photosynthesis

'Photosynthesis is a process in which light energy is captured and stored by an organism, and the stored energy is used to drive cellular processes.'

Blankenship (2002).

Photosynthesis is, in many respects, the most important process in the world. Every year, 2×10^{11} tonnes of carbon are fixed – the equivalent of 3×10^{18} KJ of chemical energy. Although photosynthesis is such a dominant process in the biosphere, only 0.1 per cent of total incident solar radiation is used. Rubisco, the enzyme responsible for carbon fixation, is the most abundant protein in the world, accounting for 50 per cent of soluble protein in most leaves.

The balance between stored and atmospheric carbon depends to a great extent on ancient and modern photosynthesis, as all fossil fuels are the products of photosynthesis. Increase in atmospheric CO_2 due to anthropogenic burning of fossil fuels is a key factor in global warming, and this has focused interest on photosynthesis and reforestation to decrease CO_2 in the atmosphere (Chapter 12, section 12.1.3).

7.4.1 Basics of photosynthesis

The equation below is often used in introductory biology classes when students are first introduced to photosynthesis:

$$6\,CO_2 + 12H_2O + \text{Light energy}$$

$$\rightarrow C_6H_{12}O_6 + 6O_2 + 6H_2O$$

This balanced equation was first recognized in the 1860s and was a major advance in understanding at the time. The equation contains some truth but it is actually an oversimplification, as it is a three-carbon sugar that is produced in photosynthesis, not glucose. It can be reduced to a very simple form:

$$CO_2 + 2H_2O \rightarrow CH_2O + H_2O + O_2$$

This again is a simplification, but it does allow us to visualize where the oxygen atoms move to in the equation. Van Niel was the first to recognize the redox nature of photosynthesis in the 1930s, and he correctly predicted that the process involved the splitting of water. Some

years later, techniques became available for the isolation and analysis of heavy isotopes, including ^{18}O, a non-radioactive isotope of oxygen. When scientists eventually used ^{18}O to label the oxygen atoms in the CO_2 and the water on the left of the equation (above), they found some interesting results:

$$CO_2 + 2\mathbf{H_2O} \rightarrow CH_2O + H_2O + \mathbf{O_2}$$

$$\mathbf{CO_2} + 2H_2O \rightarrow CH_2\mathbf{O} + H_2\mathbf{O} + O_2$$

(bold type indicates labelling with ^{18}O).

The oxygen emitted from the plants contained ^{18}O *only* if the label was supplied in water. If the label was supplied in carbon dioxide, then it did not appear in the oxygen, at least for some time. This implied that the splitting of water is an important part of photosynthesis, as Van Niel had predicted. As water is split, electrons are transferred, along with hydrogen ions from the water, to CO_2, reducing it to sugar. The electrons increase in potential energy as they move from water to sugar and the required energy boost is provided by light.

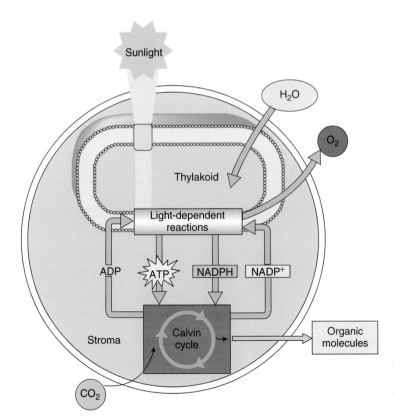

Figure 7.4 Overview of photosynthesis illustrating the interactions between the light reactions and the Calvin cycle in the chloroplast.

The simple equations above hide very much more complex processes. Figure 7.4 summarizes the reactions involved in the chloroplasts. In the light (photo) reactions, light is used to split water, producing oxygen and two energy sources, ATP and NADPH. These reactions are associated with the thylakoids of the chloroplast (see Chapter 2, section 2.5.2). Energy, as ATP and NADPH, is then used in the Calvin cycle (named after its discoverer Melvin Calvin) to fix CO_2 into carbohydrate (synthesis). These reactions occur in the stroma of the chloroplast. The Calvin cycle reactions are often rather misleadingly called 'dark reactions', implying that light is not directly involved in them. However, the Calvin cycle soon closes down in the dark, when its energy supply is turned off.

Now we need to consider the light reactions in more detail.

7.4.2 Light reactions

Visible light is just one part of the electromagnetic spectrum, which can be distinguished by different wavelengths (the distance between the crests of the waves). It is visible light (wavelengths 400 to 700 nm) that drives photosynthesis, and the most important wavelengths are in the blue and red parts of the spectrum. Ultraviolet radiation, at wavelengths below 400 nm, can be highly damaging to tissues (Chapter 9, section 9.7.2), while plants and most other organisms cannot use infrared and other wavelengths above 700 nm, as radiation of these wavelengths does not contain enough energy.

When light meets matter, it can be reflected, transmitted or absorbed. A substance that absorbs visible light is called a pigment. Different pigments absorb light of different wavelengths, and the wavelengths that are absorbed disappear. Thus white light, which is made up of all colours, is absorbed by chlorophyll in the leaf chloroplasts in the red and blue parts of the spectrum, and so reflects and transmits green. This is the reason why leaves mostly appear green.

It is relatively easy to isolate chlorophyll and other photosynthetic pigments by grinding a leaf in acetone or another suitable solvent, then separating the pigments using chromatography. A pigment's ability to absorb light of different wavelengths can be measured in a spectrophotometer (Figure 7.5). When this is done with chlorophyll a (the dominant photosynthetic pigment), the maximum absorbance is in the red and blue areas of the spectrum, with less absorbance in the green area.

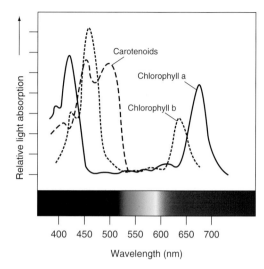

Figure 7.5 Absorbance spectra of chlorophyll a, chlorophyll b and carotenoids.

An **action spectrum** measures the performance of photosynthesis (oxygen evolution or CO_2 uptake) at the different wavelengths. Thus light of a specific wavelength is shone onto a suspension of chloroplasts and the photosynthetic rate is measured. The absorbance spectrum for chlorophyll a does not precisely match the action spectrum for photosynthesis. This is because chlorophyll a is not the only pigment involved in the absorbance of light. Chlorophyll b and various other pigments, the carotenoids, absorb light and pass the energy on to chlorophyll a. Therefore the action spectrum presents as an amalgam of all the pigment spectra, and it also has peaks in the blue and red parts of the spectrum with less absorbance at the green wavelengths.

The structure of chlorophyll a is shown in Figure 7.6. Essentially, the molecule consists of a porphyrin (tetrapyrrole) ring with a magnesium atom in its centre and a long lipid tail. In chlorophyll b, the top right CH_3 group at position R is replaced by a CHO group. This seems a small change, but chlorophyll a is blue-green and chlorophyll b is yellow-green, and the two molecules absorb light in somewhat different parts of the spectrum.

7.4.3 Photoexcitation of chlorophyll and photophosphorylation

When light is absorbed by isolated chlorophyll, its energy does not disappear. One of the molecule's electrons is

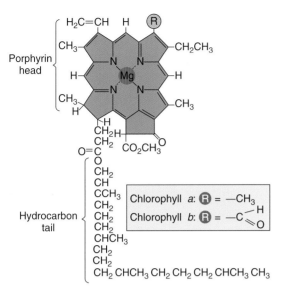

elevated to an orbital where it has more potential energy. When the electron is in its normal orbital, the molecule is in the ground state, but when it is elevated it is in an excited state. This excited state is very unstable, and generally the electron drops back in a billionth of a second, releasing excess energy as heat. A similar phenomenon is what makes the paintwork of a car hot on a sunny day (white cars are coolest as their paint reflects all wavelengths of visible light). Some pigments, including chlorophyll, emit light as well as heat when irradiated, and this is known as fluorescence.

Excitation of chlorophyll in the chloroplast produces very different results from those seen with the isolated pigment. In the thylakoid membrane of the chloroplast, the chlorophyll is arranged with other molecules into photosystems (see Chapter 2, section 2.5.3, Figure 2.10). A photosystem consists of a cluster of a few hundred chlorophyll a, chlorophyll b and carotenoid pigments. When light hits any of these molecules, its energy is transferred across the photosystem until it reaches a particular chlorophyll a molecule in a particular position called the reaction centre. In the same reaction centre is a specialized molecule called the primary electron acceptor. The chlorophyll a at the reaction centre loses an electron to the primary acceptor. This acts as a block. At this

point, isolated chlorophyll loses heat and fluoresces, but the primary acceptor prevents the immediate drop back to the ground state.

The thylakoid membrane has two types of photosystems that cooperate in the light reactions – photosystem I and photosystem II. The reaction centre of photosystem I is known as P700, as it is best at absorbing light at 700 nm (far red), while the centre of photosystem II is P680, as it absorbs light best at 680 nm (red).

There are two possible routes for electron flow in the light reactions of photosynthesis; cyclic and non-cyclic. The dominant type is non-cyclic (Figure 7.7) and occurs in the following stages:

1 Photosystem II absorbs light, causing an electron to be excited to a higher energy level, where the primary acceptor captures it. This means that the P680 chlorophyll is now an oxidizing agent with an electron 'hole' which must be filled.

2 An enzyme splits water and replaces the electrons in P680. The other products are two hydrogen ions and an oxygen atom. The latter immediately combines with another atom and O_2 is released. The mechanism of this process is still the subject of active research, but it is clear that it requires a very strong oxidizing agent to split oxygen atoms from hydrogen atoms, and that a manganese-containing enzyme complex known as the oxygen evolving complex (OEC) is involved.

3 P680*, the excited form of P680, then rapidly passes an electron to pheophytin (Phe), the primary electron acceptor of photosystem II. Pheophytin is a form of chlorophyll a, where the magnesium ion at the centre of the porphyrin ring is replaced by two hydrogen ions. Each electron then drops from photosystem II to photosystem I via an electron transport chain similar to the one used in respiration (Chapter 2, section 2.14.2). The chloroplast version consists of plastoquinone (PQ), an iron-containing cytochrome b_6/f complex, and a copper-containing protein called plastocyanin (PC).

4 As the electrons fall to a lower energy level, the energy is harnessed to produce ATP. This ATP synthesis is called non-cyclic photophosphorylation. The mechanism for the generation of ATP involves setting up a proton gradient, and we will consider this in more detail in section 7.4.4 below. The ATP is then used to drive the Calvin cycle.

5 The electron from photosystem II then fills a 'hole' in P700 that had been created when light energy excited

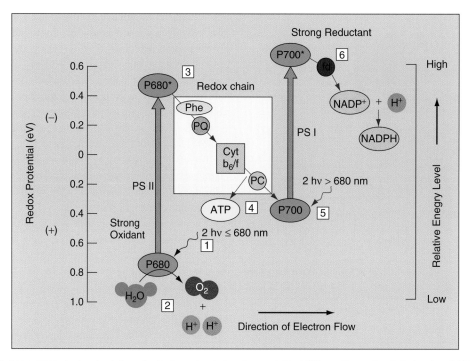

Figure 7.7 Non-cyclic photophosphorylation – the Z scheme for electron transport. Key: P680, reaction centre of photosystem II; P700, reaction centre of photosystem I; Phe, Pheophytin; PQ, plastoquinone; Cyt b6/f, a cytochrome complex; PC, plastocyanin; fd, ferredoxin. Numbers 1 to 6 refer to text in Section 7.4.3.

photosystem I, and an electron is elevated to a primary acceptor.

6 The primary acceptor then passes the electrons down another electron transport chain, which transmits them to an iron-containing protein called ferredoxin (fd). An enzyme, $NADP^+$ reductase, transfers electrons from fd to $NADP^+$. This is a redox reaction that stores high-energy electrons in NADPH, which is used as reducing power for the synthesis of sugars in the Calvin cycle.

Under some conditions excited electrons take an alternative route called cyclic electron flow (Figure 7.8). This is really a short circuit. Here there is no release of oxygen or production of NADPH, but this system does produce ATP, and it is called *cyclic photophosphorylation*.

What is the function of cyclic flow? Non-cyclic flow produces ATP and NADPH in roughly equal amounts, but the Calvin cycle requires more ATP than NADPH. So cyclic flow is believed to make up the difference.

It is worth noting at this point that we can see here in the light reactions of photosynthesis some of the essential

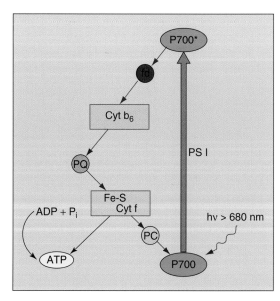

Figure 7.8 Cyclic photophosphorylation. Key as in Figure 7.7.

roles of the elements we mentioned in our plant nutrition section (Chapter 5, section 5.5). Thus magnesium is involved in chlorophyll, the OEC contains manganese, the cytochrome b_6/f complex and ferredoxin contain iron and PC is a copper-containing protein.

7.4.4 Proton gradients in photosynthesis

We have seen that cyclic and non-cyclic electron flow generate ATP; we now need to think about the mechanism. This involves setting up a charge separation or proton gradient, a process sometimes known as chemiosmosis, which we last considered in Chapter 2, section 2.14.2. Both mitochondria and chloroplasts generate ATP by a similar mechanism. In chloroplasts, the electron transport chains pump protons across the membrane as electrons pass down a series of carriers. The protons accumulate in the thylakoid space and provide a proton gradient across the membrane. In the same membrane is an ATP synthase complex (similar to that shown in Figure 2.20

in Chapter 2), which couples diffusion of the protons (H^+ ions) down their electrochemical gradient to the synthesis of ATP. Thus ATP forms in the stroma, where it is available for the Calvin cycle.

The proton gradient across the membrane can be quite large. When the lights are on, the pH in the thylakoid space can be 5, while in the stroma it increases to 8 (so the H^+ concentration is 1,000 times higher in the thylakoid). This gradient is abolished in the dark and ATP synthesis stops.

7.4.5 The Calvin cycle

The light reactions produce the high-energy compounds ATP and NADPH in the thylakoid membranes. These compounds are then used to drive the Calvin cycle (Figure 7.9).

Carbon enters the cycle as CO_2 and leaves as sugar. The sugar produced is glyceraldehyde 3-phosphate (G3P) a three-carbon sugar. To produce this sugar, the cycle has to turn three times, fixing one CO_2 each time. So, as we

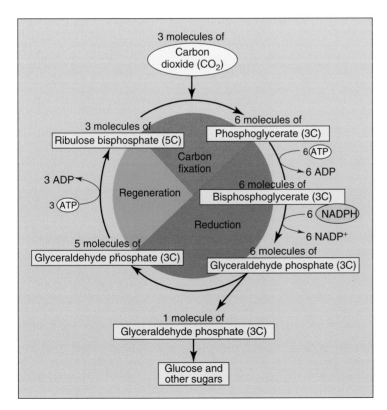

Figure 7.9 The Calvin Cycle. This can be divided into three main sections: carbon fixation; reduction; and the regeneration of the CO_2 acceptor.

go through the cycle, we are following three molecules of CO_2. The Calvin cycle can be divided into three parts:

1 *Carbon fixation*. The cycle starts by attaching each CO_2 to a five-carbon sugar called ribulose bisphosphate (RuBP). The enzyme responsible for this is the most common in the chloroplast – RuBP carboxylase or Rubisco. In vascular plants, the enzyme consists of eight large (L) 51–58 kDA subunits and eight small (S) 12–18 kDA subunits, giving an L_8S_8 quaternary structure. Two of the L subunits share the catalytic site. The assembly and structure of Rubisco was considered in Chapter 3 (section 3.3.2). The six-carbon compound produced by the carboxylation of RuBP is so unstable that it immediately breaks down into two molecules of 3-phosphoglycerate, or PGA (for each CO_2).

2 *Reduction*. The next stage is two reactions involving the use of ATP and NADPH to convert PGA to glyceraldehyde phosphate (G3P) – reactions which are the reverse of those occurring in glycolysis. At this point, one G3P molecule exits the cycle and is used as the starting material to form starch inside the chloroplast or glucose, fructose and sucrose in the cytosol (see Chapter 2, section 2.5.3 and Figure 2.12).

3 *Regeneration of the CO_2 acceptor*. In a complex series of reactions, the five molecules of G3P are then rearranged to form three molecules of RuBP.

For every one G3P molecule produced, the Calvin cycle consumes a total of nine molecules of ATP and six of NADPH. The type of photosynthesis that we have just discussed is known as C_3, as carbon is first fixed into a three-carbon compound. C_3 plants are the majority and they include many of our temperate crops, such as wheat, barley, clover and sunflower.

7.4.6 The evolution of photosynthesis

In Chapter 1, we considered the evolution of the photosynthetic eukaryote in some detail. Here we will briefly investigate what is known of the evolution of the biochemistry of photosynthesis. Undoubtedly, this has been a complex process, as it is the result of the numerous photosynthetic components each having their own evolutionary pathway.

One obvious candidate for analysis is the chlorophyll biosynthesis pathway, and particularly the genes responsible for the synthesis of the Mg porphyrin ring. A phylogenetic analysis of these genes has suggested that most anoxygenic photosynthetic organisms (i.e. those that carry out photosynthesis that does not produce oxygen) are ancestral to oxygen-evolving cyanobacteria, and that the purple bacteria have the most ancestral type of this biosynthetic pathway.

The other major target for analysis has been the apoproteins that bind the chlorophyll molecules at the reaction centres of the photosystems. This work has revealed that the reaction centre apoproteins from photosystem I and photosystem II have rather different protein sequences, but do have some similarities in the topology of transmembrane helices and the positions of the core Mg porphyrin rings. It is evident from this work that the evolution of photosynthesis cannot be explained by a linear branching pathway.

As photosynthesis was evolving at the same time as the organisms carrying it out, this had a considerable effect on the Earth's atmosphere. About 2.7 billion years ago, Earth began to be dominated by cyanobacteria, causing an early rise of O_2 in the atmosphere. These cyanobacteria increased the O_2 level to around a quarter of the present level in roughly a billion years. As we saw in Chapter 1, the cyanobacteria not only had the advantage of photosynthesis in providing energy, but also inhibited anaerobic organisms by producing O_2, which was potentially toxic.

The eukaryotes appeared 1.8 billion years ago, but it was another 0.6–0.8 billion years before cyanobacteria became included in eukaryotic algae as chloroplasts. Starting around a billion years ago the algae began to increase the oxygen concentration in the atmosphere to the present levels.

7.5 Photorespiration

Photosynthesis is not really a perfect system, as it often wastes light and it does a poor job of fixing carbon. In 1920, Otto Warburg showed that increased amounts of oxygen inhibited photosynthesis in many plants.

One might think this unlikely to have much effect in real life, as the concentrations of oxygen and CO_2 in the atmosphere are relatively constant. Although the concentration of CO_2 in the atmosphere is relatively low at about 0.039 per cent (in 2010), there is effectively an unlimited supply of it, given the sheer volume of the atmosphere.

Inside a leaf, however, the situation is more complex, and CO_2 is only available to the plant when its stomata are open. This is possible on most days, but the situation inside a leaf when the weather is dry and hot is different. Under these conditions, the stomata close (see section 7.10 below). CO_2 concentrations inside the leaf consequently drop and the oxygen concentrations produced by photosynthesis may well rise. Rubisco then stops fixing CO_2 and starts to fix oxygen. The products of this oxygenase reaction are phosphoglycolic acid, a two-carbon compound, and PGA, a three-carbon compound. The PGA stays in the Calvin cycle, but the phosphoglycolic acid leaves and enters a complex series of reactions in the mitochondria and peroxisomes that result in the loss of CO_2.

7.5.1 The photorespiratory carbon oxidation cycle

We will now work through the reactions of the photorespiratory carbon oxidation (PCO) cycle (See Figure 7.10):

1 Ribulose bisphosphate oxygenase (chloroplast).

Oxygen is reacted with ribulose 1,5 bisphosphate to give phosphoglycerate (PGA) and the two carbon compound phosphoglycolate. This oxygenation reaction is a property of all Rubiscos, regardless of their origin. Even those isolated from anaerobic, autotrophic bacteria still have potential oxygenase activity. Carbon dioxide and oxygen compete at the same active site of the enzyme. Offered equal concentrations of CO_2 and oxygen in a test tube, angiosperm Rubisco fixes CO_2 about 80 times faster than it oxygenates. However, an aqueous solution in equilibrium with air at $25°C$ has a $CO_2 : O_2$ ratio of 0.0416. So in air carboxylation outruns oxygenation by only 3:1.

2 Phosphoglycolate phosphatase (chloroplast)

The PCO cycle then acts as a scavenger system. 2-Phosphoglycolate is rapidly converted to glycolate by a phosphatase enzyme. Further metabolism then occurs in the mitochondria and peroxisomes, which are often closely associated with chloroplasts in the cytoplasm. Glycolate moves from the stroma of the chloroplast across the inner envelope membrane and diffuses towards the peroxisome, while glycerate moves in the opposite direction (Figure 7.10). It seems that a single carrier-type transporter is involved in the movement of both glycolate and glycerate across the membrane. The glycolate/glycerate transporter is flexible and does not only catalyse a coupled substrate exchange, and unidirectional influx or efflux can also occur with proton symport or hydroxyl antiport (see Chapter 5, section 5.7.3).

3 Glycolate oxidase (peroxisome)

The glycolate is then oxidized by glycolate oxidase to glyoxylate and hydrogen peroxide (H_2O_2. Glycolate oxidase is an octamer composed of identical 40 kDa subunits. The spinach enzyme has been crystallized in an octameric form and the subunit contains an eight-fold ß/α barrel motif which corresponds to a flavin mononucleotide (FMN) domain. The reaction catalysed by glycolate oxidase is irreversible and can be divided into two parts. Firstly, the flavin which is deeply buried in the barrel, oxidises the glycolate. Then FMN is re-oxidized by oxygen producing H_2O_2. The enzyme active site is formed by loops at the carboxy-terminal end of the ß-strands in the barrel.

4 Catalase (peroxisome)

Hydrogen peroxide is toxic to plant cells and is usually rapidly destroyed by catalase, an enzyme that is often found in large amounts (even as crystals) in peroxisomes (Chapter 2, section 2.10). An iron-containing haem enzyme, catalase was the first antioxidant enzyme to be found and characterized. The catalase enzymes are constructed from 50–70 kDa polypeptides that are arranged into tetramers, with the individual monomers bearing haem prosthetic groups. Catalase promotes a reaction involving the dismutation of two molecules of H_2O_2 to water and oxygen. The enzyme is particularly important in removing H_2O_2 when plants are stressed. For details on the toxic effects of H_2O_2 and its production by plants under stress see Chapter 9, section 9.1, and for the methods by which plants deal with H_2O_2 see Chapter 10, section 10.1.

5 Glyoxylate: glutamate aminotransferase (peroxisome)

Glyoxylate is also a potentially toxic substance, and it then undergoes transamination. A transamination reaction involves the transfer of an amino group from an amino acid to the carboxyl group of a keto acid. These reactions are carried out by enzymes called aminotransferases. In this case the amino donor is glutamate, and the products are glycine and α-ketoglutarate. Glycine, the simplest of the amino acids, then moves into a mitochondrion (see Chapter 2, section 2.6).

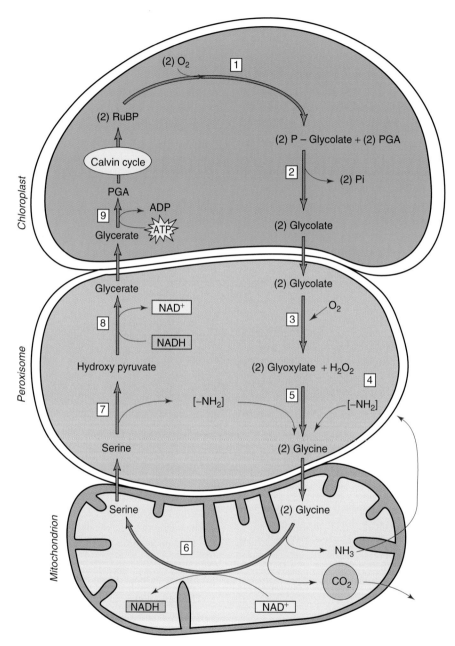

Figure 7.10 The photorespiratory carbon oxidation (PCO) cycle. Numbers refer to text in Section 7.5.1.

6 *Glycine decarboxylase and Serine hydroxymethytrans-*
ferase (mitochondrion)

In the mitochondrion, two enzymic reactions convert two molecules of glycine into one of serine. It is here that CO_2 is released and thus glycine is its immediate source. In this process considerable amounts of ammonia are released, but this is quickly assimilated by the enzymes described in Chapter 2, section 2.14.8. Serine then leaves the mitochondrion and travels back to the peroxisome.

7 *Serine aminotransferase (peroxisome)*

In the peroxisome serine is transaminated to hydrox-ypyruvate.

8 *Hydroxypyruvate reductase (peroxisome)*

Hydroxypyruvate is then reduced to glycerate, and this then leaves the peroxisome and enters the chloroplast using the transporter described in stage 2 above.

9 *Glycerate kinase (chloroplast)*

In the chloroplast glycerate is phosphorylated to give 3-phosphoglycerate (PGA), which can then rejoin the Calvin cycle, eventually regenerating RuBP (See Figure 7.9).

So, two molecules of 2-carbon phosphoglycolate (four carbons in total), which are lost from the C_3 carbon cycle by the oxygenase activity of Rubisco, are converted to one molecule of 3PGA (three carbons) and one of CO_2. Thus 75 per cent of carbon is recovered and returned to the C_3 cycle. This whole process is called photorespiration because it only occurs in the light, it consumes oxygen and releases CO_2. Unlike 'normal' respiration, photores-piration produces no ATP. In fact it actually consumes ATP and reducing power (NAD(P)H).

7.5.2 Why photorespiration?

So Rubisco only operates at about 25 per cent of its max-imal rate, and large amounts of the carbon that was fixed through the Calvin cycle are lost through photorespira-tion. This seems an incredibly wasteful process. Why has it evolved?

• Some have argued that the oxygenase reaction of Rubisco is inevitable – its active site cannot distinguish between CO_2 and oxygen.

• Another common suggestion has been that the enzyme evolved when oxygen levels in the atmosphere were low, so competition with oxygen did not matter. Green plants then produced oxygen and changed the atmosphere they grew in ('the great oxidative event' – see Chapter 1, section 1.2). The argument goes that by the time oxygen levels

had risen, the bifunctional nature of the enzyme was established. If that is the case, then photorespiration could be regarded as an unnecessary piece of evolutionary baggage.

• Others have argued that photorespiration may operate as a safety valve in situations that require dissipation of excess energy. The CO_2 generated by photorespiration is sufficient to protect the plant from photo-oxidative damage (Chapter 9, section 9.7.2) by allowing contin-ued operation of the electron transport chain. This could be very useful in conditions of high light and limited CO_2 supply (e.g. when stomata are closed due to mois-ture stress).

Plant scientists have been very interested in photores-piration because, if it could be blocked or inhibited, then crop yields of C_3 plants could be increased. They have tried to block the glycolate pathway and to use selective breeding for plants with low photorespiration. Others have surveyed large numbers of species in the hope of finding a Rubisco with a lower affinity for oxygen, but there has been little progress so far. This suggests that photorespiration may not be harmful to plants and that the idea may be a misconception.

One obvious way of decreasing photorespiration is to increase CO_2 concentrations at the active site of Rubisco. In C_3 land plants, we have seen that photorespiration is an apparently wasteful process. Particularly under warm conditions, where the stomata are closed, large amounts of carbon are lost. These conditions are prevalent where water is in short supply in arid or semi-arid conditions.

Later in this chapter we will consider C_4 and CAM plants (sections 7.7 and 7.8), which have evolved two different types of photosynthetic adaptation to dry envi-ronments. As we shall see, they have almost no photores-piration and have evolved mechanisms to concentrate CO_2. Before looking at C_4 and CAM plants, we will briefly consider the relationship between photosynthesis and transpiration.

7.6 The photosynthesis/transpiration dilemma

In Chapter 5, we saw that plants continually transpire large amounts of water into the atmosphere. A mature tree may lose more than 200 litres of water from its leaves on a hot, dry day. Of the water entering the roots of a plant, most is eventually lost in transpiration. The

transpiration ratio is the ratio of the mass of water transpired to the mass of dry matter produced. In most crops this ratio is between 200 and 1,000. Therefore, to produce 1 kg of dry matter crop, plants transpire between 200 and 1,000 kg of water.

This brings us onto the phenomenon known as the photosynthesis-transpiration dilemma. Why is so much water lost by transpiration to grow a plant? The reason is because the molecular skeletons of virtually all organic matter in plants consist of carbon atoms that must come from the atmosphere. They enter the plant as CO_2 through stomatal pores, mostly on leaf surfaces, and water exits by diffusion through these same pores, as long as they are open. Thus it could be said that the plant faces a dilemma: how to get as much CO_2 as possible from an atmosphere in which it is extremely dilute (about 0.039 per cent in 2010), while at the same time retaining as much water as possible.

Carbon dioxide has not always been in such short supply; the concentration in the Earth's early atmosphere has been estimated at about 30 per cent. At the same time, free oxygen has risen from almost zero to the present level of approximately 21 per cent. These changes suggest an evolutionary need for increased efficiency of carbon fixation relative to water loss, to cope with the diminishing carbon supply, and perhaps some adjustment to high concentrations of gaseous oxygen.

7.7 C_4 photosynthesis

So plants face a dilemma, whereby opening the stomata allows in CO_2 but leads to the loss of water. Conversely, closing the stomata cuts down water loss but means that CO_2 uptake is restricted. All plants have to deal with this problem to some extent but, for those living in arid environments, the difficulty is exacerbated. Soon after the discovery of the Calvin cycle, plant physiologists started to try to find it in many plant species. In most they were successful, finding that carbon was first fixed into C_3 compounds – but in some plants they found that fixation happened first into C_4 compounds. These C_4 plants include some of the important semitropical crops, such as maize (Figure 7.11), sorghum and sugar cane.

7.7.1 C_4 plant anatomy

One of the features of most (but not all) C_4 plants is that they have a different type of leaf anatomy from the C_3

Figure 7.11 Maize or 'corn' (*Zea mays*), the most important C_4 crop. Here the plants were photographed in the summer in Odcombe, Somerset, England.
Photo: MJH.

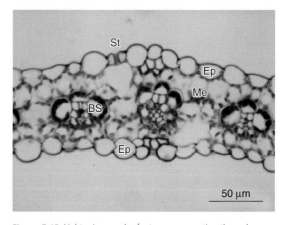

Figure 7.12 Light micrograph of a transverse section through a maize or corn (*Zea mays*) leaf. Key: Epidermis (Ep), Stoma (St) Bundle Sheath (BS), Mesophyll (Me).
Photo: Prof. Thomas Rost.

plants (Figure 7.12). In C_3 plants, the leaves have only one type of photosynthetic tissue – the mesophyll – although that is often sub-divided into palisade and spongy mesophyll. In many C_4 leaves, the vascular bundles are quite close together and each bundle is surrounded by a tightly fitting layer of cells called the *bundle sheath*. The bundle sheath chloroplasts have a different structure to those of the mesophyll, and they have no grana. This distinction between two types of photosynthetic tissues – the mesophyll and the bundle sheath – is called

Kranz anatomy, and it plays a major role in the C_4 syndrome in many plants.

7.7.2 The biochemistry of C_4 photosynthesis

In C_3 plants, three-carbon compounds were found to accumulate soon after exposure to light, while C_4 compounds only accumulated much later. In contrast, when experiments were conducted with sugar cane or maize, it was C_4 compounds that first accumulated, while C_3 compounds only appeared later. After much work, the general cycle shown in Figure 7.13 emerged.

The key enzyme in the C_4 cycle is phosphoenolpyruvate carboxylase (PEP carboxylase), which catalyses the β-carboxylation of phosphoenolpyruvate (PEP) using the bicarbonate ion, HCO_3^-, as the substrate (not CO_2, as in C_3 photosynthesis). This occurs in the mesophyll of C_4 plants. The product, oxaloacetate, being moderately unstable, is quickly reduced to a more stable C_4 acid such as malic acid, which then moves into the bundle sheath, where it is decarboxylated. The CO_2 released is then taken into the Calvin cycle, which then proceeds in the way described in Section 7.4.5. The three-carbon (C_3)

compound, pyruvic acid, then returns to the mesophyll, where PEP is regenerated at the expense of ATP.

The C_4 cycle does not *of itself* result in the fixation of carbon, as one CO_2 is fixed and one is lost in the cycle. What it does do is to concentrate CO_2 in the bundle sheath cells. Carbon dioxide concentrations in the bundle sheath cells may reach $60\,\mu M$ – about ten times higher than in C_3 plants. This suppresses photorespiration and supports higher rates of photosynthesis. Under optimal conditions, C_4 crop species can assimilate CO_2 at two to three times the rate of C_3 species. This does have a cost though and, for each CO_2 assimilated, two ATP molecules must be expended.

7.7.3 Variations in the C_4 cycle

What we have said so far hides a more complex situation and there are, in fact, three types of C_4 cycle. The type described above is known as the *NADP-Malic enzyme type*, where malate is transported to the bundle sheath, decarboxylation happens in the chloroplast of the bundle sheath, and pyruvate is the three-carbon compound translocated back to the mesophyll. In the *NAD-Malic enzyme type*, aspartate is translocated into the bundle sheath, decarboxylation occurs in the mitochondria and

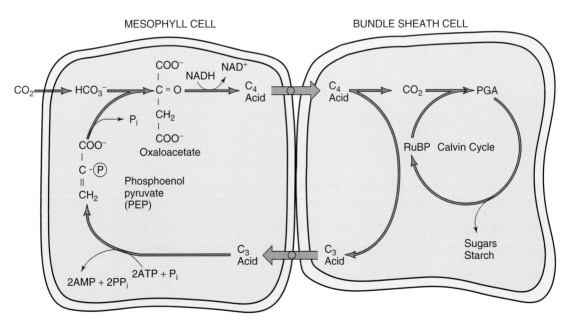

Figure 7.13 General plan of the photosynthetic C_4 cycle.

alanine is translocated back to the mesophyll. Finally, in the *PEP carboxykinase type*, aspartate is translocated into the bundle sheath, decarboxylation occurs in the chloroplast and alanine is translocated back to the mesophyll. There is no great taxonomic distinction between these three variants and, for example, the genus *Panicum* has species expressing all three types.

Thus far we have only been concerned with C_4 photosynthesis in vascular plants, but we should briefly mention that it has also been found in unicellular diatoms such as the marine diatom *Thalassiosira weissflogii*. Unicellular C_4 photosynthesis in diatoms may well have predated the appearance of multicellular C_4 plants. At least two vascular plants (*Bienertia cycloptera* and *Borszczowia aralocaspica* in the Chenopodiaceae) have been found to be 'single-celled' C_4 species, where the whole machinery for C_4 photosynthesis can be found in compartments within individual cells.

C_4 photosynthesis can also be environmentally induced and, in several species of the genera *Flaveria, Mollugo, Moricandia* and *Zea*, both types of CO_2 fixation occur within one plant. Usually, the C_3 pathway is dominant in younger plants, while in older ones the C_4 pathway is taken. The amount of C_4 photosynthesis is controlled by environmental factors. This phenomenon is even more pronounced in CAM plants (see section 7.8.3).

7.7.4 Comparing C₃ and C₄ plants

It is interesting to compare the physiology of C_3 and C_4 photosynthesis (See Table 7.1).

1 *Photorespiration*. Although C_4 plants seem to have all the machinery for photorespiration, including peroxisomes and the necessary enzymes, they have very low rates, probably due to the high concentrations of CO_2 in the bundle sheath cells. This would lead to CO_2 out-competing oxygen for binding to Rubisco.

2 *CO_2 compensation point*. This is the CO_2 concentration at which rate of CO_2 uptake by photosynthesis is balanced by rate of CO_2 evolution by respiration. In a closed environment, CO_2 concentrations at equilibrium are about $20–100\,\mu l\ CO_2\ l^{-1}$ for C_3 plants while those for C_4 plants are $0–5\,\mu l\ CO_2\ l^{-1}$. This well illustrates the ability of C_4 plants to absorb and concentrate CO_2.

3 *The temperature optimum*. Most C_4 plants have a higher temperature optimum for photosynthesis than C_3 plants. This is partly due to the greater sensitivity of C_3 enzymes to temperature. This leads to the quantum yield (moles of CO_2 per absorbed mole of photons) as a function of temperature declining in C_3 plants, while in C_4 plants this remains stable.

4 *Transpiration ratio*. C_4 plants have lower transpiration ratios than C_3 plants, and therefore they can maintain high rates of photosynthesis while conserving water. C_4 plants are able to close their stomata partially and still take in enough CO_2 as a result.

5 *Light saturation*. Even under ideal conditions, CO_2 supply limits photosynthesis in C_3 plants to the extent that light saturation occurs at about 25 per cent of full sunlight. On the other hand, C_4 plants are not saturated even at full sunlight. In C_3 plants, insufficient CO_2 is a

Table 7.1 A comparison of significant features of C_3 and C_4 plants.

	C_3	C_4
Photorespiration	Yes	No
CO_2 compensation ($\mu l\ CO_2\ l^{-1}$)	20–100	0–5
Temperature optimum	°C	
Photosynthesis	20–25	30–45
Rubisco	20–25	
PEPcase		30–35
Quantum yield as a function of temperature	Declining	Steady
Transpiration ratio	500–1000	200–350
Light saturation (μmole photons)	400–500	Does not saturate
Carbon isotope discrimination	Favours C^{12} isotope	Favours C^{13} isotope

Table modified after Fitter and Hay (2002).

common cause of suboptimal photosynthesis, especially in leaves exposed to bright light. In contrast, some C_4 plants are saturated even at present CO_2 levels.

6 *Carbon isotope discrimination.* The ratio of ^{13}C to ^{12}C is somewhat higher in plants using C_4 photosynthesis than in plants employing C_3 photosynthesis. This finding has had major applications in archaeological and palaeoecological work, as it is possible to determine whether the carbon in samples originated from C_3 or C_4 plants. This can give some information on past climates, as C_3 plants tend to grow in less arid conditions than C_4 plants.

In conclusion, given the right conditions of high temperature, high light levels and low water availability, C_4 plants have a definite advantage over C_3 plants. However, C_4 plants are not competitive in all situations and, at temperatures below about $30°C$, it is actually the C_3 plants that have a higher quantum yield.

7.7.5 Evolution of C_4 photosynthesis

The C_4 syndrome is now known to occur in around 7,500 species of flowering plants, mostly of tropical or subtropical origin, or about 3 per cent of the 250,000 total. The grasses are the dominant C_4 plants, with lesser numbers of sedges and dicots. The syndrome is spread through 19 different angiosperm families (three monocots, 16 dicots), all of which have representatives of both C_3 and C_4 plants. This would strongly suggest that C_4 has arisen recently in the evolution of the angiosperms, and on many different occasions.

How and when did C_4 photosynthesis evolve? It appears that the C_4 syndrome evolved separately at least 45 times, and it is a classic example of convergent evolution. For many years, C_4 photosynthesis has been seen as an adaptation to hot arid conditions and to low CO_2 availability. It is thought that C_4 photosynthesis first evolved in grasses, around 24–35 million years ago, during the Oligocene. The Chenopodiaceae were probably the first eudicots to have C_4 photosynthesis, about 15–21 million years ago, although many eudicots seem to have developed the C_4 trait much more recently. The simple picture that C_4 photosynthesis evolved in response to aridity and low atmospheric CO_2 concentrations has been challenged as too simplistic, and it is now thought that many other factors are involved.

In the last 150 years, atmospheric CO_2 has risen rapidly due to anthropogenic emissions – in particular, the burning of fossil fuels. This trend looks set to continue, very likely causing warmer conditions and more extreme weather events. We will consider the response of C_4 (and C_3) plants to climate change in Section 12.1.

7.7.6 Engineering C_4 photosynthesis into C_3 plants

We should also mention that there has been some interest in the idea of introducing the machinery for C_4 photosynthesis into C_3 plants, thus hopefully gaining greater efficiency in harvesting CO_2 from the atmosphere and tolerance of hot, arid conditions. Initial optimism has somewhat faded, as the difficulties of moving large suites of genes coding for anatomical and biochemical changes have become apparent.

However, an international consortium is currently working on the introduction of C_4 photosynthesis into rice. One of their approaches is to use metabolic engineering to produce a two-celled C_4 shuttle in rice by expressing the enzymes of the C_4 cycle in the correct cells. Rubisco expression will be restricted to the bundle sheath cells of rice by down-regulating its expression in the mesophyll cells.

There have been some suggestions that it might be easier to transfer 'single-celled' C_4 photosynthesis, of the type found in *Borszczowia*. However, there are concerns that the single-cell system is not efficient and only appears to be useful in extreme conditions.

7.8 Crassulacean acid metabolism (CAM)

Some plants living in extremely arid environments have developed a special type of photosynthesis known as *crassulacean acid metabolism* (CAM).[i] CAM is more common than C_4 and occurs in 20 families, including monocots, dicots and more primitive plants (e.g. *Welwitschia mirabilis*, a member of the Gnetales, a gymnosperm family, has CAM but appears not to use it very much – see Chapter 10, Figure 10.9). About 10 per cent of plants are CAM.

CAM is believed to have had polyphyletic origins in the Miocene period and, like C_4, it is held to be the consequence of reduced CO_2 levels. Most CAM plants are succulents (Figure 7.14), but there are exceptions.

---

[i]For a detailed review of the history of CAM research, see Black & Osmond (2003).

Figure 7.14 Common ice plant (*Mesembryanthemum crystallinum*), a succulent with CAM photosynthesis. Here photographed near the coast of the Alvor Estuary, the Algarve, Portugal. Photo: MJH.

For example, pineapple is CAM, but not succulent, while many halophytes are succulent, but not CAM. Although CAM is relatively common, the only plants of agricultural significance are pineapple and *Agave americana* (which is used to make tequila and as a source of fibres).

7.8.1 Biochemistry of CAM

The key problem for CAM plants is loss of water during the day, and they have developed a peculiar adaptation of closing their stomata in daylight hours. In the dark, CO_2 dissolved as HCO_3^-, is first fixed by PEP carboxylase (phosphoenyl pyruvate carboxylase). PEP is converted to the C4 acid oxaloacetic acid and then to malic acid. The malic acid is then transported to the cell vacuoles and stored there until daylight, resulting in a marked acidification of the leaves at night. During the day, the stomata close and malate is released from the vacuole and broken down to yield CO_2. The CO_2 is rapidly incorporated by the Calvin cycle.

Thus, in CAM plants, all the reactions occur in the same cells and the separation is temporal (Figure 7.15), but in almost all C_4 plants, the initial fixation of CO_2 occurs in different cells to the Calvin cycle and the separation is spatial.

7.8.2 Advantages and disadvantages of CAM

The advantage of CAM is that stomata open in the dark, leading to less water loss. CAM is an adaptation to extreme drought, and CO_2 uptake happens when water loss is minimal. CAM plants have a low transpiration ratio of 50–100 (compare this with Table 7.1 above, where we saw C_4 plants at 200–350 and C_3 plants at 500–1000). Therefore, they are very efficient in their use of water. They also retain respired CO_2, so they do not lose carbon that has been previously fixed. In the light, Rubisco is saturated and there is no photorespiration.

CAM does have some disadvantages, though. Rates for daily carbon assimilation are about half those of C_3 plants and about one-third those of C_4 plants. This suggests that they will show slow growth even when adequately watered, and this is usually the case. Overall, CAM plants are very good at surviving in hot arid environments, but they have slow growth and low yields, which is probably why so few are grown as crops.

7.8.3 Variations in CAM

The description of CAM above is an oversimplification, and its biochemistry varies quite considerably between taxa. CAM can be constitutive and can be expressed whatever the environmental conditions or it can be facultative. Thus, in some species like the ice plant (*Mesembryanthemum crystallinum*), CAM can be induced by drought or salinity. The ice plant operates as a C_3 plant under conditions of high water availability, but switches to CAM when drought conditions click in (see Chapter 10, section 10.4.2).

7.9 Sources and sinks

In Chapter 6, section 6.9 we noted that mature leaves were a 'source' of sucrose that was synthesized in the cytosol as a major product of photosynthesis (after export of triose phosphate from the chloroplast), and that this was transported in the phloem to roots, developing seeds and

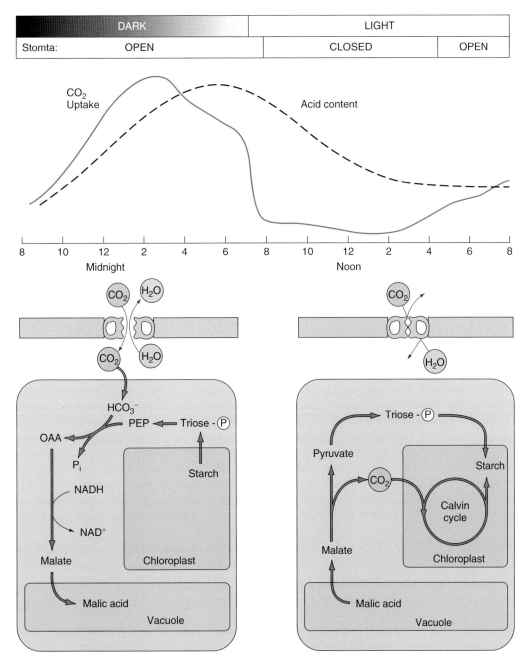

Figure 7.15 The biochemistry of crassulacean acid metabolism. The curves above show carbon dioxide uptake and acidity of the cell sap over 24 hours. Below, left: the stomata are open at night to allow carbon dioxide entry, then this is then fixed into malic acid. Below, right: the stomata are closed to conserve water in the day and stored carbon dioxide is released to be assimilated via the Calvin Cycle.
From Hopkins, 1999. Used with the permission of John Wiley publishers.

young leaves, which were 'sinks'. It is now recognized that the source and the sinks do not operate independently and that there is much interaction between them.

As long ago as 1868, Boussingault proposed that the accumulation of assimilates in the leaf has a role in the regulation of photosynthetic rate, but it is only fairly recently that the mechanisms behind this effect have been elucidated. Obviously, any factor that affects the concentration of assimilates in the leaves will affect photosynthesis. Thus, plants growing in shaded environments tend to change their photosynthetic apparatus to optimize light usage (Chapter 10, section 10.7.1).

When high levels of assimilates accumulate in the leaf, two responses are possible: feedback inhibition of photosynthesis, or the development of new sinks to consume the excess sugar. Both short- and long-term feedback responses are involved. Over a short time period, increased sugar concentrations can result in redox signals, due to changes in the redox state of plastoquinone, which quickly reduce the transcription of genes encoding for photosystems I and II. Another short-term mechanism involves the incorporation of inorganic phosphate (Pi) into the triose phosphate end-products of photosynthesis (e.g. glyceraldehyde 3-phosphate, G3P), and Pi has to be recycled back to the chloroplasts and the photosynthetic reactions. Anything limiting this recycling will decrease photosynthesis, and this will occur if less Pi is released from the triose phosphates due to decreased breakdown.

Over the longer term, imbalance between source and sink in the plant can lead to sugar accumulation in the leaves, causing a decreased expression of photosynthetic genes.

7.10 Stomata

Stomata are pores bordered by a pair of guard cells in the leaf and stem epidermis. Each stoma can regulate the size of the pore opening, and they are responsible for controlling gaseous exchange. Stomata take a vast number of shapes and sizes. The typical eudicot has kidney-shaped guard cells (Figure 7.16), while monocots have dumbbell-shaped cells (Figure 7.12 shows a maize stoma in transverse section).

The sporophyte generation of all land plants, except the liverworts, have stomata. Most eudicots have more stomata on the lower than on the upper epidermis.

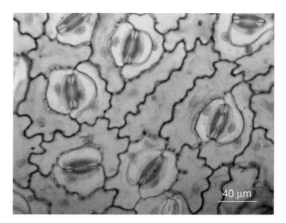

Figure 7.16 *Magnolia* sp. leaf peel showing kidney-shaped stomata.
Photo: Prof. Thomas Rost.

However, monocots usually have similar numbers of stomata on both surfaces. In hydrophytes with floating leaves, such as the water lily, stomata are often present only on the upper epidermis, while submerged leaves have no stomata (Chapter 10, section 10.3.2).

The stomata are responsible for controlling the amount of water lost by the leaf in transpiration, and CO_2 enters the plant through these openings before being used in photosynthesis. Oxygen produced by photosynthesis in the mesophyll exits through these stomata. Plants may also exert control over gaseous exchange by varying stomatal density when new leaves are produced; the greater the stomatal density, the more CO_2 can be taken up and the more water is lost. A number of environmental factors affect stomatal density, including water availability, light and atmospheric CO_2 concentration.

Water movement through the plant to the atmosphere is often said to travel through a *soil-plant-atmosphere continuum* (SPAC). As we saw in Chapter 6, section 6.8.1, the water potential of the atmosphere is approximately −50 MPa at 70 per cent relative humidity. Soil water potential is approximately zero at saturation and below −10 MPa in dry soil. Water moves down a gradient of water potential from the soil to the roots, to the stem, to the leaves, and on out into the atmosphere.

Control of water movement largely relies on the external layers of the plant and is affected by means of the stomatal, cuticular and boundary layer resistances. The boundary layer can be defined as the distance in the atmosphere through which water molecules must diffuse before their concentration reaches that of the atmosphere as a whole. This layer acts as a blanket over the surface of the leaf, reducing the amount of water loss.

The primary mechanism to reduce transpiration is stomatal closure, which occurs during drought (Chapter 9, section 9.4). As we have seen, however, the problem is that this inhibits CO_2 uptake and photosynthesis. When the stomata are open, transpiration may be appreciably reduced by the boundary layer. Adaptations that increase its thickness or protect it from wind are beneficial (Chapter 10, section 10.4), especially to plants growing in exposed situations. When the stomata are closed, water conservation will depend primarily on the permeability of the waxy cuticle, although the presence of a boundary layer should also reduce cuticular transpiration.

Stomatal aperture responds to several environmental factors. In most plants, the stomata open at sunrise and close in the darkness, allowing the entry of carbon dioxide used in photosynthesis during the daytime. The key exceptions to this are the CAM plants (see section 7.8). Day and night are not the only factors to influence stomatal aperture; high irradiance often causes wider opening. In almost all plants, low CO_2 concentrations cause the stomata to open, while a high CO_2 level causes closure. However, the amount of closure depends on the type of plant analysed, with forb, shrub and tree species being less affected than other species. Water stress, high temperature and increased wind also generally lead to stomatal closure, and we will return to these topics in Chapters 9 and 10.

7.10.1 Solutes and the control of stomatal aperture

The guard cells control the diameter of the stomatal pore by changing shape, widening or narrowing the gap between the two cells. When guard cells take in water by osmosis, they become more turgid and swell. In most eudicots, the cell walls of the guard cells are not uniformly thick, and the cellulose microfibrils are oriented in a direction that causes the guard cells to buckle outwards when they are turgid. This increases the size of the gap between the cells. When the cells lose water and become flaccid, they sag together and close the space between them.

So stomatal opening involves water movement into the guard cell, but what brings that about? The earliest idea

was proposed by Lloyd in 1908, who observed that starch decreased in guard cells during the day when the stomata were open, and increased at night when they were closed. This led to the starch-sugar hypothesis, where starch was converted to sugars in the day. This would cause water to flow into the guard cells, changing cell turgor and causing stomata to open. This starch-sugar hypothesis was the dominant idea for many years.

However, in the late 1960s, using specific cytochemical stains for potassium, it was found that reversible uptake of K^+ ions into the guard cells from neighbouring epidermal cells was involved and the starch-sugar hypothesis went into retreat. The uptake of potassium causes the water potential of the guard cells to become more negative, and the cells become more turgid as water follows by osmosis. Most of the potassium and water are stored in the vacuoles of the stomatal guard cells.

Once it was clear that potassium was involved in stomatal opening, the next question was what causes potassium to move? As in most cells, the accumulation of ions in stomata is driven by an ATP-powered proton pump located at the plasma membrane (see Chapter 5, section 5.7.3). By removing positive ions, proton extrusion hyperpolarizes the plasma membrane, producing a lower electrical potential inside the cell than outside, and also creating a pH gradient (Figure 7.17). This hyperpolarization opens potassium channels in the membrane,

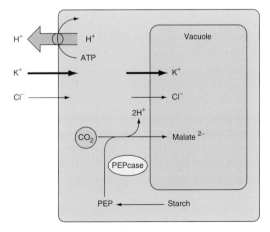

Figure 7.17 A simple model showing ion flow in guard cells that occurs during stomatal opening.
From Hopkins, 1999. Used with the permission of John Wiley publishers.

which allows the passive uptake of potassium in response to the potential difference across the membrane.

The maintenance of electrical neutrality in stomatal guard cells as large amounts of K^+ enter also needs consideration. It seems that the mechanism varies with the species, but both chloride and malate accumulation is involved. Guard cells have often been found to contain high levels of the enzyme PEP carboxylase, which is responsible for a key step in the production of malate. It is now thought that malate is the major counterion involved in most species, but there are some, such as onion (*Allium cepa*), where malate accumulation is absent, and chloride is used exclusively. However, in many species, it now appears that K^+, malate and chloride do not provide all the osmoticum required in explaining stomatal opening. It is thought that sucrose makes up the difference, so the old starch-sugar hypothesis has been (partially) revived.

7.10.2 Light and stomata

While changes in solute concentrations in guard cells are clearly involved in the mechanism of stomatal opening and closure, light (particularly red and blue light) is now recognized to play a key role in many of the processes behind solute accumulation and transport. We saw in Chapter 3 (section 3.7.5) that the phototropins (phot1, phot2), are involved in phototropic responses, and it is these blue-light receptors that also have a major part in stomatal opening. Weak blue light leads to stomatal opening, and this effect is increased by red light. In blue light, the phototropin receptor activates the plasma membrane proton ATPase in the guard cells by a mechanism that is not well characterized, but seems to be mediated by a type 1 protein phosphatase. As we noted above, this leads to K^+ uptake through voltage-gated K^+ channels.

The response of stomata to blue light seems to be strongly affected by red light, which induces stomatal opening at high intensity. Red light probably affects stomatal opening as mesophyll photosynthesis reduces the intercellular concentration of CO_2. Guard cells are unusual in that, unlike most other epidermal cells, they have chloroplasts. In many species, the effect of red light in increasing the response of stomata to blue light involves the guard cell chloroplasts. These probably provide both energy, as ATP for the proton ATPase in the plasma membrane, and triose phosphates, which

can be translocated across the chloroplast membrane and provide the precursors in the cytoplasm for the formation of malate and sucrose.

7.11 Leaf senescence and abscission

In Chapter 3 (section 3.5.2) and Chapter 6 (section 6.7.1), we considered senescence at the cellular and whole plant levels, and here we will investigate this process at the organ level. Leaf senescence is a synchronized process that responds to many developmental and environmental signals. There are essentially four ways that cells die in plants, and we will briefly consider the first three before concentrating most attention on the fourth.

In *perturbed development*, the cell dies due to a factor such as a genetic lesion interrupting growth. In *environmental stress*, cell death could be caused by physical wounding or by the attempted invasion of a pathogen (see Chapter 11). *Necrosis* may occur after severe trauma due, for example, to extreme heat or freezing stress (see Chapter 9). In *Programmed Cell Death* (PCD), sometimes known as apoptosis, death is a normal part of plant development; we covered this at the cellular level in Chapter 3 (section 3.5.2).

Leaf senescence involves age-dependent PCD, often as the days become shorter in the autumn. Organ senescence and other PCD events have universal, in addition to very characteristic, signalling pathways. Certain plant hormones have been implicated as regulators or modulators of organ senescence (e.g. auxin, cytokinin), and many of the biochemical pathways associated with the senescence syndrome have been elucidated.

In Arabidopsis, more than 800 genes have been recognized as senescence-associated genes (SAGs), reflecting the enormous changes that occur during leaf senescence. Two regulators of leaf senescence are the transcription factors WRKY53 and AtNAP. Cytokinins have long been known to delay leaf senescence and AHK3, a cytokinin receptor, has now been accepted as a key molecule controlling this phenomenon.

In Chapter 3 (section 3.4.5), we considered the proteasome pathway and the small, highly conserved protein called ubiquitin that is involved in protein degradation. Ubiquitin-dependent proteolysis is involved in the control of leaf senescence. Ubiquitin is covalently linked to substrate proteins, targeting them for degradation in the proteasome pathway. The role of ubiquitin during senescence is bulk protein degradation for nitrogen recycling.

The most well-known senescence phenomenon is the dropping of leaves from deciduous trees in the autumn by abscission. This may also occur in flowers and fruits, and in response to environmental factors such as drought (Chapter 10, section 10.4.1) and salinity (Chapter 10, section 10.5.2). The loss of leaves each autumn is an adaptation that keeps deciduous trees from desiccating during winter when roots cannot take up water from the frozen ground. Before the leaves abscise, many of their essential elements are removed back to storage tissues in the stem, and they are then re-used again the following spring.

In the autumn, the leaf stops making new chlorophyll and loses its green colour. Autumn colours are due to a combination of new pigments made at that time and those already present that were concealed by the chlorophyll. In the United Kingdom, one of the best places to see good autumn colours is Westonbirt Arboretum, and we included one photograph earlier in this chapter (Figure 7.2). The forests of the eastern parts of the USA and Canada are also famous for their fall (autumn) colours, with our example coming from West Virginia (Figure 7.18).

Abscission is a rupture along a definite line of cells across the base of a leaf petiole. The cells in the separation zone are often morphologically different before abscission. The small square-shaped parenchyma cells of this layer usually have dense cytoplasm and very thin cell walls. These cells seem to have lost the ability to expand and to undergo vacuolation as a normal part of development, unlike the cells in the surrounding area.

Species vary in the number of cells within the abscission zone; in tomato, eight to ten cell layers are involved, while in *Sambucus nigra* this number can reach 50. Just before leaf drop, the abscission layer is further weakened by enzymes such as cellulase and polygalacturonase hydrolysing the polysaccharides in the cell walls. The weight of the leaf, plus the effect of the wind, causes a separation within the abscission layer. Before this happens, a layer of cork is deposited on the twig side of the abscission layer to prevent pathogens entering through the scar.

It has been known for many years that the balance between plant hormones controls abscission. In young

Figure 7.18 Sugar maples (*Acer saccharum*) in autumn colours. Photographed at Kanawha State Forest near Charleston, West Virginia, USA.
Photo: JAB.

leaves, auxin and cytokinin levels are high, while ethylene and ABA are low. The levels of these hormones, and the sensitivity of the cells in the abscission zone to them, are both important in determining whether and when leaf drop will occur. Ageing leaves produce less auxin, and this fall in auxin concentration makes the abscission layer more sensitive to ethylene. This is reinforced by the cells producing more ethylene of their own.

Much recent work on ethylene signalling pathways has shown that abscission depends on intermediates that are analogous to those used by other processes that are affected by ethylene and auxin. Homologues of the ETR1 gene (see Chapter 3, section 3.6.6) appear to be implicated in transducing the ethylene signal during the process of abscission, but how auxin interacts with these is still the subject of research.

Selected references and suggestions for further reading

Ainsworth, E.A. & Rogers, A. (2007) The response of photosynthesis and stomatal conductance to rising [CO_2]: mechanisms and environmental interactions. *Plant, Cell and Environment* **30**, 258–270.

Black, C.C. & Osmond, C.B. (2003) Crassulacean acid metabolism photosynthesis: 'working the night shift.' *Photosynthesis Research* **76**, 329–341.

Blankenship, R.E. (2002) *Molecular Mechanisms of Photosynthesis.* Blackwell Publishing, Oxford.

Campbell, N.A., Reece, J.B. & Mitchell, L.G. (1999) *Biology.* Addison Wesley Longman Inc., Menlo Park, CA.

Dodd, A.N., Borland, A.M., Haslam, R.P., Griffiths, H. & Maxwell, K. (2002) Crassulacean acid metabolism: plastic, fantastic. *Journal of Experimental Botany* **53**, 569–580.

Edwards, E.J., Osborne, C.P., Strömberg, C.A.E., Smith, S.A. & C_4 Grasses Consortium (2010) The origins of C_4 grasslands: Integrating evolutionary and ecosystem science. *Science* **328**, 587–591.

Fitter, A.H. & Hay, R.K.M. (2002) *Environmental Physiology of Plants.* Academic Press, London.

Friedman, W.E., Moore, R.C. & Purugganan, M.D. (2004) The evolution of plant development. *American Journal of Botany* **91**, 1726–1741.

Hopkins, W.G. (1999) *Introduction to Plant Physiology*, 2nd Ed. John Wiley & Sons, New York.

Kajala, K., Covshoff, S., Karki, S., Woodfield, H., Tolley, B.J., Dionora, M.J.A., Mogul, R.T., Mabilangan, A.E., Danila, F.R.,

Hibberd, J.M. & Quick, W.P. (2011) Strategies for engineering a two-celled C4 photosynthetic pathway into rice. *Journal of Experimental Botany* **62**, 3001–3010.

Lawson, T. (2009) Guard cell photosynthesis and stomatal function. *New Phytologist* **181**, 13–34.

Lim, P.O., Kim, H.J. & Nam, H.G. (2007) Leaf senescence. *Annual Review of Plant Biology* **58**, 115–136.

Paul, M.J. & Foyer, C.H. (2001) Sink regulation of photosynthesis. *Journal of Experimental Botany* **52**, 1383–1400.

Raven, P.H. & Johnson, G.B. (2002) *Biology*, 6th ed. McGraw Hill, New York.

Reinfelder, J.R., Kraepiel, A.M.L. & Morel, F.M.M. (2000) Unicellular C–4 photosynthesis in a marine diatom. *Nature* **407**, 996–999.

Sage, R.F. (2004) The evolution of C4 photosynthesis. *New Phytologist* **161**, 341–370.

Shimazaki, K-I., Doi, M., Assmann, S.M. & Kinoshita, T. (2007) Light regulation of stomatal movement. *Annual Review of Plant Biology* **58**, 219–247.

Taylor, J.E. & Whitelaw, C.A. (2001) Signals in abscission. *New Phytologist* **151**, 323–339.

Tsiantis, M. & Hay, A. (2003) Comparative plant development: the time of the leaf? *Nature Reviews Genetics* **4**, 169–180.

Van Volkenburgh, E. (1999) Leaf expansion – an integrating plant behaviour. *Plant, Cell and Environment* **22**, 1463–1473.

Xiong, J. & Bauer, C.E. (2002) Complex evolution of photosynthesis. *Annual Review of Plant Biology* **53**, 503–521.

CHAPTER 8

Flowers

8.1 Introduction

In Chapter 4 we discussed the earliest stages in the life of a plant, starting with the one-cell embryo and moving from there via embryogenesis, seed development and germination to seedling establishment. In passing, it was noted that the way that seeds are borne on the parent plant gives rise to the taxonomic name angiosperms – 'vessel seeds'. The seeds are contained in a 'vessel' – the *carpel* – which is part of the flower, the unique sexual reproductive structure of the angiosperms. This leads to the completion of the developmental 'circle' that we broke into in Chapter 4, with the carpel providing a link to the feature that gives rise to the angiosperms' common name, *flowering plants*.

The huge array of growth forms within the angiosperms, from the tiny plants in genus *Wolffia* (in the duckweed family) to huge trees, has already been noted in Chapter 1 and there is an equal range in floral characters. *Wolffia* species (Chapter 1, section 1.7) have the smallest flowers known; these consist of one female reproductive organ (carpel) and one or two (depending on species) male reproductive organs (stamens), situated in a groove or depression on a plant that may be less than one millimetre in diameter (Chapter 1, Figure 1.8). At the other end of the scale, the flowers of the parasitic plant *Rafflesia arnoldii* (Figure 8.1) are up to a metre in diameter and weigh up to 11 kg! In between, there is an astonishing range of both floral anatomy and floral colour, often related to pollination mechanisms.

8.2 What is a flower?

Despite the wide range of flower types, we can state that all flowers have the same basic developmental history.

Essentially, they are all shoots with limited growth – that is, shoots in which the shoot apical meristem is converted into a floral meristem. This involves changes in the phyllotaxis of the primordia arising on the flanks of the meristem (see Chapter 6, section 6.3), leading to the formation of the whorls of floral organs (see below). Cell division then ceases, so the floral meristem is a determinate structure.

Botanists often state that the original apical meristem 'is consumed in the formation of the flower'. During this process, the shoot tip itself becomes swollen, forming the receptacle. Thus, in the fully formed flower, the whorls of floral organs appear as outgrowths from the receptacle. The stem which terminates in the flower is the peduncle (for a single flower) or pedicel (for flowers that are part of more complex **inflorescences**–see below).

In a typical flower, four whorls of organs may be distinguished (Figure 8.2). The outermost (and therefore basal) whorl consists of the *sepals* (known collectively as the **calyx**). In the formation of the flower, these are the first organs to appear and remain closed over the developing flower during the subsequent stages of floral development and maturation. Thus, when we speak of a 'flower bud', we are referring to the developing flower enclosed by the whorl of (usually) green sepals. Inside the sepals are the *petals* (the complete whorl being called the **corolla**). Petals are generally more delicate than the sepals and their inner surface often feels velvety to the touch. This is caused by the presence of dome-shaped cells called papillae, which increase colour reflection and may also secrete chemical attractants (see section 8.6). In the majority of eudicots and many monocot species, petals are not green. Indeed, it is the huge range of colours and colour combinations that make the angiosperms so attractive to many members of the human species – although, of course,

Functional Biology of Plants, First Edition. Martin J. Hodson and John A. Bryant.
© 2012 John Wiley & Sons, Ltd. Published 2012 by John Wiley & Sons, Ltd.

Figure 8.1 Flower of *Rafflesia arnoldii*, the largest flower in the world.
Photographed at Kuching, Sarawak, Malaysia. Photo: Chim Chee Kong.

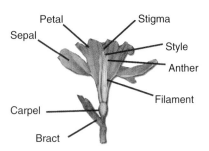

Figure 8.2 Diagrammatic longitudinal section of typical insect-pollinated flower.
From Scott, P.(2008) Physiology and Behaviour of Plants, Wiley, Chichester.

their primary function relates to pollination mechanisms (section 8.6).

The outer two whorls just described are the sterile parts of the flower, although of course they play key roles in floral function. By contrast, the inner two whorls are the reproductive organs in which the gametes are formed. The outer of these two consists of the male organs, the *stamens*, known collectively as the **androecium** (which may be translated as the *house of man*). Stamen numbers are very variable between species, ranging from just one to several hundred. The largest arrays of stamens generally occur in wind-pollinated species which, of necessity, produce large amounts of pollen. The usual form of an individual stamen is a 'stem' or filament which bears at its tip the two-lobed *anther* in which the pollen is formed (section 8.5).

The innermost floral organs are the *carpels*, the female reproductive structures known collectively as the **gynoecium** (*house of woman*). The key feature of a carpel is that it contains one or more ovules; beyond that, however, there is extensive variation in the arrangement of carpels. Flowers of many species contain more than one carpel, and in different species these may be separate or partly fused or completely fused.

Whether individual or fused, carpels consist of three regions. The basal region is the *ovary*, in which the ovules develop and are contained. Each ovary of a carpel develops a number of ovules usually ranging between one and a hundred (but sometimes even more). The number of ovules per ovary is partly species-specific.[i] Each ovule has the potential to develop after fertilization into a seed, so those species in which there are many ovules have a high potential for seed production. In many of the species that possess fused carpels, the common ovary is divided into **locules**, in which the ovules are actually located. The tomato is a good example of this.

The outermost part of the carpel is the *stigma*, which is where pollen lands. Between the stigma and the ovary, most plants possess a stalk-like structure, the style, through which pollen tubes must grow in order to achieve fertilization (section 8.6). In a few species, however, the stigma more or less sits on the ovary.

Within the general framework of flower structure just described, there is a beautiful and almost bewildering variety of floral forms between different species. However,

---

[i]In species that produce multiple ovules, there is variation between individuals in respect of the number of ovules per ovary; environmental conditions during ovary development may also influence ovule numbers.

Figure 8.3 *Euphorbia pulcherrima* (Poinsettia) with its brightly coloured bracts. The common name derives from Joel Roberts Poinsett, an American statesman who introduced the plant to the USA from Mexico in the early 19th century. It rapidly gained popularity as a decorative plant, especially at Christmas time. Photo: JAB.

we must also note that there are flowers in which there are fewer than four whorls of organs, *Wolffia* being the most extreme example of this (section 8.1 above).

In some plants, including species in the genera *Anemone*, *Helleborus* and *Tulipa*, the sepals and petals are fused to form **tepals**, which often at first glance resemble petals. There are species in which sepals and/or petals are rudimentary and, in some of these, the function of acting as a colourful attractive organ has been 'taken over' in evolution by the *bract(s)*, modified leaves that subtend the flower. Examples of this include *Euphorbia pulcherrima* ('Poinsettia'), which has (usually) bright red bracts that are nevertheless very similar to conventional leaves in structure (Figure 8.3), and plants in the genus *Bougainvillea*, with bracts that are often deep pink or purple (although among different varieties and species there is wide range of colours).

Concerning the reproductive organs, in the many species in which both stamens and ovaries are present, the flowers are described as **perfect**; in **imperfect** flowers, by contrast, one of these whorls is missing. Imperfect flowers are thus unisexual, either male (*staminate*) or female (*carpellate*[ii]). In some species with unisexual flowers, both

types of flower are carried on the same plant. Such plants are described as **monoecious**[iii], whereas species in which the two types of flower are borne on separate plants are described as **dioecious**.

8.3 Organization of flowers and flowering – inflorescences and life-styles

Bringing to mind a range of flowering plants will also bring to mind a large variety of ways in which flowers may be organized. The simplest arrangement is where a shoot ends in a single flower, thus terminating the growth of that shoot. In some small annuals, there is a single flower borne at the end of the single aerial shoot; more commonly, several shoots each end in a single flower. At the other end of the scale are the various types of inflorescence, in which multiple flowers are formed in a branched array of multiple floral axes; some of these are represented in Figure 8.4.

To complicate matters further, there are species in which an inflorescence such as a spike or raceme may retain a vegetative apical meristem. Thus the spike or raceme continues to elongate in an indeterminate manner, while initiating floral shoots that are obviously determinate. This arrangement occurs especially in species with long flowering seasons.

In addition, we need to consider the organization of flowers and flowering within the life of the plant. Annual plants, living for only one growing season, generally flower only once. They are termed *monocarpic* plants, as mentioned in Chapter 6 (section 6.7.1). Most biennials are also monocarpic; they become established in one growing season and flower in the next. Surprisingly, some perennials are also monocarpic; such plants flower once, often after many years of purely vegetative growth, and then die. Many bamboos are monocarpic, as are several *Agave* species. Thus *Agave americana*, from desert regions of the southwest USA, may live for 90 years or more before its one and only flowering event. Most perennials are, however, *polycarpic* and, once they have passed the juvenile-to-adult transition (section 8.4.1), they flower in most years during their lifetime.

[ii]Or pistillate. Some authors refer to female reproductive structures, especially when carpels are fused, as *pistils*.

[iii]This means 'a single house'; the word is clearly related to oikos (Greek for house), from which ecology (originally oecology) is derived.

Figure 8.4 Different types of inflorescence.
From Scott, P. (2008) Physiology and Behaviour of Plants. Wiley, Chichester.

8.4 Formation of flowers

8.4.1 Environment and ecology

The transition from a purely vegetative state to a reproductive state is clearly a major developmental shift. It is not surprising, therefore, that it is regulated by a network of interacting control mechanisms, some of which are responses to external, environmental signals, while others are internal.

The commonest environmental cue is day length (see Table 8.1). In regions of higher latitude, plants that flower in late spring and early summer are often *long-day* plants – that is, they require a number of days (variable according to species) of a minimum length before flowering is induced. Thus, *Sinapis alba* (white mustard) may be induced to flower after a single day of at least ten hours in length, with a maximal effect being seen with 18 hours of light. There is also an interaction with temperature: low temperatures inhibit the transition to flowering.

By contrast, many species that flower in high summer and autumn (i.e. after the summer solstice) are *short-day* plants, requiring one or more shorter days before flowering. The Japanese morning glory, *Ipomea nil* (previously known as *Pharbitis nil*), for example, requires only one short day for floral induction. Again there is an interaction with temperature; for this species, lower temperatures may induce flowering even in plants kept in long days.

In addition to plants with a specific requirement for day length, there are many that are facultative; although

induced by long or short days, they will eventually flower under non-inductive conditions. Finally, there are *day-neutral* plants, for which there is no specific day length requirement.

Although we refer to 'long-day' and 'short-day' plants, the plants of both groups are actually responding to the length of the dark period. Thus, in *Ipomea nil*, interruption of the night by a flash of white light prior to the first short day inhibits floral induction.

Plants therefore have a mechanism for measuring the length of the night. The activation of photoreceptors, especially phytochromes but also including cryptochromes (see Chapter 3, sections 3.7.2, 3.7.3 and 3.7.4) in daylight is a major factor in entraining the plant's internal circadian clock (see Chapter 6, section 6.10.2). Thus, the spectral composition of the light (in relation to the action spectra of the photoreceptors) and the quantity of light (photon flux density) are also important. It is this interaction between the photoreceptors and the clock that is involved in 'counting' the hours of darkness.

Further, the differing requirements, in terms of the number of inductive nights required *and* the increasing effectiveness of extra nights in plants with a 'one-night' requirement both suggest that a build-up of a factor or factors is involved in floral induction. This is discussed further in section 8.4.4.

Exposure to a cold period, known as *vernalization*, is another common requirement for flowering, often seen in plants which over-winter in cooler latitudes and then flower in the spring. Many of these are day-neutral, but

Table 8.1 Examples of plants with different day length requirements for flowering.

Long day	Short day	Day-neutral
Lactuca sativa, lettuce	*Chrysanthemum*	*Cucumis sativus*, cucumber
Lolium perenne, ryegrass	*Coffea arabica*, coffee	*Oryza sativa*, rice
Sinapis alba, white mustard	*Euphorbia pulcherrima*, Poinsettia	*Rosa*, rose
Spinacea oleracea, spinach	*Ipomea nil*, morning glory	*Petunia*
Trifolium repens, clover	*Lemna minor*, duckweed	*Solanum lycopersicum*, tomato

some long-day and short-day plants also have a need for vernalization. This demonstrates that whatever happens in vernalization may be 'remembered' by the plant for several weeks (see sections 8.4.3 and 8.4.4).

Finally, various stresses, including drought and/or high temperature, nutrient deficiency, excessive ultraviolet light and overcrowding, may also induce flowering. Thus, horticulturalists and gardeners alike will have experienced 'bolting' or 'running to seed' of crops such as lettuce during warm, dry periods. The former term is generally applied to those plants whose vegetative life-form is a rosette or globe-like structure; induction of flowering then involves dramatic elongation of the main axis. Indeed, although it was originally a horticultural term, scientists working with *Arabidopsis thaliana* now refer to the elongation of the stem prior to flowering as 'bolting' even when flowering is induced by day length or developmentally (Figure 8.5).

Although environmental factors have a major role, internal developmental cues, such as plant size or the number of vegetative nodes, are also important. Indeed, in day-neutral plants, such cues may be the major factor in determining when flowering is initiated. Even in plants with specific photoperiod requirements, the response to photoperiod interacts with developmental signals such that plants which are too immature will not respond. More specifically, the inability to respond appears to reside in the apical meristem, which needs to go from a 'juvenile' to an 'adult' state before the floral transition can occur.

The acquisition of the competent or responsive state is very obvious in some species, because at this time they go through a vegetative phase change in which features such as leaf shape change. Phase changes are particularly obvious in many woody perennials, but they also occur in a range of non-woody plants, including maize (*Zea mays*) and *Arabidopsis*. However, even these apparently 'built-in' developmental requirements may be overridden by environmental stresses, causing plants to flower and set seed when they are still very small. In terms of natural selection, early or rapid flowering in response to stress produces structures – seeds – that are more likely than the plant itself to survive the stress.

Figure 8.5 Flowering *Arabidopsis*.
Photo by Marco Roepers, from Wikimedia Commons, published under the GNU Free Documentation Licence.

8.4.2 Changes at the apex

At the apex, the transition to flowering – as has already been noted – involves both a change in phyllotaxis and changes in function of the organs that arise from the new primordia initiated on the flanks of the apex. Thus, instead of leaves, there are (usually) bracts (section 8.2) and the whorls of floral organs. The developmental origin of these organs indicates that they have evolved from leaves, and we can therefore think of the stamens and carpels as sporophylls or spore-bearing leaves.

Those plants that require a specific number of long or short days provide excellent experimental models for studying the changes that occur at the apex when flowering is induced under controlled conditions. In a range of species, one of the first indications of the transition from a vegetative to a floral apex is a partial synchronization of the cell division cycle among the dividing cells. There is also an increase in the growth fraction (the proportion of cells in the apex that are actually going through the cell division cycle) and a decrease in cell cycle length. The latter is caused by a shortening of the G1 and S phases of the cycle (and in some species G2).

In some species, the decrease in the length of S-phase is known to involve an increase in the number of active replicon origins (Chapter 2, section 2.13.3) and an increase in the rate of DNA synthesis (as seen in the rate at which the replication forks move along the template). All this points to a burst of cell division activity, as the meristem is reorganized to lay down floral organs which exhibit both a different pattern of phyllotaxis and a shortened plastochron (see Chapter 6, section 6.3) compared with the initiation of normal leaves. As the pattern of initiation becomes defined, the apical dome itself tends to become flattened and the rate of cell division declines. Thus, at this stage of floral development, cells on the flanks of the apex, where floral organs are being initiated, cycle faster than cells on the dome of the apex.

8.4.3 Patterns of gene expression

Great progress in the understanding of the genetic regulation of flowering has been made in recent years. This is typified by the work on floral development in *Arabidopsis* and *Antirrhinum*. Both these are eudicots but with very different floral morphologies (Figure 8.6) – *Arabidopsis* has radial symmetry (i.e. is **actinomorphic**), while *Antirrhinum* has bilateral symmetry (i.e. is **zygomorphic**).

(a)

(b)

Figure 8.6 (a) *Arabidopsis* flower. Photo by Will George; (b) *Antirrhinum* flower being visited by a bee. Photo by David Lev, Nirim, Israel.

Study during the 1980s, in the laboratories of Enrico Coen and Elliot Meyerowitz[iv], of a series of homoeotic mutants affecting the whorls of floral organs (see Coen & Meyerowitz (1991) and Chapter 4, section 4.2) led to discovery of the genes involved in controlling the development of those organs. Essentially, three classes of genes specify the correct positioning and development of sepals, petals, stamens and carpels. These three classes of genes act in an overlapping manner in different radial zones on the floral meristem, as shown diagrammatically in Figure 8.7. A-class genes work in the outer two zones and C-class genes work in the inner two zones, while

---

[iv]EC at the John Innes Centre, Norwich, UK; EM at Caltech, Pasadena, USA.

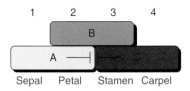

Figure 8.7 The original 'ABC' model for the control of development of the four whorls of floral organs.

the activity of B-class genes overlaps with the activities of both the A and the C-class genes.

The effects of mutations in these gene classes are thus as follows:
• Loss of A-class function leads to lack of sepals and petals; stamens and carpels are produced instead.
• Loss of B-class function leads to lack of petals and stamens; sepals and carpels are produced instead.
• Loss of C-class function leads to lack of stamens and carpels; sepals and petals are produced instead.

The genes identified in *Arabidopsis* and *Antirrhinum* that fall into these classes are shown in Table 8.2.

All of these genes encode transcription factors which are (apart from those encoded by *AP2* and *LIPLESS1 & 2*), in the MADS box[v] family (see Chapter 4, section 4.9). They have been described as the 'molecular architects' of floral morphogenesis.

This pattern of gene activity is now well-established as the 'classic ABC-model' for floral development. However, more recent work has established that a further class of MADS-box genes, the E-Class[vi], is also required. The E-class genes, which exhibit a high degree of functional redundancy, are *SEPALLATA1 (SEP1), 2, 3* and *4*. The proteins encoded by these genes interact with the proteins encoded by the ABC genes in the specification of floral organ identity. Thus, the original model as shown in Figure 8.7 is modified as follows:
• A and E are needed for sepals (see also section 8.4.4).
• A, B and E are needed for petals.
• B, C and E are needed for stamens.
• C and E are needed for carpels.

Thus the ABC model has evolved into the ABCE model. It is generally applicable across the range of

angiosperms, both eudicot and monocot, although there is some variation in detail between different plant families. Homologues of these genes are also found in basal angiosperms, but only the E-class genes appear to act in the same way as in higher level angiosperms.

The expression of the ABC genes shows much less clear-cut boundaries in basal angiosperms; it is thought that the specific roles in organ formation seen in higher-level angiosperms have arisen by evolution of gene function. Thus, in the words of Douglas Soltis and his colleagues at Gainesville, Florida (Soltis *et al.*, 2007):

'Homologues of the ABCE floral organ identity genes are also present in basal angiosperm lineages; however, C-, E- and particularly B-function genes are more broadly expressed in basal lineages. There is no single model of floral organ identity that applies to all angiosperms; there are multiple models that apply depending on the phylogenetic position and floral structure of the group in question. The classic ABC (or ABCE) model may work well for most eudicots. However, modifications are needed for basal eudicots and . . . basal angiosperms.'

At this point, two comments need to be made. First, since the ABCE genes are transcription factors, it is presumed that they regulate the activity of other genes. This presumption is correct, and it is currently thought that the targets of the ABCE genes are those directly involved in the development, growth and function of the relevant organs.

The second comment relates to the place of the ABCE genes within the overall developmental pathway that transforms a vegetative apex into a flower or that transforms a shoot into an inflorescence. It is clear that the ABCE genes act on primordia arising on an apex that has already responded to the floral stimulus. In other words, it is already or is on its way to becoming a floral apex; its identity has been established and the ABCE genes work in the context of that identity.

So, are there floral meristem identity genes equivalent to those genes that specify the identity of vegetative meristems (as discussed in Chapter 4, section 4.2)? The answer, as the reader may expect, is 'yes'. There are several genes that act before the formation of the whorls of floral organs. Some of these are active in specific induction pathways, while others seem to be involved as more general regulators of floral meristem identity.

---

[v]Thus, although these genes have a homoeotic function, they are not homeobox genes.
[vi]There is also a D-class, but D-class genes are involved in formation of ovules rather than with formation of the whorl of carpels (see section 8.5).

Table 8.2 Genes regulating floral organ formation.

	Class A	Class B	Class C
Arabidopsis	*APETALA1 (AP1)* and *AP2*	*AP3* and *PISTILLATA (PI)*	*AGAMOUS (AG)*
Antirrhinum	*SQUAMOSA, LIPLESS1* and *LIPLESS2*	*DEFICIENS (DEF)* and *GLOBOSA (GLO)*	*PLENA (PLE)*

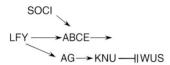

Figure 8.8 Simplified 'flow-chart' of the gene activities that lead to flowering.

Thus, mutations in the *LFY (LEAFY)* gene prevent the expression of the *ABCE* genes, giving rise, as the gene's name implies, to a leafy apex.

LFY is expressed throughout the apex and yet appears to regulate the localized expression of the *ABCE* in specific regions within the apex. Also involved in establishment of floral meristem identity is *SOC1 (SUPPRESSOR OF OVEREXPRESSION CONSTANS)*. Another function of *LFY* is the suppression of the genes that maintain the vegetative shoot apical meristem; it up-regulates *AGA-MOUS* (AG). The AG protein, in turn, down-regulates *WUS (WUCHSEL)*[vii] by activating a repressor protein, KNUCKLES (KNU), thus 'terminating' the vegetative meristem (Figure 8.8).

In relation to the requirement for phase change from juvenile to adult, it was thought until quite recently that at least some of the genes involved in floral induction would also be involved here. However, this is not the case. Indeed, for several years, attempts to identify genetic changes associated with phase change came up with nothing. Only very recently has a 'phase-change' gene (*JAT – JUVENILE-TO-ADULT-TRANSITION*) been identified in olive (*Olea europaea*) and in *Arabidopsis*. In olive, it is highly expressed in the juvenile phase and expression declines as phase-change occurs; in *Arabidopsis*, *JAT* mutant lines exhibit delayed flowering. However, there is currently no clear idea of the mode of action of *JAT*.

Finally, it is necessary to consider negative regulation. There are many circumstances in which flowering might

occur but does not do so until particular conditions have been met. An example of this is the requirement for exposure to cold conditions (vernalization). A gene that is widely involved in prevention of flowering is *FLC* which is a MADS-box transcription factor that represses several flowering genes, including *SOC1*, *FD* and *FT* (see section 8.4.4). Several induction pathways converge on *FLC*, leading to its down-regulation, thus allowing flowering to occur. Down-regulation is achieved by epigenetic mechanisms, mediated by different proteins according to the specific induction pathway.

Thus, in the vernalization pathway, *FLC* is down-regulated initially by inhibitory RNA (Chapter 3, section 3.4.4), then histone H3 in the vicinity of the gene is hyper-acetylated by POLYCOMB REPRESSOR COMPLEX2 (PRC2) (see section 4.3 in Chapter 4), which strongly represses the gene. Further, it has been discovered very recently that a long, non-coding RNA molecule, derived from a spliced out intron (Chapter 3, section 3.2.1) is also necessary. The RNA, named COLDAIR, accumulates during exposure to low temperatures and then interacts with the PRC2 protein, enabling it to interact with chromatin. The mechanism by which lower temperatures lead to the accumulation of COLDAIR remains unknown.

8.4.4 Signalling

The role of environmental signals such as day length and temperature, and the interaction of these signals with internal developmental cues such as the juvenile/adult transition, lead to the concept of multiple signalling pathways. This idea is strengthened when we consider that some species in which there is an environmental requirement, such as long days, may eventually flower in the absence of the required factor and, furthermore, that application of hormones may overcome specifically inhibitory conditions.

What, then, is involved in the signalling pathway from exposure to flower-inducing conditions and the onset of flowering? Taking long-day plants as a specific example, it has already been noted that the plant 'counts' the hours of

---

[vii]As discussed in Chapter 4, *WUS* is an essential gene in the maintenance of shoot apical meristem identity and activity.

darkness through an interaction between photoreceptors and the components of the circadian clock. The major sites for location of the photoreceptors (and therefore of perception of day length) are the leaves.

It has been a major discussion point over the years as to how the leaves then communicate with the shot apex to bring about the formation of a floral apex. In the early years of the 20th century, grafting experiments showed that a shoot apex that had not been exposed to inducing conditions will undergo the change from vegetative to floral if grafted onto a shoot that has been exposed to inducing conditions.

This led the Armenian-Russian plant physiologist, Mikhail Chailakhyan, in 1936, to propose the existence of 'florigen'. He conceived of it as a specific flowering hormone that moves from leaves to the shoot apex. But florigen remained elusive for over 70 years.

In this long search, other, already known hormones were implicated. For example, in *Sinapis alba*, a long-day plant, application of cytokinin (Chapter 3, section 3.6.3) to the shoot apex causes the same changes in DNA replication that occur after floral induction. However, cytokinin on its own does not induce flower formation, and similar things can be said about several other molecules that have from time to time been put forward as florigen. That is not to say that 'conventional' hormones and signalling molecules are not involved in flowering. Indeed, as indicated below, there is good evidence from several species of a role for GA in flowering. Nevertheless, what is clear is that the transmissible signal, florigen, is not one of the well-known hormones or signalling molecules.

The breakthrough came in a study of mutants in *Arabidopsis thaliana*, and in particular of the *CONSTANS* (*CO*) gene. This gene encodes a 'zinc-finger' transcription factor and is expressed in leaves. It is involved in the entrainment of the circadian clock, which, as already mentioned, is part of the mechanism whereby plants can 'measure' the hours of darkness. When *Arabidopsis* is exposed to long days, expression of *CO* is up-regulated, leading to the induction, in the leaves, of the *FLOWERING LOCUS T* (*FT*) gene. The discovery of what happens next – the result of work in several laboratories – was one of those events in science that we can rightly call a major breakthrough. It is the *FT protein* itself that travels via the phloem from the leaves to the shoot apex. Further, over-expression of *FT* in leaves under non-inducing conditions actually induces flowering and can, in fact, cause flowering

while plants are still in the juvenile phase (see earlier). Thus, 70 years after florigen was first postulated, it was shown to be not a 'conventional' hormone but a small protein. Chailakhyan, who died aged 90 in 1991, would have been proud to see his hypothesis thus vindicated.

So, FT protein arrives at the apex, and there it combines with a transcription factor, the product of the flowering-time gene *FD*. The next steps are still the subject of discussion, but currently it is thought that the FT-FD complex up-regulates *SOC1*, the floral apex identity gene, and also *AP1*. The latter has already been discussed in relation to establishment of the outer two whorls of the flower, but this up-regulation early in the events in the apex indicates a more over-arching role in specifying floral meristem identity and, perhaps, also in coordinating the other genes involved in whorl formation. Some evidence for this comes from the effect of the SOC1 protein; following up-regulation of its gene by FT-FD, SOC1 in turn up-regulates *LFY* and it appears that an interaction between the LFY and AP1 proteins is involved in turning on the other whorl identity genes.

All this appears quite complex – and indeed, it is. However, there is yet one more layer of regulation. Acting above *LFY* and *SOC1* are two homeobox genes, known rather poetically as *PENNYWISE* (*PNY*) and *POUND-FOOLISH* (*PNF*). There is extensive overlap of function between these two genes but, in *Arabidopsis* double mutants in an otherwise wild-type background, flowering does not occur under inducing conditions. One or other of the PNY or PNF proteins is essential for the competence of *LFY* and *SOC1* to respond to FT-FD.

Based on the discussion in this, and the immediately previous section, it is possible to construct a genetic pathway for the regulation of day length-induced flowering, as is shown in Figure 8.9.

But this is not the whole story. Although we can no longer equate florigen with any of the conventional hormones or signalling molecules, it is clear that certain hormones do influence flowering. Thus, in *Arabidopsis*, there is a GA-induced pathway to flowering which bypasses the requirement for long days. Indeed, GA can induce flowering in plants kept under short days.

As described in Chapter 3 (section 3.6.4), GA exerts its effect via destruction of transcriptional regulators, the DELLA proteins. In general, DELLAs are growth inhibitors, and thus their destruction allows growth. In the shoot apex, under non-inducing conditions,

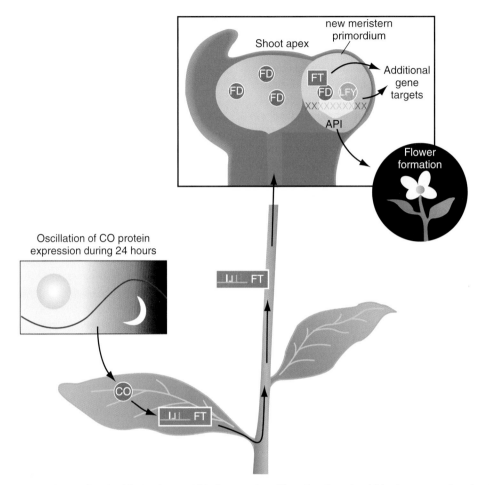

Figure 8.9 Cartoon showing the role of florigen (FT protein) in the promotion of flowering. The main additional gene target (mentioned at top right) is *SOC1* (see Figure 8.8 and text). Key CO = Constans; FT = Flowering locus T; FT protein is the transmissable signal, 'florigen'; FD is a flowering-time gene; AP = Apetala; LFY = Leafy; SOC is the floral apex identity gene.
Reproduced by permission from Blázquez MA (2005) *Science* **309**, 1024–1025, © AAAS.

DELLAs act as a further group of repressors of *LFY* and *SOC1* (see above); GA relieves this repression, allowing flowering to occur. However, this GA-induced flowering is delayed if plants are exposed to the gaseous hormone ethylene (Chapter 3, section 3.6.6). Ethylene does this by inhibiting significantly the accumulation of GA, in turn elevating the levels of the DELLA proteins. There is thus an interaction between the ethylene and GA signalling pathways which provides adaptability in response to environmental conditions.

However, one puzzle remains. Ethylene can also delay flowering in plants induced by long days (although the inhibition is not as great as in the GA pathway). This

inhibition also occurs via interaction with the GA-DELLA signalling pathway, which implies that GA may have a minor, perhaps enhancing, role in 'normal' flowering induced by long days.

8.5 Gametogenesis

8.5.1 Introduction

The function of the angiosperm flower is sexual reproduction[viii], and a key part of reproduction is the

---

[viii]But see also the discussion of apomixis in Chapter 4, section 4.8.

Box 8.1 Meiosis

With the evolution of sexual reproduction came the need for a type of cell division in which the number of chromosomes is halved, otherwise the chromosome number would double with every generation. This reduction division is known as meiosis, and in angiosperms it occurs during formation of the gametes. The first phase of meiosis resembles mitosis: chromosomes are replicated and thus each consists of two chromatids. However, the chromatids do not separate; instead, homologous chromosomes 'find' each other (e.g. the two replicated copies of chromosome 1 find each other).

The chromatids align themselves to form a bivalent and, at this stage, crossing over can occur between the chromatids of the homologous chromosomes. That is, equivalent tracts of DNA can be swapped between the chromatids of homologous chromosomes. This type of recombination is another effective way of increasing genetic variation.

The homologous chromosomes then separate and nuclear division occurs, giving two nuclei, each with a complete set of replicated but unseparated chromosomes. In the next division, the two chromatids separate and are partitioned to daughter nuclei. Thus, one diploid cell gives rise to four haploid spores.

---

formation of gametes. It is here that we see the extreme reduction of the gametophyte generation that was mentioned in Chapter 1 (section 1.8). For both female and male gametes, the first phase of gametogenesis is the formation of a spore or spores, and the second phase is the formation of the gametophyte containing the gametes themselves.

8.5.2 Development of female gametes

The first stage in female gametogenesis is ovule development. Ovule primordia arise by periclinal divisions within subdermal layers, accompanied by anticlinal divisions of the epidermis (Figure 8.10a). The developing ovule enlarges to form a **funiculus** (the stalk that anchors it within the ovary) and a tissue known as the **nucellus**. Later in development, one or two outer layers differentiate to form **integuments**. These will later form the seed coat or testa (Chapter 4, section 4.6). There is an opening in the integument layer, the **micropyle**, which will be important in the uptake of water in seed germination (Chapter 4, section 4.2). Inside the ovule, prior to the formation of the egg cell (the female gamete), *megasporogenesis* and *megagametogenesis* occur. There are some variations in these processes across the range of angiosperms, but here we describe what happens in the majority of species.

The first indication of megasporogenesis, which often occurs before integument formation (see above) is the differentiation of a megasporocyte within the nucellus near the apex of the ovule. This cell undergoes meiosis (see Box 8.1), thus generating four *haploid* megaspores. In about 70 per cent of angiosperms, only one of these – the one located furthest from the micropyle – is retained; the other three disintegrate. The majority of angiosperms are thus *monosporic*, while the remaining 30 per cent are

either bisporic (two of the megaspores disintegrate) or tetrasporic (all four megaspores are retained).

Focusing on monosporic species, the megaspore enlarges, displacing the nucellus and absorbing its contents[ix]. It then undergoes three rounds of mitosis (nuclear division) without cytokinesis (cell division), thus producing a cell with eight haploid nuclei. As shown in Figure 8.10, these nuclei become arranged in two groups of four, one near the micropylar end of what is now the megagametophyte and one at the opposite end (known as the chalazal end – the region of the ovule opposite the micropyle is named the **chalaza**). One nucleus from each group then migrates to the centre of the cell. These are the **polar nuclei**.

Cellularization then occurs around the three nuclei at the micropylar end and around the three nuclei at the chalazal end (Figure 8.10). The former make up the egg apparatus, consisting of the *egg cell* (ovum) itself and two **synergids**. The three cells at the chalazal end are the **antipodal cells**. The entire structure, consisting of just seven cells (one of which has two nuclei) and known as the *embryo sac*, is the mature megagametophyte, an extreme case in the reduction of the 'plant' that carries the 'mega-gametes'.

8.5.3 Development of male gametes

As with female gametes, two main phases are detected in the formation of male gametes, namely *microsporogenesis* and *microgametogenesis*. These processes occur in the anthers, borne at the tips of the stamens (see Section 8.2).

[ix]Except in the few species in which the nucellus becomes the perisperm (Chapter 4, section 4.4).

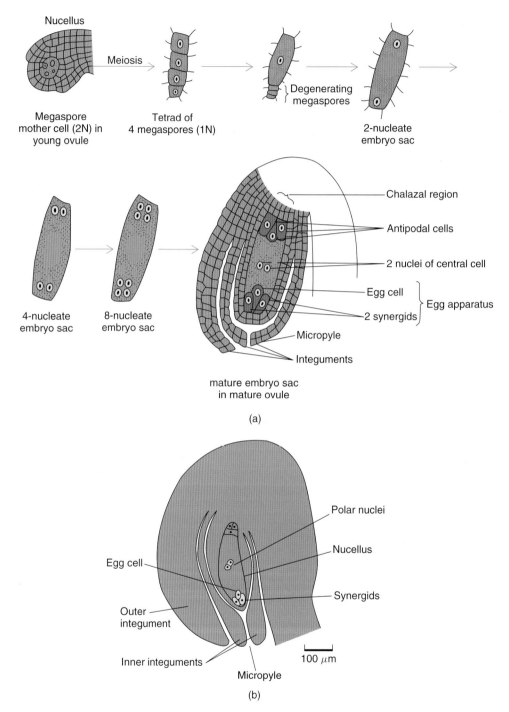

Figure 8.10 (a) Diagram of ovule formation and megagametogenesis. (b) Embryo sac of a typical angiosperm.
Part a is reproduced by permission from Berry, SJ and Mascarenhas, JP "Gametogenesis," in AccessScience, ©McGraw-Hill
Companies, 2008, http://www.accessscience.com.

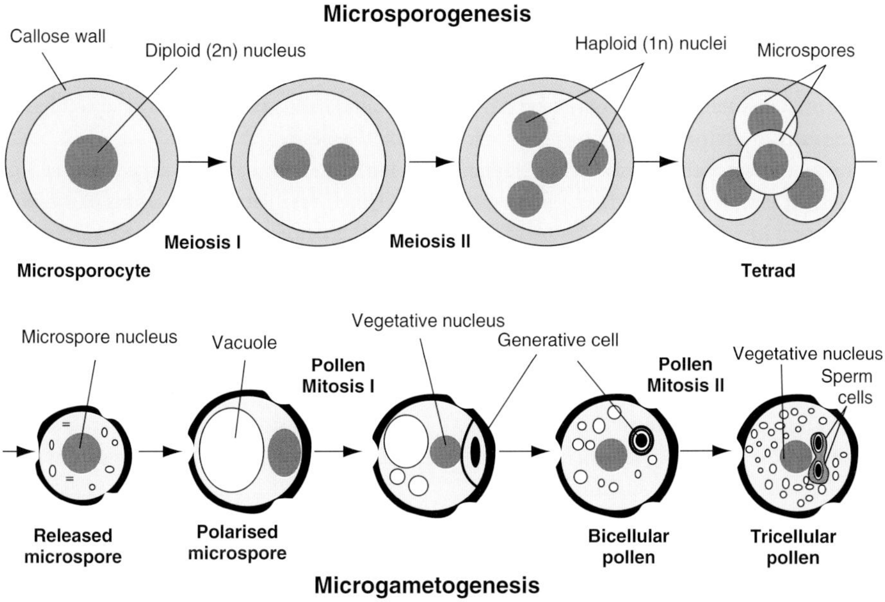

Figure 8.11 Diagrams showing first the formation of the microsporophyte, then microgametogenesis (pollen formation). Diagrams provided by Professor David Twell, University of Leicester, UK (http://www2.le.ac.uk/departments/biology/people/twell/lab/pollenis/development).

Early in development, the anther shows little evidence of differentiation except for the presence of an epidermis.

As the anther matures, four groups of sporogenous (i.e. fertile) cells develop within a ground tissue of sterile **parietal cells**. Each sporogenous cell develops into a pollen mother cell or microsporocyte, which undergoes meiosis, giving rise to four haploid microspores from each microsporocyte. Cellularization may be delayed until after the second meiotic division, as occurs in most eudicots, or it may occur after both nuclear divisions, as in most monocots. Either way, the end result is the formation of four pollen grains, each of which is a single-celled *microspore*. In eudicots, prior to cellularization, the tetrad of microspores is enclosed in a sac that is bounded by a wall of callose[x]. The four microspore protoplasts are released by the action of a callase enzyme, produced by the tapetum (see below) so that each microspore can start to grow its own cell wall (Figure 8.11).

Meanwhile, the parietal cells have developed into the wall of the pollen sac, the innermost layer of which, the **tapetum**, consists of nutritive cells (in addition to its role

in secreting callase). These cells deliver nutrients to the developing microsporocytes. In the later stages of pollen formation the tapetum also deposits a lipid-rich coat around each microspore.

As the microspores/pollen grains mature, a two-layered wall is formed. The inner layer, the **intine**, is deposited from the cytoplasm of the microspore in a manner similar to the deposition of the primary walls of cells in the plant body (see Chapter 2, section 2.2.2). Indeed, the composition of the intine is similar to normal primary walls, consisting of cellulose and pectic polysaccharides. The outer wall laid down by the tapetum, or **exine**, is much more complex, consisting initially mainly of lipids and then of a very resistant polymer called *sporopollenin*[xi], which is built from carotenoid units.

In the final stages of pollen wall formation, the tapetal cells disintegrate. Some of the cellular contents, including lipids and proteins, are added to the pollen coat. Many of these pollen coat compounds are human allergens, causing 'hay fever', which can be very severe in people who are allergic to these compounds.

---

[x]Callose is a polymer of glucose, mainly joined by β-1,3 linkages but sometimes with a few β-1,6-branches.

[xi]This polymer has a long evolutionary history. It is found in the spore walls of all plants.

The next stage in formation of the microgametophyte is the mitotic division of the microspore. This is an asymmetric division, giving rise to a small *generative cell* and a larger *vegetative cell* (sometimes known as the tube cell). Both cells are retained in the original microspore wall, with the generative cell nearer the middle of the pollen grain. This two-celled entity is the immature microgametophyte. Maturation involves a further division of the generative cell to give two male gametes or sperm cells.

In about 33 per cent of species, including *Arabidopsis thaliana*, this division occurs prior to the release of the pollen from the anther but, in the majority, the final division is delayed, usually until pollination has occurred (see section 8.6). Thus, pollen represents an even more dramatic reduction of the gametophyte generation than is seen in egg cell development – the microgametophyte contains just three cells.

8.5.4 Genetic control of gametogenesis

It will be clear already that gametogenesis is a complex process, and it thus not surprising that its genetic control is equally complex. In this brief account, we focus on the most important features of genetic regulation.

Examination of the range of genes being expressed shows that there are many genes whose expression is specific for gametogenesis. For example, in male gametogenesis, the number of specific protein-coding genes has been estimated at over 300. Many of these are sex-organ-specific homologues of genes with much wider functions in the plant – for example, those involved in development of the cytoskeleton and those that encode proteasome components (Chapter 3, section 3.4.5). The latter are exemplified by *RPN5*, the gametogenesis equivalent of the more widely expressed *RPN1*. Alongside these are widely expressed genes which have a particular role in gametogenesis, of which *SLOWWALKER1* (*SLA1*) is a good example. The protein encoded by *SLA1* has a role in integrating cell division with cellular protein synthesis, probably acting via regulation of ribosomal RNA synthesis. Equally, chromatin remodelling and, thus, epigenetic programming, is a key part of embryogenesis; thus, not surprisingly, genes encoding chromatin-modifying enzymes are important.

There are also proteins with metabolic functions that are essential for gametogenesis. Examples include the vacuolar H^+ ATPase, involved in Golgi function, which is essential for male gametogenesis and partly essential for female gametogenesis. Similarly, a glucose-6-phosphate/P_i translocator which mediates transport of G6P into non-green plastids is absolutely required for gametogenesis.

Turning now to gene expression specific for each sex, genetic regulation of megagametogenesis starts with the activity of a MADS-box transcription factor, AGL11 (AGAMOUS-LIKE11). This was originally identified as the 'D-class' gene involved in floral determination, and it is now known that it specifies ovule identity. At a level below this are the genes that regulate patterning within the ovule, coupled with the integration of meiosis and mitosis into this patterning process. A key gene in patterning is NOZZLE (NZL), also known as SPORO-CYTELESS (SPL). This is described as the reproductive organ-building gene and it is negatively regulated by genes encoding transcription factors AINTEGUMENTA and INNER NO OUTER, which are thus upstream of NZL/SPL. However, there is a feedback loop in that NZL/SPL antagonizes the action of AINTEGUMENTA and INNER NO OUTER (and also of another transcription factor, BELL).

The embryo sac is thus established and, within it, a gradient of auxin (high at the micropyle end, low at the chalazal end) has a major role in establishing the specific patterns of gene expression that lead to the development of specific cell types within the gametophyte (see above). The patterns of expression of the genes *YUCCA1* and *YUCCA2* (*YUC1* and *YUC2*), which code for proteins in the auxin biosynthesis pathway, are involved in setting up this auxin gradient. There is also evidence for the involvement of cytokinins and GA in the patterning of the female gametophyte, but their specific functions have not been determined.

As with female gametogenesis, male gametogenesis involves a series of interacting transcription factors, finally leading to the expression of genes specific for each cell type. *NZL/SPL* again participates in tissue patterning (to form the sporogenous and the parietal cells) but, in microgametogenesis, its upstream regulatory gene is *AGAMOUS* (*AG*). A balance between *NZL/SPL* and *BAM1/2* (*BARELY ANY MERISTEM 1* and *2*) maintains in turn the balance between sporogenous and non-sporogenous cells.

Within the latter, the formation of the tapetum is again regulated by a cascade of gene activity in which,

Box 8.2 Pollen from the past

Pollen grains vary greatly between species. There are differences in the size and shape of the grains themselves, in the number, shape and location of apertures (pores) and surface troughs (colpi) and in the sculpting of the pollen coat (Figure 8.12). It is thus possible, with many angiosperms (and indeed, gymnosperms), to identify the genus, and sometimes the actual species, from which a pollen grain has been released. Furthermore, the resistance of sporopollenin to degradation means that pollen grains are often very long-lasting in the environment.

Figure 8.12 Pollen from (left to right) Daisy (*Bellis perennis*), Lily (*Lilium*) and Willow *(Salix)*. Scanning electron micrographs by Barry Martin, Oxford Brookes University.

This has led to the science of palynology, in which environmentally preserved and 'sub-fossil' (and even fossil) pollen grains are identified in order to gain information on vegetation types, floral assemblages and past land use. Applications include studies of vegetation changes during the glaciations and inter-glacial periods of the Quaternary era and of the advent of agriculture as early *Homo sapiens* settled in different regions of the planet. Forensic investigations also frequently make use of palynological data.

very surprisingly, two genes, *ROXY1* and *ROXY2*, encoding *glutaredoxins*[xii] acting downstream of *NZL/SPL* and *BAM1/2*, are needed for pollen formation. There is redundancy of function between the two *ROXY* genes, but double mutants are male-sterile. The mode of action of the ROXY proteins is to modify cysteine residues in the TGA9 and TGA10 transcription factors, thus facilitating activation of the next 'level' of genes, including *DYSFUNCTIONAL TAPETUM 1* (*DYT1*). As with so many of the key regulatory genes, *DYT1* encodes a transcription factor, and the genes acting downstream act in the formation of a properly working tapetum, in the absence of which microsporogenesis is also greatly impaired.

As noted earlier, the very last stage in forming two sperm cells is the division of the generative cell. Recent research suggests that this is under the control of a specific set of genes that can act partly independently of the normal cell cycle regulators. These are the *DUO POLLEN*

genes, *DUO 1, 2* and *3*. Although they have partly overlapping functions, mutations in any of these lead to a failure of the generative cell division. The DUO1 protein, a Myb transcription factor, brings about the G2/M transition (Chapter 2, section 2.13.3) by up-regulation of the regulatory cyclin CycB1:1, whereas the DUO3 protein appears to be able to facilitate the transition independently of CycB1:1. Less is known about DUO2, except that its timing of action is early in metaphase; *DUO2* mutants arrest at this point during mitosis.

Hormones are also involved in the control of gene expression in pollen formation. Mutants in brassinosteroid synthesis, or in the brassinosteroid signalling pathway, exhibit male sub-fertility or infertility. The transcription factor BES1, a component of the signalling pathway, interacts with the promoters of several genes acting early in pollen development, including *NZL/SPL*. Similarly, mutants defective in auxin biosynthesis or in perception and signalling or in auxin polar transport have aberrant patterns of both anther development and pollen maturation. Jasmonic acid (JA) and GA (the latter acting, as usual, via an effect on DELLA proteins) are involved in

[xii] Small oxidoreductase proteins involved in, among other things, stress responses.

pollen maturation, while JA and ethylene act with ABA to regulate the expression of genes involved in pollen grain separation and in anther dehiscence.

The final point to make in this section is that some of the genes involved in gametogenesis (and especially in pollen formation) are subject to fine tuning by micro-RNAs (Chapter 3, section 3.4.4).

8.6 Pollination and fertilization

8.6.1 General features

In mature stamens there is in most species, a line of dehiscence that forms between the two pollen sacs. The opening formed along this line, which is shared between the two sacs, is called the **stomium**, through which the mature pollen grains are released. Released pollen grains are carried on the wind or by an animal vector (according to plant species), and some reach a receptive stigma (see section 8.6.3).

At the time of release from the anthers, the pollen grains in most species are dehydrated (water content of between 2 per cent and 30 per cent), although there is a range of species in which pollen water content varies between 30 per cent and 70 per cent (e.g. *Zea mays* – maize, 37 per cent; *Humulus* – hop, 67 per cent). Rehydration occurs when the pollen lands on a receptive stigma, and then the vegetative cell 'germinates' to form the pollen tube. In pollen that is shed with a higher water content, pollen germination can occur more quickly. However there is also a disadvantage in that, between release from the anther and landing on a stigma, the pollen grains may lose water to the atmosphere. Uncontrolled water loss leads to shrinkage, damage and loss of viability.

8.6.2 Pollination mechanisms

Male gametes in all but one species of seed plant are non-motile. The exception is *Gingko biloba*, a primitive gymnosperm that was discussed in Chapter 1 (see Box 1.2). The non-motility of male gametes has great advantages for land plants, since it avoids the need for water as a medium through which the sperm need to move in order to reach the egg cell. Inevitably, then, the sperm cells (as part of the pollen grain) must make use of other means to reach their targets.

In about 80 per cent of angiosperms, animals transfer pollen from the anther to the stigma (in contrast to

gymnosperms, none of which have animal pollinators). About 200,000 different animal species have been listed as pollinators, but the vast majority of those animals are invertebrates, especially insects. In most of the remaining 20 per cent of angiosperms, wind is the main vector but, in some aquatic plants (comprising about 2 per cent of angiosperm species), pollen is released into the water to be carried passively (because the male gametes are non-motile) to a receptive stigma. Finally, there are some species (about 5 per cent of the total, and included in the 80 per cent mentioned above) that employ both wind and an animal vector.

The first angiosperms were pollinated by insects, most probably beetles (see section 8.7), and the radiation from that initial state in a relatively short period of geological time has been remarkable.

It is not possible here to discuss in detail all the different pollinators and pollination mechanisms that are found amongst angiosperms. Instead, we present a summary of the main mechanisms, with mention of the more unusual examples. Among animals, in addition to the huge array of insect species, many species of birds[xiii], over 60 species of mammals and even geckos and lizards have been recorded as pollinators.

In respect of pollination by animals, we can discern three main types. First, there are mutualistic arrangements in which both the plant and the animal benefit (see also Chapter 11, section 11.2.4) – the main benefit for the plant being, obviously, the transmission of pollen. Second, there are relationships of stricter interdependence, where the plant and its specialist pollinator absolutely depend on each other. Third, there are one-sided relationships in which one side or the other gains no benefit at all. Examples include pollination by deceit in orchids, in which the flower resembles a female insect, and flower robbery, in which an insect obtains nectar but does not pick up pollen.

• **Bee pollination.** A majority of specifically bee-pollinated flowers (Figure 8.13) are yellow or blue[xiv]. Many have ultraviolet nectar guides on the inner surfaces of the petals, which direct the bee to the nectaries. The flowers are often sweet-scented (the scent is sometimes

---

[xiii]The common names of some of these birds reflect their interaction with flowers: honey eaters, honey peckers, flower creepers, etc.

[xiv]Despite poor colour vision at the red end of the spectrum, bees will pollinate pink or red flowers.

Figure 8.13 Oilseed rape (*Brassica napus*) flowers. These are bee-pollinated as is normally the case for yellow flowers. Photo: Margot Hodson.

very delicate). Pollen is collected on the body of the bee or in 'pollen baskets' – specific pollen collection areas on the legs of the insect, as seen in honey bees and bumble bees. Bee-pollination is probably the most widespread form of pollination but, within this type, we can discern more generalist plants that are also pollinated by other insects and, at the other end of the scale, plants that are pollinated by specific types of bee.

• **Butterfly pollination**. Butterflies have good vision, covering the whole visible spectrum, and thus butterfly-pollinated flowers are usually brightly coloured. They also have ultraviolet nectar guides. The flowers are not scented (because butterflies have no sense of smell) and they offer nectar which, in many species, is 'fortified' with amino acids. Typical butterfly flowers are tubular and are often arranged in tightly packed inflorescences such as the flower heads of the Compositae or the spikes of *Buddleja* (Figure 8.14). Indeed, the latter is one of several shrubs that are known as 'butterfly bushes.'

• **Moth pollination**. Many moths are active at night and, in turn, many moth-pollinated flowers produce scent in the evening and during the night, while being much less fragrant in daylight hours. The flowers are often tubular in structure and pale in colour (although some horticultural varieties may have been bred for more showy colours). Unlike butterflies, which stand and walk on the flowers, moths generally feed by hovering above or next to the flowers and inserting their long tongues in order to harvest the nectar (which again may be fortified with amino acids).

• **Fly pollination**. Fly-pollinated flowers exhibit a range of morphologies. Many of them smell of dung or carrion and some, typified by the Dead-horse Arum Lily (*Helicodiceros muscivorus*) have a very strong and unpleasant odour (that may be experienced in Crete and other Mediterranean islands in spring). In common with other members of the Araceae (including *Arum maculatum*, 'Lords and Ladies') *H. muscivorus* is able to raise the temperature of its flowers a few degrees Celsius above the ambient temperature. Flies and midges, attracted by the scent and by the heat, visit *Arum* flowers to feed and/or lay eggs and, in some species (e.g. *Arum maculatum*), they are trapped temporarily inside the flower (Figure 8.15).

• **Beetle pollination**. Pollination by beetles is believed to be the primitive state in angiosperms (see section 8.7). Beetle-pollinated flowers are typically simple and dull. They usually emit an aroma but not a sweet scent. The flowers do not offer nectar, and the pollinator's reward consists of some of the pollen that it picks up.

• **Mimicry and deceit**. Orchids in the genus *Ophrys* are the best known of the plants whose flowers closely resemble females of particular insect species (Figure 8.16). In fact, the resemblance is so close that males attempt to copulate with the flowers. Obviously, the mating is unsuccessful, so the male seeks out another female; if the plants are close enough together, that 'female' is likely to be another orchid flower. The male insect has received no kind of reward, but pollen has been transferred from one plant to another.

• **Obligate mutualism**. There are several examples in which the lives of a plant and its pollinator have become very closely interdependent. Examples include the figs (plants in the genus *Ficus*) with fig wasps (members of the family Agaonidae) and *Yucca* with Yucca moths. Here we focus on the latter. *Yucca* (Figure 8.17) is a New World genus occurring in South and Central America,

Figure 8.14 *Buddleja*, a typical 'butterfly bush' (left; photo by JAB), as beautifully illustrated by the visiting Small Tortoiseshell butterfly, *Aglais urticae* (right; photo by MJH).

the southern part of the USA and in the Caribbean region. All *Yucca* species are outbreeding, and all except one (*Y. aloifolia*) rely for pollination on *Yucca* moths (in the genera *Parategiticula* and *Tegiticula*); individual *Yucca* species are pollinated by just one or two species of *Yucca* moth. It is a remarkable system. Adult moths emerge at the time when *Yucca* is in flower. Moths visit the flowers, landing initially on the stamens. Pollen is transferred to the moth's body and the moth collects it into a sticky ball, which it carries on an appendage underneath its neck. Only moths in the two *Yucca* moth genera possess this appendage. On visiting the next plant, the moth lays an egg or eggs within each ovary, then climbs the style to deposit the pollen mass into a depression on the stigmatic surface. So precise is this phase of the system, that it appears that the moth 'deliberately' cross-pollinates the plant. The eggs laid in the ovaries hatch and the larvae feed on some of the developing seeds. Again, remarkably, the number of eggs laid usually ensures that only a proportion of the seeds are eaten. The full-grown larvae exit the ripening fruit and burrow into the ground, where they remain dormant until the next flowering season. Metamorphosis then occurs, and full-grown moths emerge to start the cycle again.

• **Bird pollination**. Several bird species have been recorded as pollinators, but the commonest are humming birds. Bird-pollinated flowers are usually brightly coloured (often red or orange), reflecting birds' good colour vision. The flowers are rich in nectar, but unscented (birds have only a poorly developed sense of smell). Humming birds, in particular, are hover feeders,

and the flower structure of specialized 'humming bird flowers' ensures that while the bird has access to nectar, its head becomes dusted with pollen (Figure 8.18).

• **Bat pollination**. Bats are the most widespread mammalian pollinators, but 59 species of non-flying mammals are also known to perform this role. Bat pollination is widespread in the tropics, with over 500 plant species being wholly or partly dependent on bats for pollen transfer (Figure 8.19). These plant species include some important crops, such as bananas, mangoes and guavas. Bat-pollinated flowers typically open at night, are white or pale coloured, and are often large and bell-shaped. The bats feed on the copious supplies of nectar and on pollen itself but, in doing so, much pollen adheres to their fur, some of which brushes onto the stigmas of the next plant that is visited. Bats are attracted by the strong, musty odour which results from secretion of a complex mixture of volatiles, including esters, alcohols and aldehydes. Some plants, including a Cuban rainforest vine, *Marcgravia evenia*, reinforce the attraction signal by having deeply cup-shaped leaves that create an echo.

• **Wind pollination**. While pollination by animals depends on the animal's foraging habits (e.g. in performing consecutive visits to plants of the same species), wind pollination is much more haphazard. It is very widespread in trees of cool-temperate regions and in grasses. Pollen is produced in very large amounts (thus exacerbating the misery of pollen-allergy sufferers), and the stigmas are prominent and readily accessible to pollen grains blown in their direction. There is no

(a)

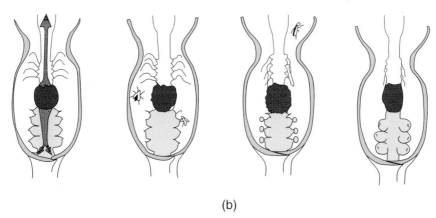

(b)

Figure 8.15 (a) Arum lily, *Arum maculatum* growing near Odcombe, Somerset, England. Photo: MJH. (b) Cartoon showing pollination of *Arum maculatum*. Small flies, attracted by the heat of the spadix, enter the flower. They are trapped by the downward pointing hairs but the hairs later wilt. This allows the flies to escape and then to visit another *Arum* flower, thus effecting pollen transfer.
Part b is reproduced with permission from Prime CT (1960) Lords and Ladies, Collins, London.

Figure 8.16 Bee orchid, *Ophrys apifera*.
Photo: Phil Corley.

Figure 8.17 Joshua tree, *Yucca brevifolia*.
Photograph by JAB.

Figure 8.18 Pollination of *Penstemon* flowers by humming bird.
Photo: Dr Bill May, US Forest Service.

need to attract pollinators, so typical wind-pollinated flowers are small and often green, are not scented and do not produce nectar. In several wind-pollinated trees, the flowers are borne on inflorescences known as catkins (Figure 8.20).

So, by a great variety of methods, depending on plant species, pollen reaches the stigmatic surface. What happens next will determine whether pollination culminates in fertilization.

8.6.3 Pollen germination and fertilization

The initial interactions between pollen and stigma, namely pollen adhesion and rehydration, rely both on properties of the pollen coat (exine) and on the properties of the stigmatic surface. Stigma morphology shows a wide range of variation between different species,

especially in respect of the absence or presence of surface papillae and the structure of papillae (ranging from small and simple to large and branched). Nevertheless, it is possible to order all these varying morphologies into two general groups – wet and dry. In both types, successful adhesion of pollen to stigma results from interaction between the stigmatic surface and the pollen coat. Adhesion is followed by pollen rehydration. Again,

Figure 8.19 The sausage tree, *Kigelia pinnata*, a tree of the African savannah. It is mainly pollinated by bats, but also by sunbirds.
Photos: JAB.

Figure 8.20 Catkins of hazel (*Corylus avellana*).
Photos: JAB.

this involves signalling between pollen and stigma, with lipids playing a significant role in this.

Interestingly, although rehydration appears to be a more straightforward (and possibly less regulated) process on wet stigmas, it is also now clear that lipids and aquaporins (see Chapter 2, section 2.3.4) control the movement of water into the pollen grain from stigmas of either type.

Successful rehydration leads to pollen germination: the vegetative or tube cell elongates through one of the pores in the exine (see above). In those species (the majority) in which division of the generative cell has been delayed, that division now occurs.

Thus, in plants of both types, two sperm cells are now carried in the cytoplasm of the tube cell. Elongation of the tube cell is a classic example of tip growth, regulated by gradients of calcium ions. Cytoplasm remains localized at the tip of the cell and, behind the growing tip, the cell is plugged by the deposition of callose (see earlier). The pollen tube grows from the stigma, down through the style and towards the ovary itself at speeds of up to 10 mm per hour. In most monocots, the style is hollow so that the pollen tube grows through a stylar tube. In most eudicots, the style is solid and the pollen tube grows between cells. However, both types of style secrete an extracellular matrix consisting of glycoproteins, particularly of arabinogalactan proteins (AGPs). These are now increasingly recognized as signalling molecules in several aspects of sexual development. Interactions between this matrix and compounds on the surface of the pollen tube cell are important for pollen tube growth through the style (see below).

On reaching the ovary, the pollen tube locates the micropyle of an ovule by growing along a funiculus. The ultimate target of the pollen tube is an embryo sac, and there is some evidence to suggest that signalling from the embryo sac is involved in guiding the tube tip to the micropyle. Signalling from the synergids, by secretion of LUREs (cysteine-rich proteins similar to defensins), is involved after the pollen tube has entered the ovule, thus attracting the pollen tube to the synergids. Signalling both from the embryo sac and from the synergids thus involves specific proteins which act as a guidance system for the elongating pollen tube.

The tube penetrates one of the synergids, leading to its collapse, and the pollen tube tip then ruptures. Rupture requires participation of another synergid protein, ES4, which induces osmotic bursting by opening the pollen tube's potassium channels. This causes an influx of potassium ions and water, upsetting the osmotic balance and causing an explosive bursting of the tube tip. Rupture releases the tube (vegetative) nucleus and the two sperm cells. There then follows the remarkable double fertilization that is only seen in angiosperms. One sperm cell fuses with the egg cell to form the zygote or one-celled embryo (Chapter 4, sections 4.1 and 4.2), while the other migrates to the central cell, fusing with the two polar nuclei to form the initial endosperm cell. The next generation has been started.

8.6.4 Self and non-self: inbreeding and outbreeding

As noted earlier, flowers may be hermaphrodite or unisexual. Plants with unisexual flowers may bear individual flowers of both sexes on the same plant (monoecious), or may bear flowers of each sex on separate plants (dioecious). Dioecious plants are, of course, obligate outbreeders but, in both monoecious plants and plants with hermaphrodite flowers, inbreeding (fertilization by pollen from the same individual) is possible unless there are specific mechanisms to prevent it (see below). Indeed, most such species are self-fertile.

However, inbreeding depression is often seen in self-fertilized plants, such that the offspring are less vigorous or less successful than the offspring resulting from a non-self-fertilization. This is generally thought to be caused by the possession in the homozygous state of alleles with mildly deleterious effects. Inbreeding depression may become apparent at any stage of development, including embryogenesis, such that inbred embryos develop more slowly than outbred embryos and form seeds with lower germination potentials. Interestingly, the extent of inbreeding depression (or its reciprocal, the fitness of the offspring) may be strongly influenced by environmental factors.

Inbreeding for most self-fertile species is probably a failsafe mechanism to ensure that some seed is set even if successful non-self pollination is not achieved. Even in self-fertile plants, mechanisms often exist to lower the rate of self-pollination (in addition to inbreeding depression). For example, on an individual plant, anthers may mature and release their pollen before or after the stigmatic surface is (fully) receptive. This obviously reduces the probability of a self-pollination.

In some species, there are also specific mechanisms to prevent inbreeding. Essentially, these prevent the egg from being fertilized by sperm from the same plant. Referring back to section 8.6.3, it may be envisaged that there are several stages at which self-pollen may be rejected, but actually the majority of mechanisms involve either the interaction between the pollen and the stigma and/or the style. Indeed, as Simon Hiscock and Alexandra Allen at the University of Bristol put it so clearly: *'The events before fertilization (pollen-pistil interactions) comprise a series of complex cellular interactions involving a continuous exchange of signals between the haploid pollen and the diploid maternal tissue of the pistil (sporophyte)'*.

Having said that, it must also be mentioned that some mechanisms involve interactions between the gametes themselves. In terms of genetic control, self-incompatibility mechanisms are classified as either *gametophytic* or *sporophytic* (Figure 8.21). Within each group, there is a range of inter-specific variation in the detail. Here, we discuss particular examples in order to illustrate the general features.

In *gametophytic incompatibility*, the process is governed by the genotypes of the pollen grains (i.e. the male gametes – hence gametophytic) and the genotypes of the diploid stigmatic or stylar tissues. The most widespread type of gametophytic incompatibility involves an adjacent pair of tightly linked and highly polymorphic genetic loci, the *S*-loci. Indeed, they are so tightly linked that until the mid-1990s, it was thought that the pollen-expressed and the style-expressed loci were actually the same gene. The style-expressed *S*-gene encodes an antigenic arabinogalactan protein (AGP). Growth of the pollen tube

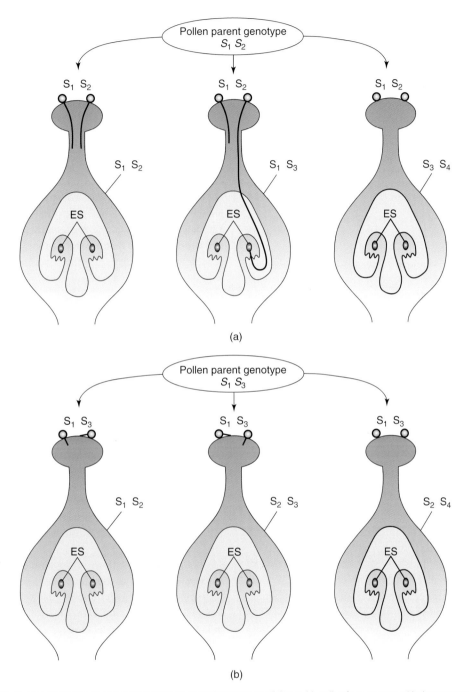

Figure 8.21 Diagram illustrating (a) gametophytic and (b) sporophytic incompatibility. In (a), pollen from a parent with the genotype S_1S_2 will be effectively fertile or infertile, according to which of the parental alleles is carried by a particular pollen grain and by the genotype of the receiving plant. In (b), the behaviour of the pollen is determined by the genotype of the parent. Thus, in the left-hand example, S_3 pollen will not germinate on an S_1S_2 stigma if the pollen parent had the genotype S_1S_3 (the S_1 allele of the pollen is incompatible with the S_1 allele of the stigma). The middle example shows a similar rejection of S_1 pollen on an S_2S_3 stigma because one of the alleles of the male parent was S_3. Key: ES = embryo sac.
Redrawn from page 306 in Lea, P and Leegood, R (1999) Plant Biochemistry and Molecular Biology, Wiley, Chichester.

is prevented or inhibited if the pollen grain carries the equivalent *S*-gene, which also encodes a glycoprotein.

The recognition process in many gametophytic species occurs early, when the pollen has landed and just started to, or is about to, germinate, as happens in *Oenothera* (evening primrose) and in grasses. In grasses, the stylar papillae produce an exudate of glycoproteins which has a role in pollen recognition. Pollen grains adhere to this and start to germinate but, with self-pollen, tube growth stops almost immediately. In *Oenothera*, the surface of the stigma is covered with a liquid secretion; pollen tubes from self-pollen fail to penetrate this.

However, in other species with gametophytic incompatibility, including many members of the family Solanaceae, the major molecular interactions which recognize self- or non-self-pollen occur while the pollen tube is growing through the stylar tissues. Inhibition of the growth of self-pollen occurs in several different ways, and which of these predominates varies, even within one species. The most widespread mechanisms are a swelling or bursting of the pollen tube tip or a complete occlusion of the tube by callose. This is a different pattern of callose deposition from that which occurs in normal pollen tube growth (see above: section 8.6.3).

Cloning and analysis of the *S*-genes in *Petunia*, *Nicotiana* (tobacco) and *Solanum lycopersicum* (tomato) has helped us to understand this incompatibility reaction in more detail. On the male side, the pollen's *S*-glycoprotein is expressed after initial germination of the pollen tube. On the female side, the stylar tissues secrete the female *S*-glycoprotein. It has been shown *in vitro* that this glycoprotein inhibits the growth of pollen that secretes the equivalent male glycoprotein. In an extension of this study, the essential role of the stylar *S*-glycoprotein was shown in genetic modification experiments. Plants of the S_1S_2 genotype (which normally accept S_3 pollen: see Figure 8.21) were transformed with the style-expressed S_3 gene; they then rejected S_3 pollen.

Biochemical and structural analysis of the style-expressed *S*-protein shows first that it is a ribonuclease with a 'classic' RNase active site. Second, there are several possible glycosylation sites for addition of a heptasaccharide. Third, there are two hypervariable regions, well separated from the active site; it is in these regions that the differences between the female *S*-alleles reside.

In incompatible (self) pollinations, the *S*-ribonuclease enters the pollen tube pollen and destroys its RNA. The RNA of compatible (non-self) pollen is not degraded, and neither is the RNA of the stylar tissue itself. Intriguingly, the stylar *S*-gene is expressed, and thus the ribonuclease is synthesized *before* the self/non-self recognition occurs. This must mean that the ribonuclease activity is regulated post-translationally (see Chapter 3, section 3.4.3). Indeed, an asparagine-rich protein, HT, is required to activate the *S*-ribonuclease.

However, the key interaction is between the *S*-ribonuclease and the pollen *S*-locus encoded F-box protein[xv], SLF. In non-self (compatible) pollinations, SLF binds strongly to the *S*-ribonuclease, tagging it for destruction by the ubiquitin-proteasome system (Chapter 3, section 3.4.5). This does not occur in self (incompatible) pollinations, so the *S*-ribonuclease remains active.

The ribonuclease self-incompatibility system described above occurs in several angiosperm families, including the Solanaceae, Rosaceae and Scrophulariaceae. However, there is another system based on programmed cell death (PCD) in self pollen; this has so far only been seen in the Papaveraceae (the poppy family).

Much of our knowledge of this system comes from the extensive work of Noni Franklin-Tong and her research group at the University of Birmingham, UK. Using *Papaver rhoeas* as their experimental model, the group has shown that an incompatible reaction between pollen and style elicits PCD in the pollen. The stylar *S*-gene product is a small secreted peptide, while the pollen component is a 20 kDa trans-membrane receptor, and their interaction in self-incompatibility initiates the first phase, a Ca^{2+} signalling cascade. This leads to actin depolymerization and, hence, a cessation of tip growth. The change in actin filament numbers initiates a further chain of signalling events, including phosphorylation of two soluble inorganic pyrophosphatases and of an MAPK (mitogen-activated protein kinase). The end result is the onset of PCD in the pollen cells. This includes the induction or activation of caspases – protease enzymes specifically associated with PCD.

In *sporophytic incompatibility*, the male side is based not on the haploid genotype of the pollen grain but

---

[xv]F-box proteins contain an essential 50-amino-acid tract that is involved in protein-protein interaction.

on the diploid genotype of the male parent. The genes involved in this system are again called *S*-genes, but they are not in any way homologous to the *S*-genes that determine gametophytic incompatibility. In sporophytic incompatibility, pollination fails if *either* of the *S*—alleles of the male parent is matched by *either* of the *S*—alleles of the female parent (see Figure 8.21).

The best-known examples of sporophytic incompatibility occur in the families Compositae and Cruciferae. Research initiated over 40 years ago by the British plant scientist Jack Heslop-Harrison showed that recognition occurs very early, almost as soon as the pollen has become rehydrated on the stigmatic surface. Recognition involves both the pollen's exine (containing the products of the male parental *S*-alleles) and the cell walls of the stigmatic papillae. If the pollen is non-self (and hence compatible), the pollen tube secretes hydrolytic enzymes that dissolve the cell walls of the stigmatic papillae and grows down into the stylar tissue (as described in section 8.6.3). If the pollen is incompatible (i.e. self-pollen), pollen tube growth ceases very quickly and callose is deposited both in the pollen tube and in the stigmatic papillae.

More recent research has shown that the antigenic substances that mediate the interaction of pollen and stigma are, on both the male and female sides, glycoproteins. In *Brassica*, several alleles of the *S*-genes have been cloned and sequenced, and their glycoprotein products have been characterized. These sporophytic system glycoproteins are very different from those involved in the gametophytic system. There are up to seven sites, clustered in the N-terminal half of the molecule, at which they may be glycosylated by the addition of quite complex branched oligosaccharides. Comparison of proteins encoded by different *S*-alleles shows that allelic variation occurs as individual amino acid changes throughout the whole molecule. The proteins encoded by different alleles also differ in the number of potential glycosylation sites that are actually glycosylated.

In gametophytic incompatibility, it was noted that there are two very tightly linked genes within the *S*-locus. In sporophytic mechanisms, there are three tightly linked *S*-genes. One of these is pollen-specific in its expression pattern; the other two are style-specific. The glycoprotein described above is encoded by one of the style-expressed genes, while the other encodes a receptor-like membrane-spanning protein kinase (SRK) which is localized in the

epidermal cells of the stigma. The pollen-expressed *S*-locus encodes a small cysteine-rich protein (SCR) located in the pollen coat. In incompatible pollinations, SCR binds to the external domain of SRK, and this binding event triggers a phosphorylation cascade which leads eventually (by mechanisms as yet not understood) to the inhibition of pollen tube growth as described above.

Two further points need to be made about this system. First, there is evidence from *Brassica* that self-incompatibility becomes less rigid as a flower ages; late in the life of an individual flower, some self pollen gets through. This may be regarded as another failsafe mechanism if there is a poor supply of non-self pollen. Second, the 'model' member of the Cruciferae, *Arabidopsis thaliana*, is self-fertile. It is thought that the initial mutation that led to loss of self-incompatibility was in the pollen-expressed *S*—locus encoding SCR. Nevertheless, full self-incompatibility can be restored by transfer from the closely related *A. lyrata* of the genes (from incompatible *S*—genotypes) encoding the SRK and SCR proteins. This exciting development, achieved in June Nasrallah's laboratory in Cornell University, will enable pollination biologists to study incompatibility in a species about which there is already extensive knowledge of genes and gene expression.

The final point to be considered in this section is inter-specific pollination. In addition to systems that prevent self-pollination, there are also barriers to inter-specific pollination. These barriers occur right across the range of angiosperms, although some species retain an ability to hybridize with other, closely related, species. Sometimes those barriers are mainly ecological, with no specific pollen rejection mechanisms but, when ecological boundaries become blurred, hybridization may occur at the boundaries. For example, in the UK, red campion (*Silene dioica*) interbreeds readily with white campion (*Silene latifolia*) to produce the fertile and often vigorous hybrid, pink campion (*Silene* x *hampeana*) shown in Figure 8.22.

However, many possible hybridizations are actually prevented by a range of general and fairly non-specific morphological or physiological barriers to pollen growth. For example, something as simple as the osmotic potential of stylar secretions being 'wrong' for foreign pollen may be enough to prevent pollen germination and tube growth. In addition to these general mechanisms, some plant families, again including the Cruciferae and Compositae,

Figure 8.22 Pink campion, *Silene* x *hampeana*, in a typical hedgerow habitat. Both parent species were growing nearby. Photo: JAB.

have more specific barriers to interbreeding. In these families, pollen from other families is rejected by the general types of mechanism mentioned above. Pollen from other species within the family elicits a reaction which is similar to that seen in the rejection of self-pollen in the self-incompatibility reaction.

Lastly, rejection of pollen from other species in the family may not be an all-or-nothing reaction. Foreign pollen may sometimes get through, leading – albeit rarely – to the formation of hybrids. This is equivalent to the occasional failure of the self-incompatibility systems mentioned above. There is also evidence that the system can be 'fooled'. Plant breeders may sometimes facilitate inter-specific breeding by providing some compatible pollen alongside the incompatible pollen. Some of the normally incompatible pollen may be accepted, leading to the formation of some hybrid seeds.

8.7 Evolution

As mentioned in Chapter 1 (section 1.7), true angiosperms emerged either in the late Jurassic era or in the early Cretaceous era. There has been extensive discussion of the origins of the flower itself. In extant angiosperms, it is readily seen that the whorls of floral organs develop in place of leaves, albeit with altered phyllotaxis. Some of the seed ferns (see Chapter 1, section 1.7) already had flower-like structures.

Fossil floral structures and the expression of floral genes in extant plant groups (including *LFY* and the *ABCE* genes – see section 8.4) suggest a feasible evolutionary route from seed fern morphology to angiosperm flower morphology. Fossil evidence (from both plant and insect fossils) suggests that the earliest angiosperms had simple hermaphroditic flowers and that they were pollinated by beetles. Consistent with this view is the observation that a significant proportion of extant 'ANITA grade' angiosperms (see Chapter 1, section 1.7) are beetle-pollinated. At the time of the emergence of the angiosperms, beetles were by far the commonest insects and, indeed, they remain so. Beetle-pollinated flowers offer no nectar reward, and it is likely that in the early angiosperms, it was the feeding by beetles on pollen that actually effected pollen transfer. Wind pollination, which seems to be a simpler system, is actually a much more advanced trait.

Extensive radiation of the insects had occurred before the arrival of true angiosperms, so that all the major groups known today were already present in the insect fauna. This represented a huge untapped resource of pollination agents, and the evolution of angiosperms to exploit that resource is a major feature of the 'Cretaceous explosion', to which reference was made in Chapter 1. The radiation involved, as is evident from the examples presented in section 8.6.2, changes in morphology, physiology, biochemistry and even lifecycle. It facilitated the invasion of a wide range of habitats.

The reciprocal of this was that particular groups of pollinators became adapted to particular providers of pollen, nectar and other rewards. Such adaptation also involved changes in morphology, physiology, biochemistry (e.g. in utilizing different food sources) and lifecycle. For this reason, many scientists speak of the *co-evolution* of angiosperms and their pollinators, especially their insect pollinators. Furthermore, co-evolution has led to some extremely close interdependent relationships, as seen in the genus *Yucca* and the Yucca moths. This leads to the final points to be made in this chapter: modern angiosperms are present in a very wide range of habitats; they have entered into relationships with an equally wide range of other organisms, in many cases in order to aid reproduction; they exhibit a very large range of vegetative and floral morphologies. Without them, we would not be here.

Selected references and suggestions for further reading

Adrian, J., Torti, S. & Turck, F. (2009) From decision to commitment: The molecular memory of flowering. *Molecular Plant* **2**, 628–642.

Bosch, M. & Franklin-Tong, V.E. (2008) Self-incompatibility in *Papaver*: signalling to trigger PCD in incompatible pollen. *Journal of Experimental Botany* **59**, 481–490.

Chandler, J.W. (2011) The hormonal regulation of flower development. *Journal of Plant Growth Regulation* **30**, 242–254.

Coen, E.S. & Meyerowitz, E.M. (1991) The war of the whorls – genetic interactions controlling flower development. *Nature* **353**, 31–37.

Franklin-Tong, N. (2010) Plant fertilization: Bursting pollen tubes! *Current Biology* **20**, R681–R683.

Ge, X.C., Chang, F. & Ma, H. (2010) Signalling and transcriptional control of reproductive development in *Arabidopsis*. *Current Biology* **20**, R988–R997.

Hiscock, S.J. & Allen, A.M. (2008) Diverse cell signalling pathways regulate pollen-stigma interactions: the search for consensus. *New Phytologist* **179**, 286–317.

Irish, V.F. (2010) The flowering of *Arabidopsis* flower development. *Plant Journal* **61**, 1014–1028.

Jaeger, K.E. & Wigge, P.A. (2007) FT protein acts as a long-range signal in *Arabidopsis*. *Current Biology* **17**, 1050–1054.

Klein, A.-M., Vaissiere, B.E., Cane, J.H., Steffan-Dewenter, I., Cunningham, S.A., Kremen, C. & Tscharntke, T. (2007) Importance of pollinators in changing landscapes for world crops. *Proceedings of the Royal Society B – Biological Sciences* **274**, 303–313.

Lifschitz, E., Eviatar, T., Rozman, A., Shalit, A., Goldshmidt, A., Amsellem, Z., Alvarez, J.P. & Eshed, Y. (2006) The tomato FT ortholog triggers systemic signals that regulate growth and flowering and substitute for diverse environmental stimuli. *Proceedings of the National Academy of Sciences, USA* **103**, 6398–6403.

Rea, A.C., Lui, P. & Nasrallah, J.B. (2010) A transgenic self-incompatible *Arabidopsis thaliana* model for evolutionary and mechanistic studies of crucifer self-incompatibility. *Journal of Experimental Botany* **61**, 1897–1906.

Scott, P. (2008) *Physiology and Behaviour of Plants*. Wiley-Blackwell, Chichester.

Soltis, D.E., Chanderbali, A.S., Kim, S., Buzgo, M. & Soltis, P.S. (2007) The ABC model and its applicability to basal angiosperms. *Annals of Botany* **100**, 155–163.

CHAPTER 9

Environmental Stresses

'Stress is what I don't think I would like if I were a buttercup'

John L. Harper, quoted in Turkington (2009).

Many years ago, one of us (MJH) had the good fortune to spend five years working in the School of Plant Biology at the University College of North Wales, Bangor. The head of the school at the beginning of my time there was the world-famous plant ecologist, Professor John Harper. He was undoubtedly one of the greatest plant scientists of his generation, and he loved a good debate. I well remember a seminar that he gave in the school, when he did a major critique of the words 'stress' and 'adaptation' as they were applied in biological science. It was soon evident that Harper did not like either word and felt that they had been misused, particularly by plant physiologists. About half of the audience were plant physiologists! As the quote above suggests, Harper's main criticism was the introduction of anthropomorphic thinking, and that it is not possible for us to know what a plant does not 'like'.

Despite the attacks of Harper and others on the word 'stress', it has continued to be widely used by plant scientists. In 2009 'stress' appeared in the title of 632 papers in the plant science subject area[i], and it is evident that the word has not gone out of use. It is, however, obvious that we need to use it with care. The eminent plant physiologist, Jacob Levitt, suggested that the biological terms 'stress' and 'strain' could be related to the concepts used in mechanics and engineering. In these terms, stress is the force applied per unit area to an object and, in response to the stress, the object is strained and changes in dimension.

Thus for Levitt, the biological sense of stress is 'an external factor acting on an organism (e.g. bars of water stress)', while strain is 'any physical or chemical change produced by a stress.' Levitt took these analogies further, considering elastic and plastic strains and their moduli, but few plant scientists have followed him in this.

Possibly the best definition of a biological stress is an 'adverse force or influence that tends to inhibit normal systems from functioning' (Jones & Jones, 1989), but even here the explanation of 'normal' is problematic. We can conclude that stress in plants is difficult to define, but that it involves inhibition of growth and other processes by some external factor.

9.1 Responses to stress

Most plants suffer from stress at some stage. In the present chapter, we will consider the effects of abiotic stresses that are caused by the physical and chemical environment in which they are living. The ways plants attain tolerance to a certain abiotic stress, adaptation and acclimation, are covered in Chapter 10. Finally, in Chapter 11, the interactions of plants with each other and with herbivorous and pathogenic organisms will be investigated under the topic of biotic stress. However, before we look at each of the stresses in turn, we will

---

[i]ISI Web of Knowledge (accessed 21 August 2010).

Functional Biology of Plants, First Edition. Martin J. Hodson and John A. Bryant.
© 2012 John Wiley & Sons, Ltd. Published 2012 by John Wiley & Sons, Ltd.

consider one generic response of plants to all stresses, the production of **Reactive Oxygen Intermediates** (ROI).

Atmospheric oxygen can be changed to much more reactive, reduced ROI forms by transfer of energy or by electron transfer reactions. In the first case, singlet oxygen ($^1O_2$) is produced, whereas electron transfer leads to a sequential reduction to a superoxide radical (O_2^-), hydrogen peroxide (H_2O_2) and a hydroxyl radical (HO^-). Unlike atmospheric oxygen, ROIs can bring about oxidation of cellular constituents, leading to oxidative damage of the cell.

Some reactions that produce ROIs in plants are involved in processes such as photosynthesis (see Chapter 7, section 7.4) and respiration (Chapter 2, section 2.14.2). Thus, some ROIs are by-products of aerobic metabolism. ROIs are also produced by pathways that are stimulated by abiotic stresses. One example of this is glycolate oxidase, a key enzyme in photorespiration (Chapter 7, section 7.5). Other sources of ROIs have more recently been found in plants, including NADPH oxidases, amine oxidases and cell wall peroxidases. Regulation of these is tightly controlled, and they play a part in the manufacture of ROIs during processes such as programmed cell death (PCD: see Chapters 3, section 3.5.2 and Chapter 7, section 7.11) and defence against pathogens (Chapter 11, section 11.3.8).

Under normal growth conditions, the production of ROIs in cells is low, and many stresses that disrupt the cellular homeostasis of cells enhance the production of ROIs. These include all of the stresses considered below, nutrient deficiency (see Chapter 5, section 5.5) and pathogen attack (Chapter 11). The increased manufacture of ROIs during stress can be toxic to cells, but they also act as secondary messengers for the activation of responses to stress through signal transduction pathways (Chapters 10 and 11).

9.2 Temperature

Different living organisms can survive over a considerable temperature range, from almost boiling point down to $-80°C$ or lower. Blue-green algae can endure temperatures as high as $98°C$ in hot springs, but these are photosynthetic cyanobacteria, not plants. Higher plants are considerably less tolerant, and probably the highest survival temperature recorded for an actively growing plant is around $45°C$. Resting tissues in the dehydrated state can survive higher temperatures (e.g. seeds at around $120°C$).

Dry and dormant tissues can also endure very low temperatures, even to within a fraction of a degree of absolute zero (seeds, spores and pollen grains). This is the basis of seed banks, which store seeds at low temperature to preserve their genetic material for future use (Chapter 12, Section 12.2.3). However, these same tissues can easily be damaged by only a slight freeze in the hydrated condition. These observations point to the critical role of water in temperature stress.

Plants are essentially **poikilotherms** and **ectotherms**, and their internal body temperatures vary with that of the ambient environmental temperature. This is due to their low heat production in relation to mass, as well as having large, not very well insulated surfaces. The inability of plants to regulate their temperature is shown by their loss of heat at night, and leaves may fall several degrees below ambient temperature. Moreover, the sunny side of a tree trunk may be $30°C$ higher than the shaded side on a cold day.

There are relatively few examples where the temperature of plants is raised above ambient by the plant's metabolism. The most famous is the fleshy inflorescence of *Arum* species, which can increase in temperature by some $25°C$ (Chapter 2, section 2.14.2); this acts as an attractant to thermophilic flies (Chapter 8, section 8.6.2 and Figure 8.15). The Eastern skunk cabbage (*Symplocarpus foetidus*) from North America is able to keep its temperature at $15°C$ when the ambient temperature is $-15°C$ and can melt the snow around it in the early spring. For the most part, however, plant temperatures reflect those of the environment.

9.2.1 Chilling

In some plants, particularly those of tropical origin (e.g. maize, rice, tomato) injury occurs when the temperature drops to some point above freezing but low enough to cause damage to tissues, cells or organs of the plants. For many sensitive plants, this happens when they are exposed to temperatures of about $10°C$ to $15°C$. Visible signs of injury can take the form of necrosis, discoloration, tissue breakdown, reduced growth or, in seeds, failure to germinate.

A number of physiological and biochemical phenomena are known to occur during chilling injury in plants.

Frequently the cells show pseudo-plasmolysis, in which the protoplasm gradually detaches itself from the cell wall. Increase in membrane permeability leads to solute loss from cells, mostly as potassium leakage. Ion uptake by roots is often decreased due to a lack of respiratory activity. A sharp drop in photosynthetic rate is frequently observed in chilled plants, and this seems to be related to damage to chloroplast thylakoids (see Chapter 2, section 2.5.2). In some chilling-sensitive plants, non-photosynthetic organs may be starved as translocation is inhibited. For example in sugar cane, translocation ceases completely at $5°C$.

Chilling can also have marked effects on respiration. Often, respiration rate will first rise during injury, then fall as tissues approach death. It seems that aerobic respiratory processes are inhibited, while anaerobic processes are either unaffected or even stimulated. Any inhibition of aerobic respiration will allow anaerobic respiration to go to completion, producing toxic compounds such as acetaldehyde and ethanol (see Chapter 2, section 2.14.2).

Protein synthesis is also inhibited, as decreased oxidative phosphorylation leads to a lower supply of ATP. In addition, sensitive plants can suffer secondary water stress injury. Sugar cane, for example, will wilt if its root temperature drops to $15°C$. It is possible to prevent this injury by covering the shoot with a plastic bag, or by severing the shoot from the root and placing it in water. The site of injury is in the roots and is almost certainly due to a decrease in membrane permeability.

What is the overall mechanism behind all of the effects described above? Abrupt changes to metabolism often occur at a temperature of about $10–15°C$. The systems most often affected are enzyme systems that are associated with membranes (e.g. mitochondria, chloroplasts). It has been found that chilling-sensitive plants have higher levels of saturated fatty acids in their membranes, so their membranes tend to solidify at higher temperatures than those of tolerant plants.

The membranes in chilling-sensitive plants appear to undergo a physical phase transition at a temperature of $10–15°C$, from a flexible liquid crystal to a solid gel structure. This explains many of the observations mentioned previously. For instance, we would expect a change in membrane state of this sort to bring about a contraction of membrane components, leading to the formation of holes and increased permeability (e.g. K^+ leakage). The phase transition could also cause changes in

the conformation of membrane-bound enzymes (e.g. H^+ ATPases), leading to interference in metabolic processes. A reduced ATP supply might also result, since many of the enzymes in mitochondria are associated with membranes.

Work with mutant and transgenic plants has confirmed the idea that damage to membranes is involved in chilling injury. For example, *Arabidopsis* plants that were transformed to increase the proportion of saturated fatty acids in their membranes showed greatly increased chilling sensitivity.

9.2.2 Freezing

Freezing stress is often associated with alpine or arctic environments, but it is also encountered in temperate zones, particularly in early spring. Figure 9.1 illustrates one particular alpine environment, Plateau Mountain in Alberta, Canada, here shown in the summer. As the name suggests, the mountain is unusual in having a gently sloping plateau at its summit, but its height is 2,519 m above sea level and the mean annual air temperature at the summit is $-4.1°C$. The periglacial features known as 'patterned ground' include active permafrost and frost polygons caused by repeated freezing and thawing. Strong winds remove most of the snow in the winter, making this a particularly difficult environment for plant growth but, as Figure 9.1 shows, the alpine flora is well able to grow here, at least in the summer. Many of the alpine plants survive the winter as seeds, as mats close to the rock, or in crevices. However, subalpine fir (*Abies lasiocarpa*)

Figure 9.1 Frost polygons caused by freezing and thawing of the rock at the summit of Plateau Mountain, near Kananaskis, Alberta, Canada. This photograph was taken in the summer and shows an alpine flora inhabits this inhospitable environment. Photo: MJH.

Figure 9.2 Subalpine fir (*Abies lasiocarpa*) growing in Krummholz formation at the summit of Plateau Mountain, near Kananaskis, Alberta, Canada.
Photo: MJH.

is a perennial tree able to survive on Plateau Mountain (see Figure 9.2). This species normally grows to 20 m in height, but here continual exposure to fierce, freezing winds causes the trees to become stunted and deformed. This is known as the *Krummholz* formation (German: *krumm*, 'crooked' and *holz*, 'wood') and it is a common feature of subarctic and subalpine tree line landscapes.

Susceptibility of plants to frost is probably the single most important factor limiting the growth and distribution of plant species, and it is significant for crop growth. Many plants are exposed to freezing stress at some time in their life cycle, and in this section we will cover the mechanisms of freezing damage.

As long ago as the early 1800s, scientists observed that tree trunks made cracking noises that sounded like a gun being fired when exposed to freezing stress. At first it was thought that these noises were due to the expansion of water within the tissues, causing the cells to burst. It is well known that the volume of frozen water is greater than that of liquid water, so it seemed a reasonable assumption. However, it was not realized at the time that plant cell walls can easily stretch to accommodate the extra volume. After much searching for burst cells, the idea was abandoned and the cracking noises were attributed instead to uneven contraction of wood. It is now known that there are two kinds of freezing in plants – intracellular and extracellular.

Even hydrated plant tissues will survive freezing if the rate of cooling is fast enough. So, if tissues are frozen by plunging them into liquid nitrogen (at $-196\,^{\circ}$C) then

they are generally not damaged, provided that subsequent warming is equally rapid. The key is that rapid freezing to the temperature of liquid nitrogen involves the vitrification of water as it forms a solid without the creation of ice crystals (formation of such intracellular crystals would cause major damage to cellular structures and certain cell death).

Under laboratory conditions, many observations have been made concerning intracellular freezing. Nearly all of these experiments have involved fairly rapid freezing of isolated tissues. The freezing rates used (10 to $100\,^{\circ}$C min^{-1}), although much slower than when tissue is frozen in liquid nitrogen, have generally been considerably faster than those normally observed in nature (normally less than $10\,^{\circ}$C hour^{-1}). Therefore it is not clear how important intracellular freezing is in the natural environment, but one possibility is 'sunscald'. On a cold winter day, the sun shining on a tree may raise its temperature on one side some 20–$30\,^{\circ}$C above that of the shaded side. If the sun then suddenly disappears behind a cloud, the temperature of the tissue may drop so rapidly that intracellular freezing does occur. This leads rapidly to the death of the tissue.

If the significance of intracellular freezing is doubtful in nature, there is no doubt that extracellular freezing is important. Ice crystals form in the intercellular spaces of plant tissues when the freezing rate is less than $10\,^{\circ}$C min^{-1} because the water in the apoplast contains less solute than the intracellular fluid. As the chemical potential of ice is lower than that of water at the same temperature, the water potential outside the cell is lowered. Thus, water is drawn out of the cell protoplasts down a potential gradient, and extracellular crystals continue to grow. Often these crystals are confined to particular areas and they can form large masses up to one thousand times larger than a cell.

On thawing, if the tissues are not injured, the water is reabsorbed by the cells. However, the major problem with extracellular freezing is that it causes dehydration of the protoplasm, as water is sucked out and cells contract or collapse. Most interest in freezing injury is now concentrated on the indirect effects of dehydration caused by extracellular freezing. This leads to mechanical stress, contraction and collapse of cells and loss of water from the protoplast. These changes may cause solution effects involving cell pH, increases in ionic concentrations and close contact of solute molecules, including proteins, which are often denatured. Above all, it seems that damage to membranes is crucial in freezing stress, and thus stabilization of membranes is essential in cold acclimation (see Chapter 10, section 10.2.2).

9.2.3 High temperature

Typically, high-temperature stress in plants is associated with desert environments (Figure 9.3), but it is now also a major factor in future climate change predictions. Here we will consider the physiological and biochemical effects of heat stress, while Chapter 10 (section 10.2.3) covers adaptation and acclimation and Chapter 12 (section 12.1.2) will put this material in the context of climate change. It is often difficult to separate out the effect of high temperature from the effects of drought (section 9.4) and high light levels (section 9.7.2), as these phenomena frequently accompany each other.

With freezing stress, the time of exposure is of relatively little importance and, once frozen, the damage to tissues is

Figure 9.3 The saguaro (*Carnegiea gigantea*) is a tree-sized cactus species. It is native to the southwestern USA and part of Mexico, and is here photographed north of Phoenix, Arizona.
Photo: Margot Hodson.

either done or it is not. With heat injury, time of exposure is of great significance, and heat-killing temperature varies inversely with exposure time. Thus, *Tradescantia discolor* cells are killed at $65°C$ by an exposure for 1.8 minutes, at $60°C$ for seven minutes, at $50°C$ for four hours and at $40°C$ for 22 hours. Plants can only survive a few seconds at even higher temperatures, but some plants are more tolerant than others. Among crop plants, squash, maize and oilseed rape are killed by exposure to $50°C$ for ten minutes. It is worth mentioning here that temperatures exceeded $50°C$ in several countries in 2010.

There are two types of heat injury: *indirect* at the lower temperature range, and *direct* at the upper temperature range. Indirect injury is essentially injury by desiccation or starvation. Transpiration rate increases with increase in temperature, because of the effect on the diffusion coefficient of water and an increase in the water potential gradient between the plant and the surrounding air. This increased water loss could result in dehydration, resulting in stomatal closure, but then the situation is aggravated because the cooling effect of evaporation is decreased. At 70 per cent relative humidity, a $5°C$ rise in leaf temperature above that of the atmosphere doubles the transpiration rate.

The temperature maximum for photosynthetic assimilation is $3–12°C$ below the heat-killing temperature. Starvation can occur, however, before the temperature maximum for photosynthesis is reached, because respiration has a higher temperature optimum than photosynthesis. For example, in potato leaves the optimum for respiration is $50°C$, while the optimum for photosynthesis is only $30°C$.

The point at which respiration rate and photosynthesis are equally rapid is called the **temperature compensation point**. If the temperature rises above this point, then the plant will begin to use up its reserves. A sufficient time above the compensation point will exhaust these reserves and the plant will starve to death. This deficit increases particularly rapidly in C_3 plants with an active photorespiration rate (Chapter 7, section 7.5) in addition to normal respiration.

The indirect effects of high temperature stress occur quite slowly, but it can also have direct and rapid effects on cytoplasmic or membrane proteins. The sensitivity of photosynthesis to high temperature seems to be mainly due to chloroplast damage. In particular, the electron transport system of photosystem II seems to be more sensitive than other systems. This may well be related to the phase changes observed in membranes under heat stress.

An early idea for heat stress injury was that it liquefied membranes, increasing permeability, and the mobility of lipids must increase with increased temperature. Under these conditions, lipid vesicles may form, and this would almost certainly lead to death. Protein denaturation was proposed by Belehradek in 1935 as an explanation for heat injury and it is still accepted as important today, although greater emphasis is now placed on the inhibition of photosynthesis. Cytoplasmic streaming is halted at a lower temperature than other cell processes and it also appears to be sensitive to direct heat stress.

9.2.4 Fire

Stress caused by exposure to fire is related to that caused by heat and, in many ways, fire is the ultimate in heat stress. Fires can be caused naturally by phenomena such as lightning strikes, but they are often also caused by careless humans. Some environments are prone to wildfires, and Mediterranean areas, with cool, wet winters and dry, hot summers, are particularly so. Figure 9.4 shows the effects of a wildfire on woodland in southern Portugal. Temperatures reached during wildfires are usually well in excess of the heat-killing temperatures noted in Section

Figure 9.4 A woodland in the Monchique mountains in Southern Portugal. There had been a fire in the previous year, and this shows the first shoots of the recovery. Most of the trees are cork oak (*Quercus suber*), but there are also a few maritime pine present (*Pinus pinaster*).
Photo: Will Simonson.

9.2.3 and may reach 150°C at the soil surface. Although exposure to fire is often disastrous for individual plants, many plant communities are adapted to being frequently burned, and these adaptations will be covered in Chapter 10, section 10.2.4.

9.3 Waterlogging

Marshes, mires and swamps are habitats that are regularly waterlogged and the plants growing in them (hydrophytes) are frequently adapted specifically to these environments (see Chapter 10, section 10.3.2). Waterlogging can be a major problem after heavy rainfall for arable crops or, indeed, for any non-adapted plant. Here we will consider the effects of waterlogging, first on soil, and then on plants.

9.3.1 Soil waterlogging

As we saw in Chapter 5 (section 5.4.1), approximately half of the total volume of a mineral soil is occupied by solid material. The rest is taken up by water and gas, and one cannot be affected without the other as they are interlinked. The two most important biological reactions involving gases in soils are the respiration of higher plants and the decomposition of organic compounds by microorganisms. Both of these processes use oxygen and generate carbon dioxide. The composition of soil air is dependent on the amount of air space available, together with the rates of biochemical reactions and gaseous interchange.

A sufficiently well-aerated soil must have two characteristics: spaces free of solids and water, and opportunity for the movement of gases. The total porosity of a soil is determined by its bulk density, the weight of soil per unit volume (g cm^{-1}). The greater the bulk density, the less easy it is for water and air to travel through the soil and the greater the chance of waterlogging. For this reason, soil compaction is a major cause of waterlogging in agricultural contexts.

There are two main kinds of naturally waterlogged soils. Organic soils include peat and fen soils, and these have a low content of mineral matter. Gleys are mineral soils that are saturated with water annually for a sufficiently long time to give the soil the distinctive bluish-green gley horizons resulting from reduction-oxidation (redox) processes.

The major effect of waterlogging is to decrease gaseous interchange between the soil and the atmosphere. This depends on the rate of biochemical reactions (affected by temperature and the concentration of organic residues) and the speed at which each gas is moving. The mechanisms of mass flow and diffusion facilitate this movement. Oxygen diffuses in the gas phase about 10,000 times faster than it does in solution, and the concentration of oxygen is 30 times less in water than in air. Thus concentration also decreases with depth.

There are three ways of characterizing the degree of soil aeration: direct measurement of gaseous oxygen in the soil; estimating oxygen diffusion rates; and determining the soil redox potential (E_h). The redox potential is an important chemical characteristic of a soil. If a soil is well aerated, oxidized states dominate (e.g. ferric iron, manganic manganese, nitrate and sulphate). In poorly aerated soils, reduced forms are found (e.g. ferrous iron, manganous manganese, ammonium, and sulphides).

The E_h value at which redox reactions occur varies with the chemical to be oxidized or reduced (see Table 9.1). In a well-drained soil, E_h is between $+400$ mV and $+700$ mV. At these E_h values, dissolved oxygen is used by plant roots and aerobic microbes in aerobic respiration. Oxygen concentration is decreased by this activity and, in waterlogged soils, aerobic bacteria are replaced by anaerobic bacteria, which use electron acceptors other than oxygen. As the soil E_h value drops, nitrates, sulphates and ferric oxides are reduced and the energy gained is utilized for anaerobic microbial metabolism.

Under fully waterlogged conditions, E_h is about -400 mV. At -240 mV, carbon dioxide is reduced to methane by methanogenic bacteria. This is important in both naturally occurring anaerobic soils and in rice paddies, and increasingly so as the methane produced is a powerful greenhouse gas (See Chapter 12, section 12.1.1). Furthermore, under waterlogged conditions, incomplete breakdown of organic compounds allows the build up of toxic substances. For example, organic acids, ethylene (a plant growth regulator), acetic acid, butyric acid and alcohols can all accumulate under waterlogging.

9.3.2 Effects of waterlogging on plants

Flooding is more harmful to plants in warm summer temperatures, as metabolism is greater, while flooding in cool conditions may not cause much damage. The major problem for plants growing in waterlogged soils is anaerobiosis and, at greater than $20°C$, plant roots, soil fauna and microbes can remove all of the oxygen from the soil water within one day. The activity of roots is usually hampered when the oxygen content of soil falls below 10 per cent, and flooding-sensitive plants are often badly damaged by an anaerobic environment for 24 hours.

Anaerobic conditions affect roots to varying extents. Root apices of cotton are killed after three hours, and in soybean after five hours, but in maize only after 70 hours. For maize root tips, the critical oxygen pressure (COP) at which the root first shows a decrease in respiration rate is 20 per cent oxygen by volume (almost ambient). When oxygen concentrations are below the COP, the centre of the root becomes anoxic or hypoxic (partially lacking in oxygen).

Lowered oxygen leads to changes in respiratory metabolism from aerobic to anaerobic. In the absence of oxygen, the Krebs cycle cannot operate due to a lack of a terminal electron acceptor for the oxidation of

Table 9.1 Reactions occurring in soils at differing redox potentials.

Redox couple	Microbial process	Redox potential mV (pH 7)
$O_2 \rightarrow H_2O$	Aerobic respiration	$+820$
$NO_3^- \rightarrow N_2, N_2O$	Denitrification	$+420$
$Mn^{4+} \rightarrow Mn^{2+}$	Manganese reduction	$+410$
Organic matter $\rightarrow$ organic acids	Fermentation	$+400$
$Fe^{3+} \rightarrow Fe^{2+}$	Iron reduction	-180
$NO_3^- \rightarrow NH_4^+$	Dissimilatory nitrate reduction	-200
$SO_4^{2-} \rightarrow H_2S$	Sulphate reduction	-220
$CO_2 \rightarrow CH_4$	Methanogenesis	-240

NADH (see Chapter 2, section 2.14.2). ATP can then only be produced by fermentation, where pyruvate is first converted to lactate. However, this does not last for long, as lowered cytoplasmic pH causes inhibition of lactate dehydrogenase and a switch to ethanol fermentation. Under anaerobic conditions, leakage of protons from the vacuole to the cytoplasm adds to the acidity caused by lactate dehydrogenase. The lowered pH, termed cytosolic acidosis, is a major cause of damage, and it frequently leads to cell death in sensitive plants.

Under anaerobic conditions, only two molecules of ATP are produced per molecule of glucose respired, compared with 34–36[ii] molecules with aerobic respiration. The growth of roots is curtailed due to lack of energy, leading eventually to decreased shoot growth. Lowered ATP availability causes ion uptake to be decreased. Waterlogging also decreases the permeability of roots and one, perhaps unexpected, symptom of waterlogged plants is that they wilt.

Anaerobiosis also causes alterations in gene regulation in affected plants. Regulation takes place at the level of mRNA accumulation and translation in *Arabidopsis* and other species. Under hypoxia, selective mRNA translation is an important regulatory mechanism. Changes in mRNA accumulation in response to anoxia are affected by many factors, including the length and severity of the stress and developmental age. In *Arabidopsis*, the genes that are induced by anaerobiosis include those coding for proteins involved in carbohydrate metabolism, fermentation, ethylene synthesis, processes mediated by auxin, amelioration of reactive oxygen species, calcium-mediated signal transduction and gene transcription.

The concentrations of endogenous plant growth hormones change in the waterlogged plant. Gibberellins and cytokinins are produced in the root, and their transport to the shoot is known to be decreased by anaerobiosis. Ethylene concentration in susceptible plants often increases markedly. Lowered oxygen availability speeds up the manufacture of the ethylene precursor 1-aminocyclopropane-1-carboxylic acid (ACC) in roots. Phosphorylation of two isoforms of ACC synthase (ACS) by MAP kinase 6 has been observed in *Arabidopsis*, leading

to an increase in ACS protein and activity (the unphosphorylated proteins are rapidly degraded). Subsequently, ACS activity and ethylene manufacture are increased. In some sensitive plants, ACC is transported in the xylem to the shoot, where it is converted to ethylene by ACC oxidase (ACO), often within four hours of transfer to hypoxic conditions.

Ethylene biosynthesis cannot happen in totally anaerobic conditions, as conversion of ACC to ethylene by ACC oxidase needs oxygen. Ethylene is known to bring about a nastic growth response (see Chapter 6, section 6.14), termed **epinasty**. Here, ethylene causes cells on the upper adaxial surfaces of leaf petioles to expand, and thus the leaves droop.

In addition to the problems caused by lowered oxygen availability many of the organic and inorganic substances produced in waterlogged soils (see Section 9.3.1) are toxic and can cause growth reduction.

9.4 Drought

Drought will cause stress to a plant if too little water is available in a suitable thermodynamic state. The term 'drought' usually denotes a period without significant precipitation, during which the water content of the soil is reduced to such an extent that plants suffer from lack of water. This is often coupled with a period of strong evaporation, caused by dryness of the air and high levels of radiation. Drought can also occur for other reasons: lowered osmotic potential of water in saline soils (see section 9.5); lack of water availability in a frozen soil; or due to the soil being too shallow for the development of an adequate root system.

How do plants respond to drought? Whatever the cause of drought, it nearly always develops slowly over a period of time. Drought stress comes on gradually in almost all cases, and the processes affected can be considered as a sequence of events which are covered below.

As soil water potential decreases and the water content of the soil declines, the most sensitive response is a decrease in cell turgor and a decrease in cell growth in response to this. Thus, the earliest response to water stress seems to be mediated by biophysical rather than chemical changes. As the water content of the cell protoplasts decreases, they shrink and the cell walls relax. This decrease in cell volume results from a lower hydrostatic

---

[ii]There is some uncertainty about the actual number, because the oxidation of NADH entering the mitochondrion from the cytosol may generate only two, rather than three, ATP molecules.

pressure or turgor. Expansion growth requires the plant cells to be turgid and, if they are not, growth soon slows down, as seen particularly in the decrease in leaf expansion in response to water deficit.

The response to water deficit involves perception by specific receptors, and activation of a signal-transduction pathway. When cell turgor first decreases in plants, it appears that initial perception is through an osmosensor in the plasma membranes as they undergo changes in physical state. In *Arabidopsis*, the osmosensor AtHK1 has a histidine kinase domain, a receiver domain and two transmembrane domains. There is now reasonable evidence to suggest that AtHK1 is the first sensor that triggers a downstream signalling cascade, resulting in the changes in gene expression observed during water stress (Figure 9.5).

After perception of the osmotic changes during the early stages of drought, signal transduction involves the up-regulation of genes responsible for production of protein kinases and phosphatases. Several calcium-dependent and mitogen-activated protein (MAP) kinases have been found in plants, and these are involved in the transduction to the nucleus of the signal first sensed in the plasma membrane.

Once the initial perception and transduction events have happened, further responses can be divided into those that are abscisic acid (ABA)-dependent and those that are ABA-independent. The latter responses are less well understood, although ABA-deficient and ABA-insensitive mutants have a number of genes that are rapidly induced by drought, indicating that ABA is not needed for expression. ABA-dependent responses are discussed below.

As a drought sets in, several factors contribute to decreased water flow through the plant. As the root cells shrink they tend to move away from the soil particles that hold water. The outer layers of the root cortex often become suberized (suberin is a waxy substance that increases resistance to water flow). Even during quite mild drought, the xylem elements show some cavitation, and this increases as the drought continues. It can also be a continuing problem after the drought subsides.

We noted in Chapter 7 (section 7.10) that stomata tend to close when plants are exposed to drought, and now we need to consider the mechanisms behind this response. It seems that there are two:

1 **Hydropassive closure**. We have seen that leaf surfaces are protected from water loss by a heavy waxy cuticle. Stomatal guard cells are not protected in this way, possibly because they are required to open and close continually. When the atmosphere becomes dry and hot, they lose water and thus turgor. The guard cells become flaccid and close. Thus, this type of closure requires no metabolic activity in the guard cells, which just act as simple osmometers. Brodribb & McAdam (2011) have shown that lycophyte and fern stomata are not responsive to ABA and only have this passive control of cell turgor, making stomatal behaviour predictable. This suggests that the change from passive to active metabolic control of stomatal aperture occurred after the divergence of seed plants from ferns, about 360 million years ago (see Chapter 1, section 1.6). We will now consider the active responses of seed plants.

2 **Hydroactive closure**. This is metabolically dependent and essentially involves a reversal of the ion fluxes that cause opening. This type of closure is often triggered by decreasing water potential in the mesophyll and it involves ABA. ABA was discovered in the late 1960s (see Chapter 3, section 3.6.5) and was soon found to have a prominent role in the response of plants to water stress. It was shown to accumulate in water-stressed plants and, if it is applied externally, it is a powerful inhibitor of stomatal opening. Tomato mutants that fail to accumulate normal ABA levels have been observed to wilt very easily. Thus the evidence is very strong that ABA is involved in plant water stress responses. It seems that in leaves there is always enough xanthoxin, a precursor of ABA, available for biosynthesis when drought conditions begin, whereas in roots, genes for precursor synthesis need to be induced to provide the necessary substrate. Xanthoxin synthesis in both roots and leaves occurs in the plastids, but this is then converted to ABA by two enzymes in the cytoplasm. The enzyme responsible for xanthoxin synthesis, NCED (9-cis-epoxycarotenoid dioxygenase), has a promoter that is induced by water stress in roots and leaves, and this seems to be the main regulatory step. ABA is released eventually into the apoplast, where it flows in the transpiration stream as far as the stomata. Another feature of stomata is that they have no plasmodesmata, so transport to them must be apoplastic. ABA does not enter the guard cells, but it inhibits phosphorylation of the plasma membrane proton ATPase that is mediated by phototropin (see Chapter 3, section 3.7.5). ABA causes

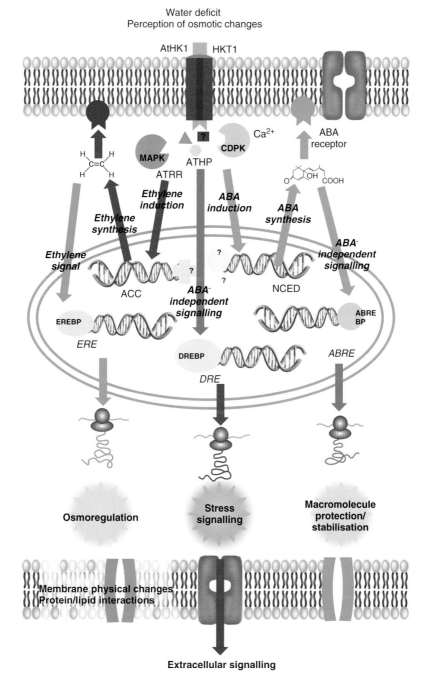

Figure 9.5 Sensing, signalling and cell-level responses to drought stress. Key: ABA-responsive element (ABRE); ABRE binding protein (ABREBP); 1- aminocyclopropane-1-carboxylic acid (ACC); *Arabidopsis thaliana* histidine kinase (AtHK1); a phospho-relay intermediate (ATHP); a response regulator (ATRR); Ca^{2+} dependent kinases (CDPK); dehydration-responsive element (DRE); DRE binding protein (DREBP); ethylene- responsive element (ERE); ERE binding protein (EREBP); a high affinity K$^+$ transporter (HKT1); mitogen-activated protein kinase (MAPK); 9-*cis*-epoxycarotenoid dioxygenase (NCED).
From Chaves *et al.*, 2003. (Reproduced by Permission of CSIRO Publishing, Australia).

hydrogen peroxide production in the guard cells, resulting in stomatal closure. Cysteine residues in the phototropins are possibly the main target of the hydrogen peroxide.

Finally, there is also quite a lot of evidence that, in some cases, the mesophyll does not need to sense water stress for stomata to close. Under some conditions and in some species (e.g. tobacco, *Nicotiana plumbaginifolia*), the roots can sense a drop in water potential, synthesize ABA and transport it to the leaves in the xylem transpiration stream. Other signals, such as xylem sap pH, cytokinins, an ethylene precursor and malate have all been suggested to be involved in root to shoot signalling under drought conditions in some species.

Soon after leaf expansion declines, protein synthesis starts to decrease, including the synthesis of the enzyme nitrate reductase, one of the most sensitive to drought. Nitrate accumulates in drought-affected plants that have been supplied with nitrate fertilizer.

Once the stomata close, photosynthesis is reduced due to lack of carbon dioxide. Photosynthesis (Chapter 7, section 7.4) is not as sensitive to water stress as leaf expansion and it is not affected by mild stress very much. As stress becomes severe, carbon dioxide levels in the leaves decline as the stomata are closed. This, plus direct effects of water stress on photosynthesis, lead to decreased photosynthetic rates.

Another critical process that might be affected by water stress is translocation of photosynthate in the phloem (Chapter 6, section 6.5.2). Experiments have shown, however, that translocation is much less affected than photosynthesis, and it is only inhibited when stress is very severe. This means, for example, that plants are able to move photosynthate to seeds at the end of the season even if there are drought conditions.

Plant respiration is one of the last processes to be affected by drought and, when this happens, the plant is usually suffering serious damage or nearing death. At first, plants undergo reversible wilting by day and recover at night – but later, if this continues, they senesce and die.

9.5 Salinity

Soil salinity is a major problem for agriculture, particularly in arid zones, and has been at least partly responsible for the collapse of previous civilizations (e.g. the Sumerian civilization in Mesopotamia in the period 2100–1700 BC).

Here we will first consider saline environments and soils, and then the effects of salinity on higher plants. In Chapter 10 (section 10.5.2), acclimation and adaptation to salinity and a model for salt tolerance in plants will be presented.

9.5.1 Saline environments and soil

The major saline environments in which plants grow are:

1 **Maritime environments**. Most higher plants are terrestrial and do not grow permanently under sea water. The exceptions are the eelgrasses such as *Zostera marina*, which is found in estuaries either underwater or partially floating. *Zostera* beds, where they occur, are important habitats and provide food for geese. The major maritime habitats for higher plants are salt marshes, sand dunes and cliffs. In salt marshes, the plants are often inundated with sea water, while sand dunes and cliffs are more frequently exposed to salt water in the form of a spray. Figure 9.6 illustrates several maritime habitats at Bradwell-on-Sea in Essex, England. Growing on the mud flats at Bradwell there is marsh samphire (*Salicornia europaea*), a succulent halophyte that can be eaten both raw and cooked (Figure 9.7).

2 **Inland saline lakes**. Examples of inland saline lakes include the Dead Sea in Israel, the Great Salt Lake in Utah and Neusiedler See in Austria. In all cases, higher plants grow around the edges of these lakes and have to cope with soil salinity as a result.

Figure 9.6 Coastal scene at Bradwell-on-Sea, Essex, England, revealing several maritime habitats. In the distance are the mud flats, and there are sand dunes in the middle foreground and a creek in the foreground.
Photo: MJH.

Figure 9.7 Marsh samphire (*Salicornia europaea*), a succulent halophyte, here growing on the salt marsh at Bradwell-on-Sea, Essex, England.
Photo: MJH.

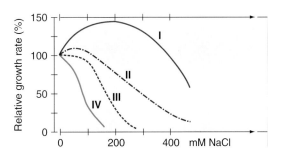

Figure 9.8 Generalized growth responses of various species to increased salinity. Key: I, Halophytes; II, crops like sugar beet that have halophytes as ancestors; III, salt tolerant crops like barley; IV, salt sensitive crops like beans.
From Greenway & Munns, 1980. (Reproduced with permission of Annual Reviews, Inc.)

3 Roadsides. Winter de-icing salt operations frequently leave roadside verges with residual salinity problems.

4 Irrigation agriculture in arid environments. Even irrigating crops with slightly saline water in arid environments can lead to an increase in soil salinity. The irrigation water often evaporates rapidly, leaving the salt to accumulate at the soil surface.

Any classification of saline environments has to be arbitrary, as they form a continual series. The US salinity laboratory definition of a saline soil is one whose solution, when extracted, has an electrical conductivity greater than 4 mmhos cm^{-1} (about 40 mM NaCl). In nearly all cases, the ions involved are sodium and chloride, but there are some instances where other ions can be important (e.g. sulphate and magnesium).

9.5.2 Effects of salinity on plants

Plants growing in saline soil, be it in a salt marsh, an inland environment or an agricultural soil in the developing world have three major problems:

1 High salt concentrations lower soil osmotic potential and, hence, overall water potential (see Chapter 5, section 5.4.2). Plants will therefore have difficulty in taking up water from the soil; thus, salinity stress is related to drought stress (see section 9.4).

2 High concentrations of sodium and chloride ions are potentially toxic to most species, including nearly all crops. Enzymes are particularly sensitive to these ions (see Chapter 10, section 10.5.2).

3 Sodium and chloride compete with nutrient ions (e.g. potassium and nitrate) for uptake by the roots. This can bring about nutrient deficiencies. Ratios of ions (e.g. Na$^+$: K$^+$, Na$^+$: Ca^{2+}) are therefore important and are often used in salt tolerance research.

Figure 9.8 shows some generalized responses to increases in salinity. Plants differ considerably in their salt tolerance. The growth of **halophytes** (salt plants) is optimal at high levels of salinity, and some have their growth increased by salinity over a fairly wide range (I). These plants include many of the common seashore plants, such as marsh samphire (*Salicornia europaea* – see Figure 9.7). Only a very few crop species have their growth stimulated by salinity and are halophytic (II). Sugar beet is an exception, as it was originally derived from the halophyte *Beta maritima*. Most crops have their growth reduced considerably by salinity; they are called **glycophytes** ('sweet plants'). The response of crop species to salinity also varies quite considerably; barley and wheat (III) are much more tolerant than pea or bean (IV). There are also considerable differences between cultivars and varieties of the same species, and this variation has led scientists to search for increased salt tolerance in crops.

The responses of plants to salinity depend on the species. The vast majority of species, including almost all crops, are excluders of salt and keep it out of their tissues as much as possible. However, no plant is a total excluder of salt; peas and beans, for example, would be regarded as *excluders*, but they still let through some salt. On the other hand, most halophytes are *includers* and take up a lot of salt.

Both of these strategies have advantages and disadvantages. The principal problem of exclusion is that the plant suffers water deficit. If there is a low external water potential, then water cannot be taken into the plant very easily. This causes several side effects that are related to those caused by drought stress (decrease in cell expansion, stomatal closure causing decreased photosynthesis and inhibition of protein synthesis).

Most of the halophytes are includers and, by taking up large amounts of ions into their tissues, they avoid water deficit as they maintain a lower water potential than their substrate. As we have already noted, however, Na^+ and Cl^- are potentially toxic and can also cause deficiencies in nutrient ions. Halophytes have several strategies for dealing with ion toxicity, and these will be covered in Chapter 10 (section 10.5.2).

9.6 Chemical stress

In this section, we will consider a group of stresses that are often, but not always, caused by humans: heavy metal toxicity, air pollution and soil acidity. In some respects, salinity could also be included here, but it is less often associated with human pollution and has a number of features that connect it with drought stress.

9.6.1 Heavy metal toxicity

The definition of the term 'heavy metal' is imprecise, but has been used to describe about a dozen elements with densities above $5\,g\,cm^{-3}$ that are metals or metalloids. Examples of heavy metals include cadmium, copper, lead, manganese, mercury and zinc.

In what environments is heavy metal toxicity most likely to occur in higher plants?
• Some soils are naturally very high in metals, due to the bedrock from which the soil was derived. One of the most famous examples is the serpentine soils, which are high in iron, magnesium, nickel, chromium and cobalt, but very low in calcium. These soils support a specialized flora that is adapted to the soil conditions.
• Metal toxicity can also be brought about as the result of human activity. Mining and smelting often leads to localized pollution, and the spoil heaps from metal mines are often very high in metals that are toxic to plant growth. A classic example is that of Parys Mountain on Anglesey in Wales, the site of a large copper mine that was worked

Figure 9.9 Parys Mountain, on the island of Anglesey in Wales, is the site of a copper mine. More than 200 years after the mine reached its peak and closed, the mine spoil and contaminated soil is still too toxic for most plants to grow.
Photo: MJH.

in the late 18th century. Even today, very little is able to grow in its toxic soils (Figure 9.9). In addition, sewage sludge often contains high concentrations of heavy metals that have been adsorbed; release onto agricultural land is limited by the metal content.
• Finally, changes in soil pH caused by human activity can increase the availability of metals to toxic levels. Mostly, heavy metals are more available for plant uptake at low pH, so acid rain (see section 9.6.3) and treatment with some fertilizers can indirectly lead to metal toxicity.

As we saw in Chapter 5 (section 5.5), some heavy metals (e.g. copper and zinc) are micronutrients that are essential for plant growth, but above a certain concentration they will become toxic. However, other heavy metals (e.g. lead and cadmium) are only known for their toxic effects. Heavy metal toxicity gives rise to many symptoms, some of which are specific to the particular metal. Decrease in growth, particularly root growth, is frequently observed, as are changes in colouration of the leaves (e.g. chlorosis and necrotic patches).

There are several mechanisms whereby metals cause toxicity problems. As has often been the case with the stresses considered so far in this chapter, damage to the plasma membrane, resulting in leakage of solutes (e.g. potassium) is frequently observed with metal toxicity. Heavy metals can also compete for uptake with nutrient ions, potentially causing deficiency symptoms (e.g. copper can cause iron deficiency). Bonding of heavy metals with proteins and other compounds because of their similarity to essential metals is another possible mechanism. Oxidative stress can also be caused because of the oxidative-redox properties of many heavy metals.

9.6.2 Air pollution

A number of common air pollutants are known to cause stress symptoms in plants. Many industrial processes produce large amounts of dust particles which can cover plants, reducing photosynthetic rates and sometimes blocking stomatal pores. Such particles can also be abrasive when blown in the wind. More serious problems for plants are caused by chemical pollutants, the major ones being sulphur dioxide (SO_2), nitrogen oxides (NO_x) and ozone. Here we will consider the effects of these pollutants in their dry forms, and in the following section the topic will be acid rain – the result of sulphur and nitrous oxides reacting with water.

The major sources for anthropogenic emissions of sulphur dioxide are coal-fired power plants and metal smelting plants. World SO_2 emissions continue to increase, particularly in China, but in North America and Western Europe SO_2 is declining due to the use of cleaner fuels in residential areas, power stations burning cleaner fuels, and flue gas desulphurization.

The Prunéřov power station, near Kadaň, is the largest coal-fired power station in the Czech Republic, with a capacity of 1,490 MW (Figure 9.10). The plant has been retrofitted with flue gas desulphurization, as were many others in the country. As a result, the Czech Republic decreased its SO_2 emissions by 87 per cent between 1990 and 2002. However, the Prunéřov power station remains the largest source of CO_2 in the Czech Republic, and one of the largest in Europe (see section 12.1.1).

Dry deposition of SO_2 can occur long distances from the source, particularly if the latter is a tall stack. For example, if the weather is dry, half of the SO_2 is still in

Figure 9.10 The Prunéřov coal-fired power station, near Kadaň, is the largest in the Czech Republic and has been retrofitted with flue gas desulphurization.
Photo: MJH.

the atmosphere one day after emission and, if the wind speed is 7 m s^{-1}, then it will have travelled 600 km. Plants can metabolize SO_2 that is absorbed through the stomata, but if SO_2 concentration is too high or if exposure is for too long a period, toxicity symptoms will be observed. Increased stomatal opening has been observed in plants treated with SO_2, as have changes in enzyme activity (e.g. glutamate dehydrogenase is increased) and degradation of chlorophyll.

Concentrations of NO_x from sources such as motor vehicles, industry, and nitrogen fertilizer application are still increasing worldwide. Dry deposition of nitrogen accounts for about half of the total in the SE of England but, in mountainous areas, wet deposition dominates ($\approx$90 per cent). It has been known for many years that NO_x as a pollutant does have some biochemical effects on plants, including decreasing photosynthesis. More recently, the role of nitric oxide (NO) as a signalling molecule in plants has become apparent (Chapter 3, section 3.6.12). However, as pollutants, NO_x species seem to be more important for their secondary effects, through the production of ozone and in the production of acid rain (see section 9.6.3 below).

Ozone in the upper atmosphere (stratosphere) is beneficial, as it absorbs much ultraviolet light (see section 9.7.2). At ground level, however, it is a serious pollutant. Nitrogen oxides from vehicle emissions react in sunlight to form ozone. This is particularly problematic in hot, sunny weather in cities; ozone is a major constituent of photochemical smog, while peroxyacyl nitrates (PANs) are another important component. Although most ozone is produced in cities, it can then be blown into agricultural areas, causing damage to crops. Crop yields are decreased and plants frequently show chlorosis.

9.6.3 Acid rain, soil acidity, forest die-back and aluminium toxicity

'"Acid rain" is a general, short-hand term used to describe the deposition of all atmospheric pollutants of an acidic, or potentially acidic nature whether they be deposited in rain or snow (wet deposition) or in the dry state as gases and small particles'

Mason (1992).

Acidic precipitation ('acid rain') is a global problem that has deleterious effects on both aquatic and terrestrial environments. In the atmosphere, SO_2 and NO_x react with water and are oxidized to produce sulphuric and nitric acids, which return to earth in rainfall. It should be remembered that 'normal' rain has a pH of 5.0–5.6. This acidity mainly comes from the dissolution of carbon dioxide. However, in areas affected by acidic precipitation, the mean annual pH of rainfall ranges from 3.0 to 5.5, with individual storms as low as pH 2.0.

Wet deposition in rain will reflect both the distance from the source and the amount of rain. For example, the areas of greatest anthropogenic sulphate wet deposition in the UK have been the East Midlands and the Pennines, close to the sources, and in the mountains of Wales and Scotland, where the concentrations are low but the rainfall is high. 'Occult deposition' is the impaction of mist and fog on surfaces such as tree canopies. This can be very important in temperate regions at high altitudes (above 450 m).

Acidic precipitation has had major ecological effects in both aquatic and terrestrial environments. We will concentrate on the latter, and particularly on forest systems. Acidic precipitation has been implicated in damage to forests in both Europe and North America, but there is a considerable debate concerning the mechanism of damage among scientists working in this field.

Direct damage to leaves (including needles) can be caused by acidic precipitation. Exposure of vegetation to high levels of acidity (pH 3.3 or lower) can lead to foliar lesions or the abscission of leaves. Much will depend on how long the leaves are exposed, their surface properties and rain events after the pollution episode. Symptoms include chlorosis, the wilting of tips and accelerated senescence. Plants vary markedly in their responses to pollutants. It is also possible for acid to increase the permeability of leaves, leading to the leaching of important nutrients from them, e.g. calcium.

The effects of acidic precipitation on the soil will vary markedly according to the soil type and the bedrock. In soils with calcareous bedrock, the effect may be very limited due to a strong buffering effect. Even in these soils, though, high acid inputs can greatly increase available calcium in the soil, which can lead to potassium deficiency in the plants growing under these conditions. Unfortunately, most soils which have high anthropogenic acid inputs are already acidic naturally, and are often podzols. For example, the Black Forest soils in Germany and the Canadian Shield soils are acidic.

What seems to have happened is that anthropogenic inputs have increased the acidity at a much faster rate than would otherwise have been the case. The problem is that changes do occur quite naturally, but these changes are slow and we frequently do not have a baseline study with any pH data from a time before significant pollution began.

Relatively few studies have investigated the changes in soil pH caused by acidic precipitation. However, one unique study involved a family of Swedish scientists called Tamm. A forest site in southern Sweden that was first sampled by the older Tamm in 1927 was re-sampled in 1983 by his son, who found that soil pH was lower by between 0.5 and 1.0 units than it had been in 1927. Soil pH in this area has continued to decline, and the effect has also been observed in other parts of Europe. Only in extremely low pH situations does soil pH itself seem to have any effect on plant growth, so research has concentrated on the effects of acidity in changing the availability of elements in the soil.

Acidic precipitation increases nitrate and sulphate deposition, causing increases in soil levels of these elements. This is thought to have a fertilizing effect and, ironically, the recent decrease in acid rain in parts of England has led farmers to reintroduce sulphur into their fertilizer formulations for the first time in many years. Increased nitrogen and sulphur may also result in nutrient imbalances in plants, particularly in trees. High nitrogen produces faster growth rates but can lead to the plants suffering deficiencies in calcium, magnesium and potassium much faster than would otherwise be the case.

One of the major effects of soil acidification is a decrease in calcium, magnesium and potassium held on the cation exchange sites due to leaching caused by the increase in proton concentration in the soil. Such decreases have been observed both experimentally and over time in field situations, and they are likely to contribute to the die-back of trees. If the soil is naturally high in heavy metals, it is likely that these will be made more available in increasingly acidic soil, leading to plant toxicity in some cases. In poorly buffered soils such as podzols, acidic precipitation can lead to increased mobilization of metals into soil solutions. Although metal toxicity from elements such as manganese can be important in certain circumstances, interest has concentrated on aluminium (Al).

It is well known that aluminium toxicity is a problem in acid soils. This is a natural phenomenon and has nothing to do with pollution. In very acid soils, aluminium is mobilized and Al or hydroxy-Al cations are released into the soil solution. Thus, in soils that were affected by acid rain, acidity gradually increased and exchangeable aluminium increased as a result.

Die-back of trees in Western Europe and North America was first noted in the mid-1970s, but it took a lot of work to find out possible causes. It was not until 1980 that Ulrich and his co-workers suggested that aluminium might be an important factor (Ulrich *et al.*, 1980). They measured soil aluminium in a beech forest over a number of years at Solling in Germany, an experimental forest station where both beech and conifer plantations have been studied over many years (Figure 9.11). Between 1966 and 1979, aluminium in the soil solution markedly increased, particularly deeper in the soil profile. Ulrich's group were quick to point out that the levels reached (1–2 mg Al/l)

Figure 9.11 An experimental plot of Norway spruce (*Picea abies*) at Solling, Germany. Rain that has fallen through the canopy and plant needles are collected.
Photo: MJH.

were potentially highly toxic to plants and that this could have been the cause of decreased tree growth and toxicity symptoms. Exchangeable aluminium was also shown to have increased as soil pH decreased in southern Sweden, and in North America. The aluminium ions, once in the soil solution, can easily be lost, and this caused toxicity problems for lakes and freshwater environments.

Although there has been general agreement that aluminium toxicity is an important factor in the die-back of trees observed in Europe and North America, it is now known that it is not the only factor involved. Most work has been conducted on the conifers, where symptoms include decrease in crown density, yellowing of needles and decrease in root growth. Laboratory work by Godbold and his co-workers on Norway spruce (*Picea abies*) showed that root growth is rapidly reduced in the aluminium treatment (Godbold *et al.*, 1988). Calcium, and in particular magnesium, levels were drastically reduced in the shoots of aluminium-treated plants, and the resulting deficiency symptoms were remarkably similar to those observed in the field.

Now a large body of data has suggested that aluminium toxicity in trees is linked to calcium and/or magnesium deficiency. Most workers consider that aluminium itself has a lesser role in toxicity, but that it causes calcium and magnesium deficiency. Using radioactively labelled calcium, Godbold's group showed that aluminium blocked root uptake, even at quite low levels. Finally, x-ray microanalysis, a specialized electron microscopy technique which enables detection of mineral elements at the cellular and sub-cellular levels, showed that aluminium displaced calcium and magnesium from root cortical cell walls. Thus, a combination of field and laboratory work suggested a series of events that was at least partly responsible for tree die-back.

Since the 1980s, legislation in both Europe and North America has been very effective in reducing emissions of sulphur dioxide and has also had some success with nitrogen oxides. Although 'acid rain' is much less in the news nowadays, recent soil analyses suggest that while recovery from previous pollution is under way in some parts of Europe, there is still quite a long way to go before the problem is totally solved. In other parts of the world (e.g. China), problems with acidic precipitation are still severe, and there is much work to do before acid rain is consigned to the history books.

9.7 Light and radiation

As we saw in Chapter 7 (section 7.4.2), visible light is only one small part of the electromagnetic spectrum. It is visible light of wavelengths between 400 and 700 nm that drives photosynthesis. Obviously, when visible light levels are reduced by shading, photosynthesis and plant growth will be affected, and we will consider this in Section 9.7.1. Conversely, too much light or ultraviolet radiation (wavelengths below 400 nm), can be highly damaging to tissues (see section 9.7.2). Finally, we will consider the problems caused by ionizing radiation, including gamma rays (below 1 nm in wavelength) and alpha and beta radiation.

9.7.1 Shading
There are a number of situations where plants are shaded. In some, the shading is a temporary effect, while in other cases, the plants grow in shaded environments throughout their life cycle. There are also times where shading happens to just part of a plant, while other parts are not shaded. This is likely in large trees, where the upper leaves are fully exposed to sunlight but the lower leaves are heavily shaded.

For example, in a typical temperate oak forest, we might expect the following layers, all of which are subject to different amounts of shading:
• a closed canopy of oak leaves
• a lower, more broken canopy of shrubs, such as hazel

• a herb layer whose density will depend on that of the upper layers
• and beneath all of these, a layer of mosses and maybe soil algae and cyanobacteria.

Light levels below the tree and shrub canopies can be as low as one per cent of levels in the open, and below the herb layer they will be even less (see Chapter 3, section 3.7.2).

In many herbaceous plant communities reduction in light levels is a seasonal phenomenon. For instance, in the oak woodland there are three distinct periods:

1 Winter, when there are no leaves on the trees but light levels are low.

2 Spring, between March and May in the Northern hemisphere, when there are still no leaves on the trees but it is brighter. This is where plants on the forest floor get a distinct peak in irradiance. One example of a species that takes advantage of this peak is the bluebell (*Hyacinthoides non-scripta*), which flowers in early spring before the canopy overhead develops (see Figure 9.12).

3 Summer, when external light levels are high, but when the canopy closes, irradiance levels at the forest floor are lowered considerably.

Whether shade is a temporary phenomenon, is seasonal, or is a permanent feature of the environment, in all cases plants must still be able to achieve net photosynthesis. The main problem with shade is that respiration may exceed photosynthesis, because there is not enough light to allow photosynthesis to occur at a sufficient rate. If respiration

Figure 9.12 Bluebells (*Hyacinthoides non-scripta*) flowering under hazel coppice (*Corylus avellana*) in early May, photographed near Fawley, Berkshire, England.
Photo: Margot Hodson.

exceeds photosynthesis for a long enough period, then the plant will slowly starve to death. Adaptation and acclimation to shading will be covered in Chapter 10 (section 10.7.1).

9.7.2 Excess light and ultraviolet radiation

Although green plants are totally dependent on light, too much light can be, and often is, harmful. First, the light-harvesting may 'out-strip' the biochemical CO_2 fixation phase of photosynthesis (Chapter 7, section 7.4). Under these circumstances, the rate of photorespiration increases, thus decreasing photosynthetic efficiency.

Nevertheless, it is important that photorespiration does increase in order to lower the oxygen tension and lessen the production of ROI (see section 9.1.1). ROI damage chlorophyll by photo-bleaching and they also damage both chloroplast and nuclear DNA. Damage to DNA includes oxidation of guanine and deamination of 5-methyl-cytosine (to form thymine) and of cytosine (to form uracil). This affects base-pairing and is potentially mutagenic.

Very bright sunlight also increases exposure to ultraviolet (UV) light. Over the past ten years or so, we have become more aware of the damaging effects of UV light on human skin. Light of these wavelengths is also harmful to plants, causing significant damage to DNA. UV has several effects, which include increasing the concentration of ROI. However, its most important effect is to generate the formation of **pyrimidine dimers**, i.e. dimers between pyrimidine residues that are adjacent to each other in the DNA molecule. Thymine dimers are the commonest of these, followed by thymine-cytosine dimers; cytosine dimers are the rarest. All three types distort the double helix and prevent the passage of RNA and DNA polymerases; they thus interfere with transcription and replication.

The combined effects of ROI, UV light and ionizing radiation (next section) are immense. Depending on the exact combination and intensity of these three factors, a single cell may suffer many thousands of DNA lesions in a day! It is therefore not surprising that there are comprehensive repair mechanisms that deal with these lesions (see Chapter 10, section 10.7.2).

9.7.3 Ionizing radiation

The final topic we will cover in this section is ionizing radiation. Radioactive mineral elements behave in the environment in a similar way to any others, the only differences being that they are unstable and that, as a by-product, they emit radiation. There are three types of radiation – alpha, beta and gamma.

Alpha radiation has the greatest potential to cause damage, but only travels a few centimetres in air, and is stopped by thin foil. Gamma radiation, on the other hand, is the least damaging, but is very difficult to stop, requiring 4–5 cm of lead. All living organisms are exposed to a low-level background of radiation, mainly in the form of gamma radiation from cosmic rays and, in areas containing radium-rich rocks, alpha radiation from radon gas.

Soils can become contaminated with radioactive isotopes from a number of sources, including weapon tests, poor waste disposal and accidents at nuclear plants. This will give local levels of ionizing radiation that are above background. However, even background levels of alpha and gamma radiation may be damaging, because of their effects on DNA. Alpha radiation causes single-strand breaks (nicks), while gamma radiation causes both single-strand and double-strand breaks. The latter are especially dangerous, with the possibility of significant chromosome damage and disruption of gene expression, DNA replication and cell division.

Of much wider importance, however, is the uptake of radioactive isotopes by plants, with the potential that these could then be passed up the food chain when the plants are eaten by animals. The Chernobyl accident in 1986 contaminated about $125,000 \, km^2$ of land in Belarus, Ukraine and Russia with radioactive caesium and radioactive strontium, while lower levels of these elements were detected in environments around the world. In 2011, a major earthquake and the resulting tsunami caused serious damage to the nuclear reactors at Fukushima in Japan, leading to the release of radioactive caesium and radioactive iodine.

Caesium-137 (Cs-137) and caesium-134 (Cs-134) are the most important radioactive isotopes of the element, with half-lives of 30 and two years, respectively. Nuclear fission is the only process that creates them. Both emit beta and gamma radiation when they decay. Caesium is chemically related to potassium, and both are taken up by plants using similar pathways. Potassium fertilizer can be used to reduce the uptake of caesium by plants. Caesium is least taken up by the roots of plants in a clay soil, being bound to the soil minerals. It is taken up most easily from acid, organic and sandy soils without clay.

Radioactive strontium (Sr) occurs in the environment as Sr-89 and Sr-90, with half-lives of 51 days and 29 years, respectively. Deposition mainly occurs with rain or other precipitation. Chemically, strontium resembles calcium, and the elements tend to follow similar routes in the environment. When radio-strontium undergoes decay, it emits beta radiation. Strontium is roughly ten times more accessible to plants than caesium, and uptake is greater from the soil. Strontium is taken up equally well from all types of soil, although a large amount of chalk in the soil reduces uptake. If plants contaminated with strontium are eaten by vertebrates, then this is absorbed via the gastrointestinal tract and accumulates in bone tissue.

The radioactive isotopes of iodine (I) are I-129 and I-131, which have half-lives of 15.7 million years and eight days respectively. Both are beta emitters. Iodine is fixed by organic materials and some soil minerals in the soil, and this slows movement in the environment. It can be taken up by plants and thus contaminate food. If contaminated plants are eaten by humans, the radioactive iodine tends to accumulate in the thyroid. Fortunately, most of the iodine released at Fukushima was I-131, which decays fairly rapidly in the environment.

Selected references and suggestions for further reading

Bailey-Serres, J. & Chang, R. (2005) Sensing and signalling in response to oxygen deprivation in plants and other organisms. *Annals of Botany* **96**, 507–518.

Brodribb, T.J. & McAdam, S.A.M. (2011) Passive origins of stomatal control in vascular plants. *Science* **331**, 582–585.

Chaves, M.M., Maroco, J.P. & Pereira, J.S. (2003) Understanding plant response to drought: from genes to the whole plant. *Functional Plant Biology* **30**, 239–264.

De Vries, W., van der Salm, C., Reinds, G.J. & Erisman, J.W. (2007) Element fluxes through European forest ecosystems and their relationships with stand and site characteristics. *Environmental Pollution* **148**, 501–513.

Fitter, A.H. & Hay, R.K.M. (2002) *Environmental Physiology of Plants. 3rd Ed.* Academic Press, London.

Godbold, D.L., Fritz, E. & Hütterman, A. (1988) Aluminum toxicity and forest decline. *Proceedings of the National Academy of Science, USA* **85**, 3888–3892.

Greenway, H. and Munns, R. (1980) Mechanisms of salt tolerance in nonhalophytes. *Annual Review of Plant Physiology* **31**, 149–190.

Hallbäcken, L. & Tamm, C.O. (1986) Changes in soil acidity from 1927 to 1982–1984 in a forest area in south-west Sweden. *Scandinavian Journal of Forest Research* **1**, 219–232.

Jones, H.G. & Jones, M.B. (1989) Introduction: some terminology and common mechanisms. In Jones, H.G., Flowers, T.J. & Jones, M.B. (eds.) *Plants under stress.* Cambridge University Press, Cambridge, UK. pp. 1–10.

Levitt, J. (1980) *Responses of plants to environmental stresses. Volume II. Water, radiation, salt and other stresses.* Academic Press, New York.

Liu, Y. & Zhang, S. (2004) Phosphorylation of 1-aminocyclopropane-1-carboxylic acid synthase by mpk6, a stress-responsive mitogen-activated protein kinase, induces ethylene biosynthesis in Arabidopsis. *The Plant Cell* **16**, 3386–3399.

Mason, B.J. (1992) *Acid rain. Its causes and its effects on inland waters.* Oxford University Press.

Mittler, R. (2002) Oxidative stress, antioxidants and stress tolerance. *Trends in Plant Science* **7**, 405–410.

Schachtman, D.P. & Goodger, J.Q.D. (2008) Chemical root to shoot signaling under drought. *Trends in Plant Science* **13**, 281–287.

Turkington, R. (2009) Professor John L. Harper FRS CBE (1925–2009). *Journal of Ecology* **97**, 835–837.

Ulrich, B., Mayer, R. & Khanna, P.K. (1980) Chemical changes due to acid precipitation in a loess-derived soil in central Europe. *Soil Science* **130**, 193–199.

CHAPTER 10

Acclimation and Adaptation to Environmental Stresses

Having considered the effects of various environmental stresses on plants in Chapter 9, here we will cover acclimation to stress and adaptation to stressful environments. As we have seen in the previous chapter, the concept of adaptation has been criticized in the past. Here we will take adaptations to be constitutive and heritable, meaning that the plant always has them, regardless of the environment in which it is growing. Adaptations have evolved to give the plant a selective advantage in a particular environment. So a cactus is 'adapted' to growing in a desert but retains these adaptations even when growing as a houseplant.

Most plants live for the most of the time in climatically moist environments with moderate temperatures. These environments are known as mesic, and the plants are known as **mesophytes**. They do not usually have obvious adaptations to stressful environments, but they can still show acclimation responses to any stress imposed. A simple example of this would be the accumulation of low molecular weight organic solutes as a response to many types of stress. Thus acclimation is non-heritable, is not constitutive and only occurs when the plant undergoes stress or is preparing to meet a stress. It is, of course, possible for a plant that is adapted to a particular environment also to show acclimation responses when stressed.

Finally, some plants escape stress by virtue of their life cycle. For instance, desert ephemerals lack any obvious morphological adaptations to arid conditions, but they pass the unfavourable season as seeds. After the spring rains, they germinate and rapidly flower and set seed, dying away only a few weeks after germination before the hot, dry weather begins.

10.1 Adaptation and acclimation responses

In this chapter, we will now consider the adaptations and acclimation responses to each of the stresses that we covered in Chapter 9. First, however, we will look at some general features of plant adaption and acclimation to stressful environments. It has become apparent that membranes are a critical component in many stress responses (Chapter 9) and that they are often affected before all other cell components. In many stresses (e.g. water stress), they are involved in the early detection and response to stress; in others, membrane damage is an early sign of stress (e.g. heavy metal toxicity). It will, therefore, come as no surprise that membrane properties are also important in both adaptation and acclimation, although there are several other common reactions to stress that are regarded as of adaptive value.

Very different environmental stresses frequently trigger related cell signalling pathways, including the production of increased concentrations of low molecular weight **compatible solutes**. The accumulation of these organic solutes is a feature of many plant responses to stress – and, indeed, animal systems are similar in this respect. These compounds include proline, sugar alcohols such as sorbitol and quaternary ammonium compounds (QACs) including glycinebetaine (Figure 10.1). They have been shown not to interfere with enzyme activity up to very high concentrations, and can also act as osmotica.

As we saw in Chapter 9 (section 9.1.1), one of the common features of all stresses is that the stressed plants

Functional Biology of Plants, First Edition. Martin J. Hodson and John A. Bryant.
© 2012 John Wiley & Sons, Ltd. Published 2012 by John Wiley & Sons, Ltd.

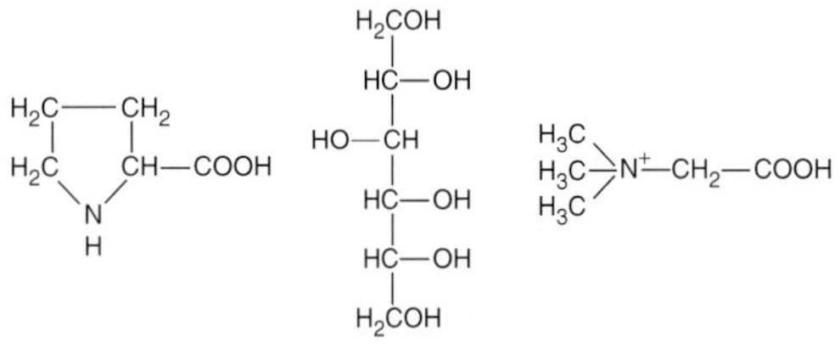

Figure 10.1 Three common low molecular weight compatible solutes: proline, sorbitol and glycinebetaine.

accumulate reactive oxygen intermediates (ROIs). If these were allowed to build up in plant cells, they would cause serious damage, so plants have evolved mechanisms to deal with excess ROIs.

As ROIs are also involved in signalling, cells need different mechanisms to regulate their ROI levels. The first concerns fine modulation of low concentrations of ROIs that are used in signalling, while the second allows detoxification of surplus ROIs. Hydrogen peroxide can be reduced to hydroxyl radicals (HO^-) by superoxide when transition metal ions, such as iron and copper, are present. Hydroxyl radicals are much more reactive than superoxide and hydrogen peroxide. Generation of HO^- occurs when superoxide and hydrogen peroxide interact. There are no known biological HO^- scavengers, so controlling the reactions leading to its production is the only means of avoiding oxidative damage. Cells therefore need to keep the levels of superoxide, hydrogen peroxide and reducing transition metals under firm control.

Plants use two methods to decrease ROI concentrations: non-enzymatic and enzymatic. Non-enzymatic antioxidants include ascorbic acid, glutathione (GSH), tocopherol, flavonoids, alkaloids, and carotenoids. Mutants showing lowered ascorbic acid or GSH concentrations have been found to be hypersensitive to stress. Enzymes responsible for scavenging ROIs comprise superoxide dismutase (SOD), ascorbate peroxidase (APX), glutathione peroxidase (GPX) and catalase (CAT). SODs convert superoxide to hydrogen peroxide, while APX, GPX and CAT then detoxify hydrogen peroxide.

Many genes and biochemical mechanisms are involved in response to abiotic stress. It seems that the molecular control mechanisms of stress tolerance involve the expression of stress-related genes of three types:

1 Those implicated in signalling cascades and in transcriptional control, including MAP kinases and various transcriptional factors.

2 Those involved in protecting membranes and proteins, including the production of heat shock proteins (HSPs) and chaperones, compatible solutes and defence against ROIs. **Pathogenesis-related (PR) proteins** are synthesized in response to several stresses, not just pathogens.

3 Those regulating aquaporins and ion transporters.

As we progress through this chapter, we will see that these responses are involved in acclimation to many different types of stress.

10.2 Temperature

In Chapter 9, we noted that plants can survive over a wide range of temperatures and can grow over a somewhat smaller range. We also considered the effects of chilling, freezing, heat and fire on plants. In this section, we will cover each of these stresses in turn, looking at adaptations and acclimation to each one.

10.2.1 Chilling resistance

We saw in Chapter 9 (section 9.2.1) that chilling sensitivity was related to high levels of saturated fatty acids in the membranes of affected plants. The membranes of sensitive plants therefore tend to solidify at higher temperatures and appear to undergo a physical phase transition from a flexible liquid crystal to a solid gel structure at temperatures of 10–15°C. This change then leads to many other biochemical lesions, which can cause decreased growth or even death of sensitive plants.

It seems reasonable to suggest that chilling resistance might be due to an ability to maintain membrane lipids in the liquid crystalline state at chilling temperatures. Variation in both the length and the degree of unsaturation (number of double bonds) influences the melting point of membrane lipids (see Chapter 2, section 2.5.2). Generally, highly unsaturated fatty acids predominate in plants that grow in cold climates. Fatty acids that are unsaturated have an important role in maintaining membrane fluidity even at low temperatures.

As is often the case nowadays, work on the model plant, *Arabidopsis*, has elucidated the changes that occur at a molecular level when a plant is chilled. Provart *et al.* (2003) compared the molecular responses of *Arabidopsis* at 22°C and 13°C (chilling). About 20 per cent of the genes analyzed were affected by chilling. Provart's team then investigated the effects of exposure to 13°C in 12 chilling-sensitive mutants and found 634 chilling-responsive genes with abnormal expression in these mutants. The genes were involved in lipid metabolism, chloroplast function, carbohydrate metabolism and ROI detoxification, suggesting that these processes are particularly important in resistance to chilling.

Chilling stress is not only important to plants growing in the field, but is also highly significant in the post-harvest physiology of sensitive tropical and semi-tropical plants. This is of considerable economic importance, as it determines the shelf life of these products. Currently, there are three ways of reducing these problems: physical; chemical and biotechnological:

• Physical methods involve thermal treatments prior to or during cold storage, and the control of gaseous exchange.
• Chemical treatment mainly involves the use of antagonists of ethylene production, as this plant hormone is important in fruit ripening and senescence processes (see Chapter 3, section 3.6.6 and Chapter 4, section 4.9).
• In the future, genetic modification techniques may allow control of ethylene production and thereby increase chilling resistance. A transgenic melon has been produced that expresses an antisense ACC oxidase gene that inhibits autocatalytic manufacture of ethylene by 99 per cent. This melon showed much greater tolerance to chilling injury during storage at 2°C than the untransformed melon.

10.2.2 Freezing resistance and tolerance

Some plants avoid the worst effects of cold temperature by hiding from it. The below-ground parts of perennial plants (e.g. roots, tubers and rhizomes) are less susceptible to freezing than above-ground structures. This is partly due to a higher inbuilt frost resistance, but also because heat losses from underground organs are minimal, while at the same time soil temperature falls less quickly than that of air. In the case of alpine and arctic plants, a layer of snow may also act as protection against frost for much of the winter.

In Chapter 9 (section 9.2.2), we saw that freezing stress was of two types – intracellular and extracellular – with the latter being more important in nature. Resistance of plant tissues to freezing involves a whole line of defences. The accumulation of soluble components in cells (e.g.

ions) will result in the depression of freezing point for cellular fluids. Even before hardening, the tissues of most temperate plants can therefore be cooled to a few degrees below zero before ice formation (−1 to −5°C). This alone can protect Mediterranean plants from the effects of frost.

Halophytes, with their high salt content (see section 10.5.2), often do not freeze until −14°C. In a similar way, the accumulation of low molecular weight organic solutes (e.g. sugars, amino acids, proline and glycinebetaine) has often been suggested to play a role in frost resistance. It is unlikely that these solutes lower the freezing point more than a few degrees, but they probably do protect enzymes against dehydration (see section 10.4.1).

Some plant tissues are able to resist freezing by the phenomenon of supercooling. Pure water melts at 0°C but, in a clean vessel, supercooling means it will only freeze at −40°C. A small speck of dust will act as a nucleus, around which crystallization and ice formation can occur. Plant tissues in hardened plants often seem to act as if they contain pure water, and they can reach as low as −38°C before ice formation. This is particularly important in the dormant buds and the xylem ray parenchyma of woody plants during the winter period. Thus, some tissues survive by deep supercooling, while the rest allow intercellular ice formation and tolerate cell dehydration.

In some areas, trees grow where temperatures drop below −40°C in the winter, so the methods of reducing freezing point mentioned above no longer operate. It seems that intracellular freezing is prevented because all of the freezable water in the cell is withdrawn to the apoplast, leaving just thin layers of water molecules to protect macromolecules.

It is a relatively rare event in nature for the temperature suddenly to drop below freezing. Usually, the ambient temperature in the autumn gradually decreases over some weeks before the first freezing event occurs. In that time, many perennial plants can anticipate future freezing and, often, only one to two days of low non-freezing temperatures will be enough to bring about acclimation. This acclimated tolerance is higher than that prior to acclimation, and it reveals the capability of plants to increase their freezing tolerance in reaction to lower but not yet freezing temperatures.

Many metabolic changes directed by alterations in gene expression are involved in cold acclimation. It appears likely that the plasma membrane may be the site of the primary sensor of cold stress and that alterations in membrane fluidity are involved, although this has yet to be confirmed. Changes in membrane fluidity probably cause the rearrangement of the actin cytoskeleton (Chapter 2, section 2.12.3). This is thought to activate Ca^{2+} channels in the membranes, increasing cytoplasmic Ca^{2+} concentrations, which in turn trigger the expression of cold-regulated (COR) genes during cold acclimation.

Cold stress induces the formation of ROI, activating a mitogen-activated protein kinase (MAPK) cascade that regulates tolerance to freezing stress. During cold acclimation, C-repeat binding transcription factors (CBF) are expressed, and these trigger many downstream genes that increase freezing tolerance in plants. As yet, the molecular link between the kinases and transcription factors is not known, but it seems that CBFs are involved in the regulation of genes that encode dehydrins (also known as late embryogenesis abundant – LEA – type COR proteins: see Chapter 4, section 4.5) and the biosynthesis of low molecular weight compatible solutes.

Tolerance of cellular dehydration during extracellular freezing is clearly an important area of research. Protection of cytoplasmic components and membranes against dehydration is thought to involve the accumulation of dehydrins. These are proteins that are distinguished by a lysine-rich amino acid motif, termed the K-segment. Dehydrins are hydrophilic, soluble and rich in polar amino acids, including glycine. They also generally have low cysteine and tryptophan contents. This unusual composition allows dehydrins to bind considerable amounts of water. They are thus able to protect proteins and membranes against unfavourable structural changes caused by dehydration, acting as emulsifiers or chaperones. The latter are responsible for protein folding, assembly, translocation and degradation in many regular cellular processes. Under stress conditions, chaperones are also involved in protein and membrane stabilization, and they often help in protein refolding.

While dehydrins have been shown to build up in the cytoplasm of plant cells during cold acclimation, antifreeze proteins (AFPs) accumulate in the apoplast. Quite a number of proteins and polypeptides, with differing structures and amino acid composition, have been isolated and shown to have antifreeze properties. Plant AFPs are powerful inhibitors of ice recrystallization, and these proteins may control the propagation of ice crystals in the cell walls.

Low temperature, ethylene and ABA are all implicated in the regulation of antifreeze activity in response to cold. Some AFPs appear to be multi-functional and are related to **pathogenesis-related (PR) proteins** (see Chapter 11, section 11.3.7). The discovery of AFPs has suggested that it might be possible to increase their production in freezing-sensitive plants using genetic modification techniques, and thereby increase their cold tolerance. Thus, transgenic tobacco, tomatoes, and potatoes have been produced which all show increased freezing resistance. This work is likely to continue in the future.

10.2.3 Heat resistance and tolerance

Many plants avoid overheating through evaporation of water from their leaves, which is a very effective way of keeping cool. The cooling effect can be quite large; for example, leaf temperatures in Death Valley, California, are sometimes $8°C$ below the air temperature. As we saw in Chapter 9 (section 9.2.3), however, any increase in water loss could lead to dehydration, and the closing of stomata (Chapter 9, section 9.4) decreases the cooling effect of evaporation.

Some plants (e.g. soybean) are able to orientate their leaves away from the sun, a phenomenon known as **paraheliotropism**, and this reduces the heating effect. Many grasses roll their leaves up, and thereby minimize the exposure of the leaf to the sun. Plants of the desert are often xerophytic (see section 10.4.2), and many conserve water by closing their stomata during the day and using crassulacean acid metabolism (CAM) as their method of photosynthesis (see Chapter 7, section 7.8). Thus, with much decreased evaporative cooling, desert plants must be able to tolerate much higher temperatures.

One way of decreasing heating up is shown by many cacti, which are covered with dense, reflective spines (Figure 10.2). Airflow is much reduced, resulting in less water loss, and much incident radiation is reflected, but the disadvantage is that light is also reflected and this can decrease photosynthesis. Rainforests, despite being another hot environment, present fewer difficulties for plants because humidity is high and shading prevents excessive heat build-up.

One of the problems of high temperature stress, as we saw in Chapter 9 (section 9.2.3), is starvation as respiration rate exceeds photosynthesis. In general, those plants that are adapted to high temperatures have much higher temperature compensation points than unadapted plants,

Figure 10.2 Golden barrel cactus (*Echinocactus grusonii*) is native to central Mexico. It is ribbed and has yellow reflective spines, and is here photographed in Peoria, Arizona, USA.
Photo: Ruth McFadden.

and can assimilate even at temperatures of 45–53°C. Adapted plants may have higher rates of photosynthesis at high temperature, due to increased thermal stability of the photosynthetic apparatus. They may also have lower respiration rates and, in many desert shrubs, rates are halved in the summer when compared with those in winter.

In a similar way to freezing resistance, if plants are exposed to moderately high temperatures beforehand, they can often endure exposure to temperatures higher than those that are optimal for growth, or can attain tolerance to lethal heat stress. This increase in tolerance is connected with the synthesis of chaperones, increases in antioxidants and the accumulation of low molecular weight organic solutes.

As a plant tissue heats up, thermodynamic effects cause changes in molecular or supramolecular structures such as membranes. These changes are usually quick, and thus all molecules can 'sense' temperature. However, to be considered a true sensor, any change induced needs to be upstream of the signalling cascade that leads to a response (in the present case, the establishment of heat tolerance). Perception can therefore be defined as *'the most upstream event(s) controlling downstream signals'* (Ruellanda & Zachowski, 2010).

There is well-documented increase in the saturation of membrane lipids with increased heat tolerance. More recent work has suggested that changes in membrane fluidity may also act as a primary heat sensor in plants. This conclusion has been reached by experiments involving dimethyl sulfoxide (DMSO), which induces membrane rigidification, and benzyl alcohol (BA), which acts by fluidizing membranes. Thus, activation by heat stress of a heat-activated MAP kinase (HAMK) and an increase of the heat shock protein HSP70 are antagonized by DMSO, but they are mimicked at room temperature by BA treatment. In addition to membrane fluidity, protein conformation, cytoskeleton assembly status and enzyme activity have all been identified as potential temperature sensors.

There has been much interest in the functions of **heat shock proteins** (HSPs) and molecular chaperones, and their roles in increasing heat tolerance. The HSPs are a special class of proteins that are involved in the prevention of aggregation, maintaining protein structure and aiding the recovery process after stress. As the name suggests, HSPs were first discovered in organisms that were undergoing heat stress, but it is now known that they are involved in tolerance to a wide range of abiotic stresses. Many HSPs act as chaperones, but there are some chaperones that are not involved as HSPs (see Chapter 3). There are at least five families of HSP/chaperones, all with somewhat different roles in the protection of plants against stress and in the restoration of homeostasis at the cell level following the stress event.

10.2.4 Adaptation to fire

In ecosystems that are affected by fire, the main plant characters that are related to post-fire survival are the capacity to re-sprout and the retention of a persistent seed bank. Re-sprouting permits individuals to survive, while a resistant seed bank allows the plant population to survive.

Those species that re-sprout after fire has destroyed their photosynthetic tissues range from small monocots to large eucalyptus trees. They generally grow again from buds hidden under a thick bark, or from underground structures such as rhizomes. One example of a re-sprouting species is cork oak (*Quercus suber*), which is adapted to arid areas in southwest Europe and northwest Africa where fires are common (see Chapter 6, Box 6.1 and Figures 6.11 and 9.4). Species that re-sprout tend to have large storage organs, and much of their photosynthetic product is partitioned into these organs.

In species that cannot re-sprout after fire, regeneration will depend on the seed bank. Dry seeds can survive temperatures up to 120°C, and survival during a fire is an important factor in allowing a species to persist at a certain site. Two other key factors are the capacity to produce seeds during the period between fires, and whether recruitment of new individuals is improved by the fire. Flowering, seed dispersal and germination are all processes that are linked to recruitment, and these may be affected by factors brought about by the fire, such as heat and smoke.

In some species, seeds only germinate shortly after the fire, while in others a percentage of the seeds remain dormant in the seed bank within the soil. The seeds of some species are dependent on fire to break seed coat dormancy (see Chapter 4, section 4.10), as the seed coats are ruptured or seed coat waxes are melted, increasing water uptake. Seed germination that is stimulated by heat seems to be an adaptive feature in habitats like shrubland and heathland, where there is a high chance of severe fires. Germination can also be promoted by smoke constituents in seeds of many species, particularly those from Mediterranean habitats that are affected by fire. It appears that nitrous oxide in the smoke may be at least partly responsible for this effect.

Seed dispersal in some shrubs and trees is considerably affected by fire. Many woody species retain viable seeds in their canopies for 30 years or more. For example, viable seeds may remain in cones on *Pinus banksiana* trees for longer than 20 years. After a fire, many of these seeds are released. In conifers, the cones usually dehydrate before opening and releasing seeds. In some species with **serotinous** (late-to-open) cones, both dehydration and heat are needed for seed dispersal. In serotinous cones, resins bind the scales, preventing opening until a fire melts the resinous material, allowing the scales to open and the seeds to disperse.

10.3 Resistance and adaptation to waterlogging

In Chapter 9 (section 9.3), we considered the effects of waterlogging on the soil and also the stress effects on unadapted plants. Essentially, the main problem of

waterlogging is anaerobiosis. In the current section, we will first cover the ways in which crop plants develop some degree of resistance to anaerobiosis, and then move on to look at hydrophytes which are higher plants that are adapted to live their entire life cycles in waterlogged conditions.

10.3.1 Acclimation to waterlogging

One of the major means by which plants adapt to lowered oxygen in the soil environment is the development of aerenchyma in roots (Figure 10.3). Aerenchyma, which occurs in many plants, is a tissue containing enlarged spaces through which gases can diffuse. It is formed as part of normal development in wetland plants through **schizogeny**, in which growth leads to cell separation. In many crop plants (e.g. barley, wheat, maize) and other non-wetland plants, it develops in response to stress induced by lack of oxygen. This happens through **lysigeny**, in which cells die to produce the gas spaces. Here we will concentrate on the development of lysigenous aerenchyma and the cell death processes implicated, which are related to programmed cell death (PCD – see Chapter 3, section 3.5.2).

We noted in Chapter 9 (section 9.3) that ethylene can be produced in waterlogged soil, and that its concentration is also increased in susceptible plants due to synthesis. Ethylene, whether made endogenously or applied exogenously, induces aerenchyma formation in a number of species. Hypoxia and ethylene bring

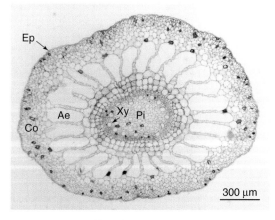

Figure 10.3 Water cabbage (*Pistia stratiotes*) root shown in cross section aerenchyma. Key: Epidermis (Ep), Cortex (Co), Aerenchyma (Ae), Xylem (Xy). Pith (Pi).
Photo: Prof. Thomas Rost.

Box 10.1 Rice

Rice is the most famous agricultural hydrophyte (Figure 10.4). Its date of domestication remains uncertain, but it has been found in archaeological sites dating back to 8,000 BCE. Two species have been domesticated: Asian rice (*Oryza sativa*) and the less well-known and less common African rice (*Oryza glaberrima*).

There are many phenotypic differences between *O. sativa* and its wild relatives. For example, domesticated rice has either short awns or no awns, and shattering of the panicle is decreased so that the number of seeds that can be harvested is maximized. However, wild rice species usually have long awns and greater shattering for increased seed dispersal. Many other traits distinguish wild and domesticated rice, including changes in seed dormancy, number, size and colour of seeds, tiller number, and mating habit (e.g. outcrossing and inbreeding).

Two different types of *O. sativa*, Hsien and Keng, were recognized in 100 CE by the Chinese Han dynasty, and these are now known as the *indica* and *japonica* subspecies. It is now thought that these two subspecies were domesticated from the wild rice, *O. rufipogon*. The first definite evidence for rice cultivation comes from the oldest known paddy fields in the lower Yangtze River Valley in China, dating from 4,000 BCE. It seems likely that the extensive use of paddies for growing rice was well underway by around 3,000 BCE in southern Asia.

According to Ruddiman, atmospheric methane concentration began to rise at that time, and rice paddies are a major source of methane emissions as they provide an anaerobic environment for methanogenic bacteria (Ruddiman, 2005). Ruddiman's 'early Anthropocene' hypothesis suggests that these methane emissions may have impacted on the global climate and were partly responsible for preventing the Earth from slipping into another ice age. Today, methane from rice paddies is still an important component in global greenhouse gas emissions (See Chapter 12, section 12.1.1).

Rice is the most important staple food for many of the world's population, particularly in Asia, the Middle East, South America and the West Indies. It is the grain crop with the second highest global production, after maize, but much of the latter is used

Figure 10.4 Aerial view of the Shonai Plains near Tsuruoka in the Shonai region of Yamagata, Japan. The plains are a major rice-producing area and in this view, almost all of the fields are rice paddies.
Photo: MJH.

as animal feed. Half of the world's population depends on rice for a major part of their nutrition. It provides 20 per cent of the world's human dietary energy supply, while wheat supplies 19 per cent and maize only 5 per cent.

Rice has also been the subject of intensive breeding programmes, most recently using genetic modification techniques. A major pathogen of rice is the blast fungus (*Magnaporthe grisea*), and much work has been carried out to decrease its effects (see Chapter 11, section 11.3.6, and Figure 11.14). Golden rice has been genetically engineered to make precursors of beta-carotene in the grain, and these have a yellow colour as a result[i]. It was developed to be used in areas where there is a shortage of vitamin A in the human diet, and it may be commercially available in some Asian countries by 2012. Golden rice has, however, had much opposition, particularly from the anti-globalization lobby.

about the development of identical forms of aerenchyma. Decreased oxygen accelerates the production of the ethylene precursor 1-aminocyclopropane-1-carboxylic acid (ACC) in roots within a few hours.

Often, as we saw in Chapter 9 (section 9.3.2), some ACC is transported to the shoot in the xylem but, in many roots, sufficient ACC remains and its product, ethylene, promotes the lysis of cells in the root cortex. The signal of low oxygen availability must be transduced into ethylene biosynthesis and thereby control aerenchyma development, but the mechanisms behind this transduction are still uncertain.

Conversion of ACC to ethylene by ACC oxidase requires oxygen (Chapter 9, section 9.3.2), so ethylene biosynthesis cannot happen in totally anoxic conditions. As a consequence aerenchyma cannot develop in anoxic roots where there is no oxygen available. It is then, perhaps unexpected that ethylene biosynthesis has been shown to increase under hypoxia. ACC oxidase is up-regulated under these conditions, and it also appears that a

---

[i]For more information about golden rice see: Bill and Melinda Gates Foundation (2011) http://www.gatesfoundation.org/agriculturaldevelopment/Pages/enriching-golden-rice.aspx

permeability barrier to ethylene is implicated in ethylene entrapment in the roots. Although there are now several known ethylene receptors in plants, that in the cortical cells that undergo lysis to create aerenchyma has yet to be found.

Maize roots, when grown under hypoxia, begin the cell death process leading to aerenchyma formation in the cortex when they are less than 12 hours old and within 10 mm of the root tip. Ultrastructural changes also start around that time, including the appearance of electron opaque deposits in the vacuole, invagination of the plasma membrane and the production of small vesicles just inside it. Furthermore, cell wall degradation occurs at an early stage in the development of aerenchyma.

Ethylene is also probably responsible for inducing a number of other adaptive physiological responses of plants to anaerobiosis, including fast underwater extension of shoots and the rapid development of adventitious roots near to the soil surface. In both cases, these adaptations increase the chances of plants being able to find areas with higher concentrations of oxygen.

10.3.2 Hydrophytes

A hydrophyte is defined in *The Penguin Dictionary of Botany* (1984) as '*a plant that is adapted to living either in waterlogged soil or partly or wholly submerged in water*'. Obtaining enough water is never a problem for hydrophytes. The two main difficulties for submerged plants are to obtain enough light for photosynthesis and to achieve gaseous exchange. As was mentioned in Chapter 9 (section 9.3.1), the diffusion of oxygen through a liquid is much slower than in a gas.

Most leaf modifications in hydrophytes enhance light absorption and gaseous exchange. Immersed leaves are protected from bright sunlight by water, and low light intensity is often the greater difficulty. Hydrophytes often have large, thin leaves with poorly developed spongy and palisade mesophyll layers. Unusually, epidermal cells are frequently photosynthetic and contain chloroplasts, while the ground tissue is modified for storage.

Hydrophyte leaves and stems also commonly contain large amounts of aerenchyma. Plant tissues are denser than water, and often the leaves and the stems of hydrophytes have gas chambers to aid buoyancy and keep the plant afloat. These chambers can also function in gaseous exchange, bringing oxygen down from aerial to submerged parts.

Figure 10.5 Water hyacinth (*Eichhornia crassipes*) is an aquatic plant from the Amazon basin, but it has been widely introduced elsewhere and is often considered a weed.
Photo: MJH. Used with the permission of Oxford Botanic Garden, Oxford, UK.

The water hyacinth (*Eichhornia crassipes*) is a completely floating plant, the petioles of which have air-filled bladders (Figure 10.5). It is the cause of many problems in the tropics and sub-tropics, where it is a notorious weed, clogging waterways and leading to considerable loss of water in reservoirs. It transpires water at eight times the rate of uncovered water.

In partially submerged plants such as rice, layers of air continuous with the atmosphere are trapped between the hydrophobic surface of the leaf and the surrounding water. Gases can diffuse from the atmosphere to submerged parts of the plant through these layers.

Hydrophytes are usually covered by a thin (or nonexistent) cuticle as underwater leaves and stems do not need a cuticle to prevent the tissues from drying out. The lack of a cuticle allows nutrients to be taken up over the whole underwater surface. However, minerals are still taken up by roots (unless the plant has no roots, as in the case of *Wolffia arrhiza*; see Chapter 1, Figure 1.8a) and, as there is no transpiration under water, ions are moved through the xylem by root pressure (see Chapter 6, section 6.8).

Unusually, hydrophytes often have a distinct endodermis in their stems and leaves. We discussed the structure and function of the endodermis, which surrounds the stele in plant roots, in Chapter 5 (section 5.7.5). In hydrophytes, the endodermis in the leaf and stem seems to be involved in channelling the flow of water and

Figure 10.6 The giant water lily (*Victoria cruziana*) is native to the Parana-Paraguay basin.
Photo: MJH. Used with the permission of Oxford Botanic Garden, Oxford, UK.

keeping it confined to the xylem. For transpiring land plants, water is pulled upwards as cells lose water to the air (Chapter 6, section 6.8) but, in submerged plants, it is pushed up by root pressure. If the submerged leaves and stems had no endodermis, then the water would squeeze out and fill up the air spaces. With an endodermis, the water is channelled and controlled.

Submerged leaves have no stomata, while floating leaves have stomata only on their upper surface. Water lilies are hydrophytes with all except the upper surfaces of their leaves and flowers submerged, and are rooted in the mud below (Figure 10.6).

Hydrophytes often show leaf dimorphism, where their submerged leaves have different shapes to the floating leaves. For example, the floating leaves of aquatic buttercup species are large and flat, while submerged leaves are highly dissected and lacelike. This means that no cell in a submerged leaf is far from water and its circulating nutrients. Leaves that are flat and have a broad lamina would be a real disadvantage in water, as even a mild current would put enormous pressure on the plant, but finely dissected compound leaves or thin cylindrical leaves offer little resistance and water moves easily around them. Most hydrophytes have relatively little xylem and supporting tissue but, even so, some fibres are often present, particularly at the edges of leaves. These provide for elastic flexibility and protect against tearing, which is a hazard, above all in flowing water.

Many wetland plants have well developed aerenchyma in their roots (see Figure 10.3). Mangroves are halophytes, adapted to living in seawater, but they are also hydrophytes, as they live in submerged conditions. In some mangroves, aerenchyma-filled roots called pneumatophores project upwards from the soil into the air and function as 'snorkels' through which gases can diffuse to submerged roots. The mangrove *Bruguiera gymnorrhiza*[ii] shows these very clearly in Figure 10.7. Pneumatophore development is most marked in sites with fluctuating water tables.

10.4 Resistance and adaptation to drought

Having considered resistance and adaptation to water-logged environments, we will now move on to acclimation to drought, and those plants that are specifically adapted to dry environments, the **xerophytes**. Drought is also a key factor limiting crop productivity, and so in the final subsection we will cover the breeding of crops for drought tolerance. This section should be read in conjunction with Chapter 9, section 9.4, which covered drought stress.

10.4.1 Acclimation to drought

As with waterlogging, acclimation to drought involves both morphological and physiological changes. As we saw in Chapter 9 (section 9.4), decreased leaf expansion is an early symptom of drought stress. It can also be regarded as the first line of defence against drought, as it means that the leaves will have smaller surface areas and thus lose less water. Total leaf area of a plant does not remain constant after all of the leaves have matured. If the plants become stressed after a substantial leaf area has developed, then some of the older leaves will senesce and eventually drop. This abscission results from the increased synthesis of ethylene (Chapter 7, section 7.11). The younger leaves may, however, remain and even become more active.

Leaf area adjustment is an important means by which plants can adapt to water stress. Some desert plants take this to the extreme and lose all of their leaves in times of drought. Another common response to drought is that increased wax deposition occurs on the leaves. This will

---

[ii]For more details concerning *Bruguiera gymnorrhiza*, see *Plants and Environments of the Marshall Islands*. http://www.hawaii.edu/cpis/MI/plants/jon.html

Figure 10.7 The mangrove *Bruguiera gymnorrhiza* (L.) Lamk., growing on Jaluit Atoll, Marshall Islands.
Numerous knee-like breathing roots (pneumatophores) are visible on the ground.
Photo: Dr. Peter Rudiak-Gould, McGill University, Montreal, Canada.

reduce transpiration a little, but it should be remembered that cuticular transpiration is only about 5–10 per cent of the total.

Increased root extension into deeper, wet soil is an important means of adaptation to drought. The relations between the root and shoot systems seem to be directed mainly by the balance between the needs for water uptake in the root and for photosynthesis in the shoot. A shoot will grow until water uptake by the root becomes limiting to growth. A root will grow until supplies of photosynthetic product from the shoot become limiting. This balance is shifted if water supply is decreased.

As we saw in Chapter 9 (section 9.4), when water uptake is reduced, then leaf expansion is affected very quickly, but photosynthesis is much less affected. If leaf expansion is lessened, this means more carbon is available for the root system. At the same time, the surface soil becomes drier and the roots in this area will tend to die off. Thus, in well-watered soil, plants tend to have shallow root systems, but these are deeper in drying soil as roots grow into moist areas at depth. Plant breeders working

on crops growing in semi-arid zones have been able to enhance rooting depth and thereby increase crop yield.

As the soil dries, its matric potential and water potential become more negative (Chapter 5, section 5.4.2). Plants can continue to absorb water only if their water potential is below that of the source of water. One way of doing this is to accumulate solutes and lower the osmotic potential of the cell sap. The increase in solute concentration is fairly small and is mostly accounted for by increases in sugars, organic acids and ions (particularly potassium).

Enzymes extracted from the cytoplasm of plants are severely inhibited by high concentrations of ions, and so most of the increase is in the vacuole. The problem with this is that the cytoplasm would become dehydrated, because of the lowered vacuolar osmotic potential. Therefore, to balance the cell's water potential equilibrium, compatible solutes accumulate in the cytoplasm. Typical compounds accumulated are proline, sorbitol and glycinebetaine. We have already had cause to mention these compounds several times in this chapter, and will return to them in greater detail when we discuss salt

Box 10.2 Papyrus

Figure 10.8 shows papyrus (*Cyperus papyrus*), a monocot belonging to the sedge family Cyperaceae, growing at the extreme northern part of its range in the Hula Valley in northern Israel. This hydrophyte usually grows 2–3 m tall, can be found throughout Africa and in some Mediterranean countries, and has escaped to become a weed species in some parts of the United States.

Papyrus grows in subtropical and tropical environments, tolerating annual temperatures of 20–30°C, but it is sensitive to frost. It thrives best in full sunlight and flowers in late summer. Papyrus is found in marshes where the water is shallow and nearly constant in level, and its water-loving nature has been known at least since Biblical times – Job 8:11 records, *'Can papyrus grow where there is no marsh? Can reeds flourish where there is no water?'*

Papyrus is also the name of a thick paper-like material produced in Egypt from the pith of the plant, at least as far back as the First Dynasty in the third millennium BCE. This was the writing surface of choice until about 100 BCE, when the use of parchment, made from animal skins, gradually superseded it. The pith of the papyrus stalk has a network of intercellular spaces that make it very buoyant, and these would have been used to make the basket in which the baby Moses was floated on the Nile (Exodus 2).

The Egyptians also used papyrus for making much larger boats. Thor Heyerdahl, the Norwegian ethnographer and adventurer, built two boats from papyrus in 1969 and 1970. He wished to prove that it was possible for the Ancient Egyptians to have sailed to the Americas, and perhaps to have conveyed the technology for building pyramids. The first boat, *Ra*, broke apart

Figure 10.8 Papyrus (*Cyperus papyrus*), growing in the Hula Valley Nature Reserve in northern Israel. Photo: MJH.

and had to be abandoned, but *Ra II* was successful and sailed from Morocco to Barbados in 1970.

tolerance (Section 10.5.2), as they also accumulate in response to salinity.

10.4.2 Xerophytes

In very dry regions, experiencing arid climates, such as deserts, annual evaporation may exceed total annual precipitation: about one-third of the Earth's land mass has a rain deficit.

A xerophyte is defined in *The Penguin Dictionary of Botany* (1984) as *'a plant that is adapted to living in dry conditions caused either by lack of soil water or by heat or wind bringing about excessive transpiration.'* The key problem for plants growing in xeric environments is how to restrict water loss without diminishing photosynthesis. This is the photosynthesis/transpiration dilemma which we covered in Chapter 7 (section 7.6), and it is of crucial importance for xerophytes. In some respects, many xerophytes use similar mechanisms in adapting to drought to more sensitive plants, but the adaptations are more prominent and have become constitutive.

Because water loss and CO_2 uptake share common pathways, the only mechanism that could retain water without reducing fixation would be a waterproof membranc that is permeable to CO_2, which seems to be a physical impossibility. Xerophytes have evolved partial solutions, however, and these will now be outlined.

Xeromorphic modifications that increase or protect the leaf boundary layer reduce transpiration while permitting passage of CO_2. These include a thicker cuticle and surface hairs, but such features have a high energy cost. Reducing leaf size or increasing leaf reflectance are common strategies in arid environments, but these decrease photosynthesis, while small leaf size also lessens evaporative cooling. Cacti store water, have low surface area-to-volume ratios, often have large root systems, and frequently their leaves are reduced to spines. The latter increase the boundary layer near the surface and reflect radiation (See Section 10.2.3), keeping the plant cooler.

Turgor pressure supports the leaves of plants growing in moist environments, but it cannot always be maintained in xerophytes and their leaves, in many cases, contain large amounts of sclerenchyma for support. Yucca is a typical example (see Chapter 8, Figure 8.17).

Succulence allows some xerophytes to store water internally and permits CO_2 assimilation during a drought. A side effect is a reduction of surface area to volume ratio. Common ice plant (*Mesembryanthemum crystallinum*) originated in South Africa but has been introduced elsewhere (See Chapter 7, Figure 7.14). It illustrates two common features of xerophytes, namely succulence and CAM photosynthesis. As we saw in Chapter 7 (section 7.8), CAM plants close stomata in the day to conserve water and open them at night to take in CO_2.

The living fossil gnetophyte, *Welwitschia mirabilis* (see Chapter 1, section 1.7), is another CAM plant and has a large taproot (Figure 10.9). It is confined to the Namib and Mossamedes deserts of southwestern Africa, where most of its moisture is derived from fog that rolls in from the Atlantic Ocean at night. Reducing shoot/root ratio to increase water acquisition relative to water loss is a widespread strategy employed by xerophytes in arid environments. A serious limitation is increased respiratory cost relative to photosynthetic potential.

Figure 10.10 Marram grass (*Ammophila arenaria*), growing here on the sand dunes at Newborough Warren near the village of Newborough, Anglesey, Wales.
Photo: MJH.

Some mesophytes exhibit xeromorphic characters even though they do not grow in arid environments such as deserts. Bare rock, thin or sandy soils subject to rapid drainage and habitats exposed to frequent high winds all tend to be xeric. Many maritime sand dune plants of high rainfall areas are ephemerals or show xeromorphic modifications. Marram grass (*Ammophila arenaria*) is found on sand dunes and is very drought-tolerant in an environment that is often dry and windy (Figure 10.10). It has a number of xeromorphic features, including in-rolled leaves, which are hairy and with stomata often sunken well into the leaf (Figure 10.11). Even tropical rainforest

Figure 10.9 The gnetophyte, *Welwitschia mirabilis*, is a CAM plant that is confined to the deserts of southwestern Africa.
Photo: MJH. Used with the permission of Oxford Botanic Garden, Oxford, UK.

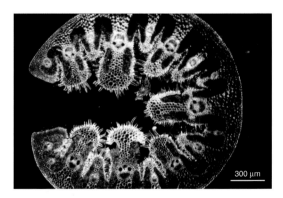

Figure 10.11 Cross-section through a marram grass (*Ammophila arenaria*) leaf. Note that the leaf is in-rolled and hairy, and the stomata are sunken deeply into it.
Photo: John Sibley.

epiphytes frequently have xeromorphic features such as succulent leaves.

10.4.3 Breeding for drought tolerance

Drought is one of the key factors limiting the growth and yield of crop plants, so increasing drought tolerance is an important aim for crop breeders. If the stress is severe, then traits that increase the chances of survival will be useful, but generally plant breeders are less interested in survival under severe stress than in increased yield under moderate stress. Tolerance to such moderate stress seems to be conferred by constitutive characters that are also expressed in the absence of stress. Improvements in three factors are involved in better crop performance under moderate stress: water use, water use efficiency (the ratio of assimilation to transpiration) and the harvest index (harvested product, such as grain, as a percentage of the total plant weight).

Thus conventional plant breeders have improved crop yield under drought conditions, taking these factors into account, for many years. Since the early 1980s, however, there has been increased interest in using genetic engineering to improve drought tolerance. One idea has been to increase the production of compatible solutes in plants undergoing drought stress, but this has largely failed for two main reasons. First, the increase in compatible solutes caused by genetic modification usually only corresponded to a small part of the total osmotic adjustment of plant cells under water stress. Second, the induced increase often resulted in impaired plant growth even when there was no stress.

Early optimism concerning the production of drought-tolerant GM crops has undoubtedly been dented as the complexity of the systems involved became more apparent. While crops that are herbicide-resistant (see Chapter 11, section 11.1.5) or insect-resistant (Chapter 11, Box 11.2) require only one or few genes to be changed, drought tolerance is vastly more complex, involving many genes that all interact with each other.

Writing in 2010, concerning the breeding of wheat for drought resistance, Fleury and her colleagues asserted that: '*The limited success of the physiological and molecular breeding approaches now suggests that a careful rethink is needed of our strategies in order to understand better and breed for drought tolerance*' (Fleury et al., 2010).

Certainly, we do appear to be some way off producing drought-tolerant crops using GM technology, but the prize is so great that efforts to do so will undoubtedly continue.

10.5 Resistance and adaptation to salinity

In Chapter 9 (section 9.5.1), we described saline environments and the problems these cause for plants in terms of salinity stress. We saw that plants could be divided into excluders (most crop plants) and includers (most halophytes). Here we will first briefly consider salt tolerance in crop plants, and then move on to halophytes in more detail. Our final topic in this section will be breeding of salt-tolerant crops.

10.5.1 Salt tolerance in crop plants

Glycophytes, including most crop plants, are sensitive to soil salinity. However, as we saw in section 9.5.2, some species are more sensitive than others and there are also considerable differences between cultivars of the same species. Almost all crop plants are excluders of Na^+ and Cl^- and, indeed, tolerance to salinity within a species is often linked to the ability to exclude these ions. Exclusion inevitably leaves plants susceptible to water stress.

There are, however, two adaptations that some excluders use to avoid internal water deficit:

• Synthesis of organic solutes (e.g. sugars) causes a decrease in shoot water potential, which allows water uptake to occur more easily.
• Succulence will cause a decrease in surface area relative to the volume of tissue, and this will lead to a decrease in water loss.

There is a large literature on the effects of salinity on crop plants and on the salt tolerance mechanisms that they possess. However, if we are to breed successfully for salt tolerance in crops, then understanding how halophytes are adapted to high internal salt concentrations is essential. We will turn to this next.

10.5.2 Halophytes

Halophytes constitute about one per cent of the world's flora, and are defined as plants that have the ability '*to complete the life cycle in a salt concentration of at least 200 mM NaCl under conditions similar to those that might be encountered in the natural environment*' (Flowers et al., 1986).

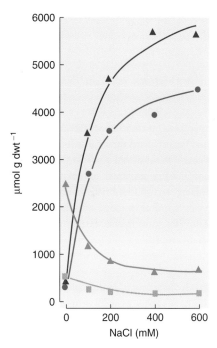

Figure 10.12 The ionic content of the halophyte *Suaeda maritima* when treated with varying salt (NaCl) concentrations in hydroponic culture. Key: sodium (red triangles); potassium (green triangles); chloride (blue circles); magnesium (orange squares). From Stewart *et al.*, 1979. (Reproduced by Permission of John Wiley Publishers).

Investigations of salt tolerance in halophytes usually begin with growing the plants at a variety of external salinities in hydroponic culture, and then assessing their growth. Some halophytes show optimal growth in saline conditions, while others grow best in the absence of salt (See Chapter 9, Figure 9.8). The next stage is usually to analyze the plant material to determine its ionic content. Figure 10.12 shows such data for the shoot of the halophyte *Suaeda maritima* when it was grown at several different salinities in hydroponic culture. As expected, Na^+ and Cl^- increase with increasing salinity, while K^+ and Mg^{2+} decrease.

As we have seen, nutrient ions are often in competition against other ions for uptake (Chapter 9, section 9.5.2). On a molar basis, seawater has about fifty times higher concentration of Na^+ than K^+. However, and as we noted in Chapter 5 (section 5.5), it is K^+ rather than Na^+ that is the essential nutrient for almost all plants and, even in the few where Na^+ is essential, it is only a micronutrient.

Thus, halophytes have evolved mechanisms for selective uptake of K^+ in an environment with a high Na^+/K^+ ratio.

It also appears that monocot halophytes show more selectivity in favour of K^+ than eudicot halophytes. Halophytes also seem to be able to use Na^+ to partially replace the roles of K^+ in the cell, particularly in osmotic adjustment. The transporters involved in the uptake of Na^+ have been investigated but they show much variation, and as yet there is considerable uncertainty over the mechanisms by which Na^+ is taken up by plant cells.

Chloride is also accumulated in significant amounts by halophytes, but generally to a lesser extent than Na^+. The mechanisms underlying these differences between Na^+ and Cl^- uptake are still obscure.

Once ions are taken up, halophytes also have several ways of decreasing internal concentrations. In some species, succulence leads to an increase in water content and thus to a dilution of the ions taken up. Retranslocation of Na^+ and Cl^- from the shoot to the root and then to the external solution is probably a fairly minor mechanism, as phloem is a living tissue that is susceptible to high concentrations of ions. Salt glands or external bladders are present on the leaves of some species, including many mangroves – salt-tolerant trees from the tropics. Where they are present, glands and bladders are used to secrete ions and modulate tissue ion concentrations, but most halophytes (e.g. *Suaeda maritima*) are able to regulate ion concentrations without resort to secretion. Finally, some species lose excess salt by dropping their older leaves, as we saw in drought stress (section 10.4.1).

The most physiologically meaningful way of expressing Na^+ and Cl^- concentrations in leaves is on a tissue water basis. When concentrations in the tissue water of halophytes are calculated in this way, they often exceed 450 mM. This leads us directly to a major paradox, as their enzymes are very salt-sensitive. Throughout the 1970s, scientists extracted enzymes from many halophytic and glycophytic plants. For example, Figure 10.13 shows the effect of NaCl on the enzyme activity of malate dehydrogenase in leaf extracts from the halophytes *Atriplex spongiosa* and *Salicornia australis* compared with that from the glycophyte, common bean (*Phaseolus vulgaris*). Almost without exception, the enzymes of halophytes proved to be just as sensitive as those extracted from glycophytes. Moreover, the results indicated that there would be very little enzyme activity at salt concentrations

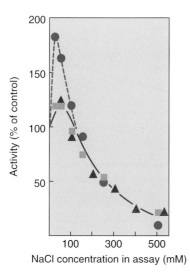

Figure 10.13 The effect of NaCl during enzyme assay on the activity of malate dehydrogenase in leaf extracts. Key: *Atriplex spongiosa* (blue circles); common bean (*Phaseolus vulgaris*) (orange squares); *Salicornia australis* (red triangles).
From Greenway & Osmond, 1972 (Reproduced by Permission of American Society of Plant Biologists).

Table 10.1 Mean ion concentrations in *Suaeda maritima* leaf mesophyll cells, as determined using freeze substitution and x-ray microanalysis.

	Sodium	Potassium	Chloride
Vacuole	565	24	388
Chloroplast	93	16	85
Cytoplasm/cell wall	109	16	21
Cell wall/intercellular space	132	13	36

Plants were grown in sand culture irrigated with a culture solution containing 340 mM NaCl. All leaf cell concentrations in mol m^{-3}. Data from Harvey *et al.* (1981).

often present in the leaves of the halophytes, and clearly this cannot be the case.

It was soon suggested that salt is compartmented away from the enzymes in higher plant cells. Most higher plant cells have large vacuoles comprising 90–95 per cent of the cell volume, and in some succulent halophytes this figure rises to 99 per cent. Most Na$^+$ and Cl$^-$ ions were suggested to be sequestered in the vacuole in halophyte cells.

Proof that this was, indeed, the case involved the use of a variety of techniques, the most important of which was x-ray microanalysis. This is a method of locating elements at the cellular or sub-cellular level using an electron microscope. It was not until the early 1980s that the problems of quantifying x-ray microanalysis and keeping ions in their original *in vivo* locations were solved. Harvey and her co-workers used a technique known as freeze substitution to study ion distribution in the leaf mesophyll cells of the halophyte *Suaeda maritima* (Harvey *et al.*, 1981; see Table 10.1). The major point to note is that the results confirm that halophytes store most of their toxic Na$^+$ and Cl$^-$ ions in their vacuoles.

Further work on salt-treated glycophytes suggested that there was less evidence of compartmentation in vacuoles, and that Na$^+$ and Cl$^-$ were present in far higher concentrations in the cytoplasm. The cytoplasmic concentrations observed in glycophytes would almost certainly cause enzyme inhibition and reduction in growth.

A question that immediately arises is how halophytes achieve the compartmentation noted above. It now appears that H$^+$-ATPases in the plasma membrane and tonoplast, and the H$^+$-translocating inorganic pyrophosphatase (H$^+$-PPiase) present in the tonoplast, supply the transmembrane proton motive force utilized by a number of secondary transporters (see Chapter 5, section 5.7.3). Exchange of Na$^+$ with H$^+$ seems to increase when plants are grown in saline media. At the same time, the activity of several enzymes that are involved in the generation of proton motive force also increases. Tonoplast Na$^+$/H$^+$ antiporters, such as those belonging to the Na$^+$/H$^+$ exchanger (NHX) family in *Arabidopsis*, are involved in the accumulation of Na$^+$ across the tonoplast into the vacuole.

The evidence strongly suggests that Na$^+$ and Cl$^-$ are sequestered in the vacuoles of halophytic higher plants. This creates a further research question: if most of the salt were in the vacuole, what would happen to the cytoplasm? A high salt concentration in the vacuole would create a low osmotic potential, water would tend to flow down the water potential gradient from the cytoplasm to the vacuole, and the former would desiccate. This obviously cannot be the case, and Stewart & Lee (1974) were the first to suggest a solution. They investigated proline concentration in a variety of halophytes and compared the levels with other groups of plants collected from the field. Most halophytes had much higher proline concentrations than other wild plants. Figure 10.14 shows the results of some laboratory experiments on proline accumulation

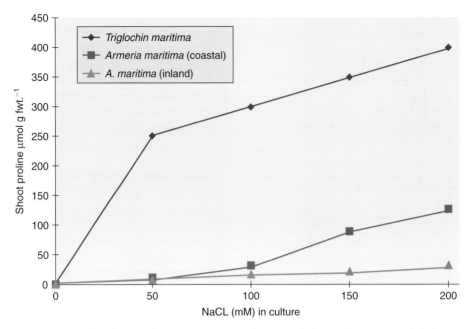

Figure 10.14 The effect of salt (NaCl) stress on proline accumulation in *Triglochin maritima*, an extreme halophyte, and sea pink (*Armeria maritima*), salt marsh and inland mountain populations.
Data from Stewart & Lee, 1974 (Reproduced by permission of Springer-Verlag Publishers).

in salt-stressed halophytes. Proline concentration in the shoots increases under salt stress. In *Triglochin maritima*, an extreme halophyte, proline increases to levels that represent ten per cent of the total dry weight. Sea pink (*Armeria maritima*) is interesting as it has coastal (salt marsh and cliff top- see Figure 10.15) populations and inland populations that grow on mountains. The coastal population accumulates more proline under saline conditions than the inland population – an expected result if proline is involved in salt tolerance.

Stewart and Lee also found that some halophytes did not accumulate very much proline, or none at all. Further investigations led to the isolation of a number of other compounds from various halophytes, all of which accumulated under salt stress in one species or another. The quaternary ammonium compound (QAC) glycinebetaine is known to accumulate in many members of the Chenopodiaceae (beet family) and in the grasses. Glycinebetaine-accumulating species appear to have little proline, and the reverse is also the case. Four related QACs, choline-0-sulphate and the betaines of ß-alanine, proline, and hydroxyproline, supplement or replace glycinebetaine in the Plumbaginaceae.

Dimethylsulphoniopropionate (DMSP), a sulphonium analogue of the QACs, accumulates in a few species like *Spartina* spp. The sugar alcohol, sorbitol, is accumulated by the Plantaginaceae (e.g. *Plantago maritima*), but inositol and pinitol are present in numerous genera.

Figure 10.15 Sea pink (*Armeria maritima*) has salt tolerant coastal populations. Here it is growing on a cliff top at Golden Cap near Bridport, Dorset, England.
Photo: MJH.

The pathways responsible for solute accumulation are now understood in many cases, and the enzymes involved in the synthesis of proline and glycinebetaine are known to increase as external salinity is increased.

Low molecular weight 'compatible' solutes have been mentioned several times already in this chapter, but what is the evidence for their compatibility with metabolic processes? Stewart and Lee investigated the effects of proline on enzyme activity and found that, in contrast to the effect of NaCl, proline had almost no effect on enzyme activity. Many other investigators have tried this with the other solutes and other enzymes, and in general terms these solutes do not inhibit to enzymes up to very high concentrations. In a number of cases, proline and glycinebetaine have also been shown to partially protect enzymes against NaCl.

There has been a little work on the sub-cellular localization of compatible solutes using either cytochemical procedures or preparation of pure vacuoles, and subsequent comparison of vacuolar concentrations with those from intact tissues. Both approaches have suggested that cytoplasmic concentrations can be considerably higher that those in the vacuoles. Compatible solute concentrations are often quite low on a whole tissue basis but, if they are restricted to the smaller volume of the cytoplasm, then cytoplasmic concentrations can be much higher. In some species, solute concentration may increase markedly under salt stress but, in others, production appears to be constitutive, not increasing in response to salinity. In such cases, there may be some solute redistribution across the tonoplast as external salt concentration increases.

A model for salt tolerance in halophytes was thus developed by the early 1980s (Figure 10.16), and work

since then has tended to confirm it. There is good evidence that most Na^+ and Cl^- in halophytes is sequestered in cell vacuoles. In glycophytes, this compartmentation does not occur to the same extent, resulting in severe stress and decreased plant growth. The cytoplasm of halophytes has low Na^+ and Cl^- concentrations, and Na^+/K^+ ratios have been calculated to be 2.5 to 21 times greater in the vacuole than the cytoplasm. To balance these high Na^+ and Cl^- concentrations, compatible solutes accumulate in the cytoplasm.

Considerable energy must be required to maintain salt tolerance in halophytes. This would be used in the ion transport processes that regulate total uptake and compartmentation of Na^+ and Cl^- at the cell level, and also in the production of compatible solutes.

10.5.3 Breeding for salt tolerance

As with drought tolerance (section 10.4.3), there has been much interest in breeding salt-tolerant crops, but there has been limited success insofar as the systems involved are very complex and not fully understood. Three methods are being used at present to increase salt tolerance.

• Direct selection in saline environments, exploiting natural genetic variation within a crop species, is the classical plant breeding approach.

• Alternatively, mapping quantitative trait loci (regions of a genome that are associated with variation in a quantitative trait, e.g. fruit yield, ion transport), followed by marker-assisted selection, is a more precise approach that improves selection efficiency, particularly for traits that are controlled by many genes and that are affected by environmental factors.

• Finally, genetic modification techniques involve the introduction of novel genes or the alteration of expression levels of existing genes to increase salt tolerance in the transgenic plant produced.

Superior compartmentation of Na^+ and Cl^- into the vacuoles of halophytes suggests that this might be a target for genetic modification in non-halophytes. Increased salinity tolerance of a range of plant species over-expressing NHX genes that code for tonoplast Na^+/H^+ exchangers indicates that this is a real possibility. As yet, however, these successes in the laboratory have not been transferred to the field.

The introduction of genes to increase the synthesis of compatible solutes in plants that do not normally contain

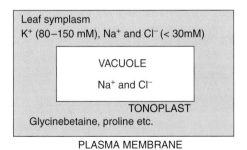

Figure 10.16 A model for salt tolerance in halophytes. Based on Figure 15.6 in Wyn Jones, 1981 (Reproduced by permission of Elsevier Publishers).

the solute has been another target for work involving transgenics. One problem with increasing glycinebetaine synthesis has been that, in some target plants (e.g. tobacco, rice), the precursor, choline, has low availability. This limits glycinebetaine synthesis, resulting in the solute not being produced at high enough concentrations.

Islam *et al.* (2007) used a somewhat different approach. They crossed salt-sensitive bread wheat (*Triticum aestivum*) with the salt-tolerant *Hordeum marinum* to produce a *H. marinum-T. aestivum* amphiploid. The amphiploid was intermediate in salt tolerance between its two progenitors and also had increased glycinebetaine and proline in its expanding leaves – a feature introduced from *H. marinum*. This kind of experiment suggests that increasing the synthesis of compatible solutes requires the introduction of whole metabolic networks, not changes in just a few enzymes.

10.6 Tolerance and adaptation to toxic metals

In Chapter 9 (section 9.6), we considered the stress caused to plants by high concentrations of chemicals. Here we will investigate what is known about tolerance to heavy metals and aluminium in plants. We will first cover ecotypes that are adapted to toxic metals, then the physiology and biochemistry of resistance to metals, and finally plants that accumulate large amounts of metal – the hyperaccumulators.

10.6.1 Metal-tolerant edaphic ecotypes

We saw in Chapter 9 (section 9.6.1) that old metal mine tips are frequently contaminated with heavy metals. Often, the soil is so toxic to plant growth that almost no plants are able to survive in it. Sometimes, however, plants have evolved races or clones that are tolerant to the toxic soil; these are known as **edaphic ecotypes** and they are plants from the same species that are more or less tolerant to a soil factor. They arise by the process of natural selection working on non-tolerant progenitors of the same species that grow in the non-toxic soils surrounding the mine site. Tolerance to heavy metals is particularly common in mine populations of grasses such as common bent grass (*Agrostis capillaris*).

If we wish to assess the tolerance of plants to an edaphic factor, there are many approaches. However, the

technique that has proved most useful, particularly with grasses, is called the rooting index of tolerance. In this procedure, plants are grown in water culture, with or without added toxic metal. As might be expected, toxic concentrations of metal in solution cause root growth to decrease. Rooting index of tolerance can be calculated as follows:

$$\text{Tolerance index} = \frac{\text{Mean length in metal solution}}{\text{Mean length in control solution}} \times 100$$

This index has been used very widely by those interested in metal tolerance, and it has also been employed to assess salt tolerance and tolerance to anaerobiosis.

A good example of the use of the rooting index of tolerance comes from the work of Cox & Hutchinson (1980) on several clones of tufted hair-grass (*Deschampsia cespitosa*) from around the Coniston smelter at Sudbury in Ontario, Canada (Figure 10.17). This area had been

Figure 10.17 Tufted hair-grass (*Deschampsia cespitosa*) growing in contaminated soil near the Coniston smelter at Sudbury in Ontario, Canada.
Photo: MJH.

Figure 10.18 Seriously contaminated land resulting from past smelting activity at Sudbury in Ontario, Canada. Note the tree stumps and the clumps of tufted hair-grass (*Deschampsia cespitosa*).
Photo: MJH.

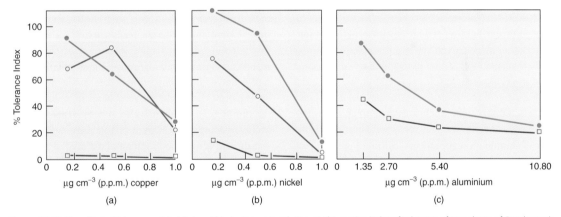

Figure 10.19 The effect of (a) copper, (b) nickel and (c) aluminium in solution on the rooting index of tolerance of two clones of *Deschampsia cespitosa* (green and blue lines) sampled near the Coniston copper/nickel smelter at Sudbury in Ontario, Canada. The results were compared with those from a control clone (red line) collected from an uncontaminated pasture at Hay Bay.
From Cox & Hutchinson, 1980 (Reproduced by permission of John Wiley Publishers).

heavily contaminated with copper and nickel from the smelter and acidified from sulphur dioxide emissions, which resulted in elevated free aluminium concentrations in the soil solution (Figure 10.18).

They also collected clones from Hay Bay, on uncontaminated pasture, 150 km away. The plants were then tested for tolerance to copper, nickel and aluminium. Not surprisingly, the Coniston population was much more tolerant to all of the metals than the Hay Bay population (Figure 10.19).

Metals differ in their toxicities, with copper the most and aluminium the least toxic. Fairly frequently, tolerant plants have been observed to have their growth stimulated by low concentrations of toxic metals, as is the case here. This phenomenon is little understood. The Coniston plants are tolerant to more than one metal – again a common observation where more than one metal is present in the soil solution at toxic concentrations. However, tolerance is usually metal-specific, so tolerance to a particular metal does not confer tolerance to another (e.g. copper and zinc). The evolution of separate tolerance mechanisms to three different metals, as with the Coniston clone, is relatively unusual, and the probability of this happening from the non-tolerant progenitor plants in the vicinity must be extremely low.

10.6.2 Tolerance at the cellular level

Some algae appear to exclude toxic metals completely from their tissues, but this has not been observed in higher plants. In a few species, exclusion from the shoot occurs in tolerant ecotypes (e.g. *Agrostis capillaris*). Other metal-tolerant species, such as *Becium homblei*, accumulate metals in their older leaves and then drop them, removing toxic ions in a similar way to some halophytes. Most interest has, however, concerned tolerance at the cellular level, to which we will now turn our attention.

Most of the tolerance mechanisms seem to involve preventing the build-up of toxic quantities of ions at sensitive sites. There are relatively few examples of enzymes that are adapted to function in the presence of high concentrations of metal. In some species, binding to the cell walls leading to accumulation in the apoplast does occur, but it is unlikely to be whole answer to tolerance as this compartment soon becomes saturated with ions. In the case of aluminium it appears that silicon, if present, may form non-toxic hydroxyaluminosilicate complexes in the apoplast of plant roots. In conifer needles, co-deposition of Al with Si in cell walls may remove toxic Al from the apoplast.

Many metals, because of their relatively large atomic radius and valency (e.g. aluminium is trivalent), show restricted movement across the plasma membrane. There is also evidence for active efflux of toxic metals from the cytoplasm of some plants. Increased resistance of membranes to metal-induced damage has been observed in a number of cases; for example, copper-tolerant clones of *Agrostis capillaris* showed less potassium leakage (a measure of membrane damage) from root tips than non-tolerant clones.

Secretion of organic acids into the rhizosphere and chelation at the cell wall-plasma membrane interface is a well-established method of decreasing metal toxicity. For example, exudation of malate by wheat roots is known to be involved in aluminium tolerance, and tolerant cultivars secrete more organic acid than sensitive cultivars. Buckwheat both secretes oxalic acid into the rhizosphere and accumulates non-toxic Al-oxalate complexes in the vacuoles of the leaf cells.

Within the cytoplasm, chelation by specific metal-complexing peptides called **phytochelatins** (PCs) is an important mechanism of decreasing metal availability. Production of PCs is induced by metal treatment, and they have the general structure $(\gamma\text{-Glu Cys})_n\text{-Gly}$, where $n = 2\text{--}10$. Most work on PCs has concerned cadmium (Cd) tolerance, and the enzyme PC synthase is activated in the presence of metal. Mutants of *Arabidopsis* that have reduced levels of PC synthase are more sensitive to cadmium.

There is no doubt that PCs have a role in cadmium tolerance, but evidence is less conclusive for some other heavy metals, and they are certainly not the only tolerance mechanism in some plants. **Metallothioneins** are another type of cysteine-enriched metal-binding peptide and are also involved in tolerance to some heavy metals. It is now evident that different plant species use a variety of compounds for decreasing the availability of specific metals, giving an apparently confusing overall picture.

We noted that compartmentation in the vacuole is an important salt tolerance mechanism in halophytes (section 10.5.2), and there is also some evidence for accumulation being involved in metal tolerance. Cadmium is transported across the tonoplast and stored as a Cd-PC complex in the vacuole. Zinc is also localized in the vacuole in some plants, and zinc-tolerant ecotypes seem to be more effective in this respect than are sensitive ecotypes.

In Chapter 9 (section 9.6.3), we saw that aluminium toxicity is a factor in the die-back of trees caused by acid rain. It is also significant in reducing crop production on naturally occurring acid soils. Therefore, improving the aluminium tolerance of crop plants is an important area of research, and Ryan *et al.* (2011) reviewed progress in this area. The efflux of organic acids (e.g. malate, citrate) from roots of plants such as wheat is controlled by members of the ALMT (aluminium activated malate transporter) and MATE (multidrug and toxic compound exudation) gene families. These genes encode membrane proteins that aid the efflux of organic acids across the plasma membrane. Attempts to use genetic engineering to increase aluminium tolerance have thus mainly focused on increasing organic acid efflux from sensitive plants. This can be done by enhancing organic acid synthesis or the transport of these substances across the plasma membrane. It appears that the latter has the greater chance of increasing tolerance to aluminium, and thus improving crop growth in acid soils.

10.6.3 Metal hyperaccumulators and the phytoremediation of heavy metal contaminated land

Certain plant species have the ability to hyperaccumulate metals, often to extreme levels – sometimes a hundred or a thousand fold higher than the levels normally found. **Hyperaccumulators** are present in at least 45 plant families, accumulating different metals including arsenic, cadmium, cobalt, copper, nickel, selenium and zinc, or sometimes combinations of these metals. Spring sandwort or leadwort (*Minuartia verna*) is now recognized as a hyperaccumulator of lead (Figure 10.20).

The two brassicaceous zinc hyperaccumulators, *Thlaspi caerulescens* and *Arabidopsis halleri*, have the advantage of being closely related to *Arabidopsis thaliana*, the model plant of molecular biology, facilitating molecular approaches. Aluminium hyperaccumulators include tea (*Camellia sinensis*), where the element accumulates in the leaves (Figure 10.21). There has been continued interest

Figure 10.21 Tea (*Camellia sinensis*) plantation, Korakuen gardens, Okayama, Japan.
Photo: MJH.

in Al in tea because of the potential human toxicity effects, and the possibility that Al could be a factor in the development of some types of Alzheimer's disease. Much depends on the bioavailability of Al in tea infusions, and most workers do not consider this to be particularly high.

In 1998, the European Environment Agency estimated that there were a total of 1,400,000 contaminated sites in Western Europe. A number of remediation techniques are available to clean up contaminated land and there has been considerable interest in using plants to help in this, leading to the creation of the whole new science of **phytoremediation**.

Here we will concentrate on the use of hyperaccumulator plants to clean up land contaminated with heavy metals, termed phytoaccumulation. The idea is to grow these plants on contaminated soil, allow them to take up metal (e.g. cadmium or lead) from the substrate, and then harvest the shoots that have accumulated the metal. It has even been suggested that the metal might be extracted from the harvested shoots in commercially viable amounts – so-called 'phytomining'. If these hyperaccumulators are grown for long enough, they will eventually decontaminate the soil of the metal.

The major problem with this idea is that that most hyperaccumulators are small plants, with low biomass, that grow very slowly. One idea to overcome this difficulty is to genetically modify faster-growing plants with high biomass to become hyperaccumulators. However, transformed plants often show slow growth, and hyperaccumulation is probably a highly energy intensive process. Moreover, an even larger problem at the moment is that the economics do not favour phytoremediation, a slow

Figure 10.20 Spring sandwort or leadwort (*Minuartia verna*), photographed near Surrender Bridge in Swaledale, Yorkshire, England. The area has many lead mines and leadwort is a hyperaccumulator of the element.
Photo: Margot Hodson.

method with many uncertainties. Land contamination consultants tend to prefer quick clean-up operations with reliable, known methods. Thus, the idea of using hyper-accumulators to clean up contaminated land is still a long way from commercial reality.

10.7 Adaptations to light and radiation

In Chapter 9 (section 9.7), we considered stress caused by shading when visible light levels are reduced (section 9.7.1), as well as that caused by too much light or ultraviolet radiation (section 9.7.2). Here we will cover adaptations to those stresses.

10.7.1 Shade plants

In the case of temporary shade, such as that under deciduous woodland, there is a definite peak of irradiance on the forest floor in early spring before the tree leaves above emerge (see also Chapter 3, section 3.7.2). The trees are deciduous as an adaptation to avoiding damage to leaves from frost in the winter, so any forest floor plants that take advantage of the early spring window must be frost-resistant (section 10.2.2).

If these plants are to utilize this radiation peak, then they must have fully expanded leaves by April in the temperate Northern hemisphere. This means that leaf growth has to occur in the low temperatures of February and March, when photosynthesis is low because of reduced light levels. To grow at this time of year necessitates that the plants must have reserves, so nearly all of these plants have underground storage organs such as bulbs, tubers, corms and rhizomes. These supply the leaves in early spring so that they can grow. Then, at the radiation peak in later spring, the underground organs are re-charged. To survive in this particular niche means considerable modifications to the whole plant life cycle. Some plants, like the bluebell (Figure 9.12), complete their carbon fixation before the overhead canopy closes. Others remain photosynthetically active throughout the shade phase, but at reduced rates.

If a plant grows in a long-term shaded environment and is still active outside the brief period of high illumination in the spring, then the selection pressure will be more on the photosynthetic process than on the life cycle. In some situations, leaves will always be in the shade, and this is typical of lower leaves of multi-layer canopies.

The photon flux density at which a positive carbon balance is attained is called the **light compensation point** (Figure 10.22). Below this point, a leaf uses more energy for respiration than it can fix in photosynthesis, and

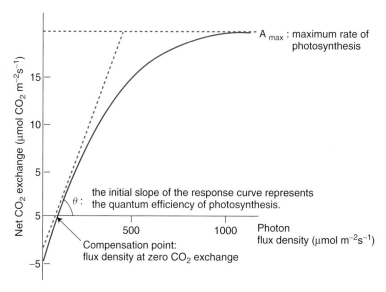

Figure 10.22 Photosynthesis response curve, where net CO_2 exchange is plotted against photon flux density. It illustrates the quantum efficiency of photosynthesis, the light compensation point and the maximum rate of photosynthesis. From Fitter & Hay, 2002 (Reproduced by permission of Elsevier Publishers).

evidently this situation cannot be maintained for very long. As light declines and shade increases, photosynthetic rate decreases for a whole variety of reasons. Stomata tend to close in darker conditions, so CO_2 absorption is decreased. There is also less energy available to regenerate RuBP (Chapter 7, section 7.4.5). At a certain flux density, CO_2 coming into the plant for photosynthesis equals that going out because of respiration and photorespiration.

There are three variables that determine the overall shape of the light response curve for photosynthesis:

1 An increase in the leaf area per unit mass provides a greater area for light absorption. The main effect of shading leaves is to increase their area but decrease their thickness. For more on light and shade leaves, see Chapter 3 (section 3.7.2) and Figure 3.25.

2 A reduction in respiration and photorespiration lowers the compensation point. The difficulty is that reduction in respiration is also likely to reduce growth, as it is needed for energy. Thus, a reduction in respiration rate is only likely to be a useful strategy in severe shade, where growth rates are so slow that competition between plants is not likely to be significant. Plants growing in deep shade generally grow very slowly.

3 The capacity for photosynthesis is mostly determined by the amount of Rubisco available for carbon fixation, and thus an increase in the responsiveness of the photosynthetic system to light levels. Generally, sun-adapted plants have high A_{max} (the saturation point above which no increase in photosynthesis occurs) and high compensation points, while shade plants have the reverse.

10.7.2 Tolerance to high light intensity, UV and ionizing radiation

As indicated in Chapter 9, section 9.7.2, high light intensity involves two main damaging factors. One is the generation of ROI and the other is the increased level of UV light. The latter is also involved in generating ROI, which are potent DNA damaging agents via oxidation of guanine and deamination of 5-methyl-cytosine and cytosine. In addition, UV causes other types of damage to DNA, the commonest of which is the formation of dimers between adjacent pyrimidine moieties in the DNA, particularly between adjacent thymines.

In respect of the effects of ROI, photorespiration must be seen as an adaptation in C_3 plants (Chapter 7, section 7.5.2) while, in all species, anti-oxidants are among the compounds synthesized in response to light stress (see also section 10.1). Another general adaption to high light (or more specifically high UV) is the synthesis of 'sun-screen' flavonoid compounds that absorb UV.

We noted in Chapter 9 (section 9.7.3) that, in addition to UV light, ionizing radiation is also a cause of DNA damage. The exposure for most plants is simply a matter of the background of gamma radiation from cosmic rays and, depending on habitat and geographical location, of alpha radiation from naturally occurring radioactive substances such as radon gas. Both cause single-strand breaks (nicks) in DNA strands (Figure 10.23), but gamma radiation also causes double-strand breaks which are potentially very serious.

The best known repair mechanism is **photoreactivation**, which is dependent on blue and/or UV light. An enzyme called **photolyase** chemically reverses dimer formation, thus restoring the integrity of the double helix. Photolyase is a member of the cryptochrome group of photoreceptors (see Chapter 3, section 3.7.4) but, unlike cryptochromes themselves, photolyases are not involved in maintaining circadian rhythms or in other signalling pathways.

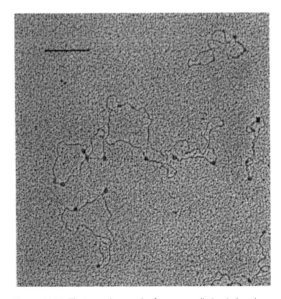

Figure 10.23 Electron micrograph of gamma radiation-induced nicks in DNA bound by protein that recognizes nicks and ss-ds junctions. Scale bar 200 nm.
From Burton *et al.*, 1997 (Reproduced by permission of John Wiley Publishers).

Transcription of the photolyase gene is strongly up-regulated by UV-B. Different rice cultivars have different sensitivities to UV-B, related to different efficiencies of their photolyases. Sequence analysis suggests that the photolyase genes in the different cultivars have very slightly different base compositions, raising the possibility of increasing the UV-resistance of susceptible varieties of crop plants by selective breeding or by genetic modification.

However, dimers can also be repaired (in the 'dark repair' mechanism) by removal mediated by a combination of DNA glycosylase and endonuclease, creating a short single-strand gap in the DNA. Single-strand gaps are also caused by various excision pathways, which remove mismatched nucleotides (the results of oxidation and/or deamination of bases). The gaps may be elongated by the action of exonuclease but, whether or not this is so, the free 3'-OH group is recognized by specific proteins and the repair DNA polymerase (polymerase-β or polymerase-λ) fills the gap. In the last stage of the process, the newly synthesized section is ligated to the 'waiting' 5' end by DNA ligase.

The repair of single-strand breaks depends on whether there has been any 'fraying' (loss of nucleotides) of the DNA strand in the region of the break. If not, it is simply a matter of recognition of the nick and its sealing by DNA ligase. If a section of the DNA chain has been lost, then the repair polymerase(s) will fill the gap prior to ligation.

Double-stranded breaks are repaired by non-homologous end joining (NHEJ). The breaks are recognized by two Ku proteins, which then recruit the DNA-dependent protein kinase catalytic subunit to form the trimeric DNA-dependent protein kinase. This forms a 'platform' onto which the other proteins involved in NHEJ are recruited. If nucleotides have been lost from one strand to create an overhang, the gap is filled by a specific DNA polymerase (polymerase-μ) and, finally, the strands are joined by DNA ligase IV. Transcription of the ligase IV gene is up-regulated by gamma radiation, but not by exposure to UV.

There is great interest in NHEJ, not just because of its importance in maintaining genome integrity, but also because it is the mechanism by which T-DNA is inserted by *Agrobacterium* (see Chapter 3, textbox 3.1). It would be very advantageous if, in GM procedures, genes could be inserted into specific sites in plant chromosomes. There is extensive research in this area, including the demonstration that that the joining of the T-DNA into the host genome is mediated by the host's enzymes, including the appropriate DNA ligase.

Selected references and suggestions for further reading

Apel, K. & Hirt, H. (2004) Reactive oxygen species: Metabolism, oxidative stress, and signal transduction. *Annual Review of Plant Biology* **55**, 373–399.

Araus, J.L., Slafer, G.A., Reynolds, M.P. & Royo, C. (2002) Plant breeding and drought in C3 cereals: what should we breed for? *Annals of Botany* **89**, 925–940.

Bray, C.M., Sunderland, P.A., Waterworth, W.M. & West, C.E. (2008) DNA ligase – a means to an end-joining. In Bryant, J.A. & Francis, D. (eds.) *The Eukaryotic Cell Cycle*. Taylor and Francis, Abingdon.

Burton, S.K., Vant Hof, J. & Bryant, J.A. (1997) Novel DNA-binding characteristics of a protein associated with DNA polymerase-alpha in pea. *Plant Journal* **12**, 357–365.

Chaves, M.M., Maroco, J.P. & Pereira, J.S. (2003) Understanding plant response to drought: from genes to the whole plant. *Functional Plant Biology* **30**, 239–264.

Chinnusamy, V., Zhu, J. & Zhu, J-K. (2006) Gene regulation during cold acclimation in plants. *Physiologia Plantarum* **126**, 52–61.

Cox, R.M. & Hutchinson, T.C. (1980) Multiple metal tolerances in the grass *Deschampsia cespitosa* (L.) Beauv. from the Sudbury smelting area. *New Phytologist* **84**, 631–647.

Evans, D.E. (2003) Aerenchyma formation. *New Phytologist* **161**, 35–49.

Fitter, A.H. & Hay, R.K.M. (2002) *Environmental Physiology of Plants*. 3rd ed. Academic Press, London.

Fleury, D., Jefferies, S., Kuchel, H. & Langridge, P. (2010) Genetic and genomic tools to improve drought tolerance in wheat. *Journal of Experimental Botany* **61**, 3211–3222.

Flowers, T.J. & Colmer, T.D. (2008) Salinity tolerance in halophytes. *New Phytologist* **179**, 945–963.

Flowers, T.J., Hajibagheri, M.A. & Clipson, N.J.W. (1986) Halophytes. *Quarterly Review of Biology* **61**, 313–337.

Greenway, H. & Osmond, C.B. (1972) Salt responses of enzymes from species differing in salt tolerance. *Plant Physiology* **49**, 256–259.

Hall, J.L. (2002) Cellular mechanisms for heavy metal detoxification and tolerance. *Journal of Experimental Botany* **53**, 1–11.

Harvey, D.M.R., Hall, J.L., Flowers, T.J. & Kent, B. (1981) Quantitative ion localization within *Suaeda maritima* leaf mesophyll cells. *Planta* **151**, 555–560.

Hopkins, W.G. (1999) *Introduction to Plant Physiology*, 2nd ed. John Wiley & Sons, New York.

Islam, S., Malik, A.I., Islam, A.K.M.R. & Colmer, T.D. (2007) Salt tolerance in a *Hordeum marinum-Triticum aestivum* amphiploid, and its parents. *Journal of Experimental Botany* **58**, 1219–1229.

Jansen, S., Broadley, M.R., Robbrecht, E. & Smets, E. (2002) Aluminum hyperaccumulation in angiosperms: A review of its phylogenetic significance. *Botanical Review* **68**, 235–269.

Kosová, K., Vítámvás, P. & Prášil, I.T. (2007) The role of dehydrins in plant response to cold. *Biologia Plantarum* **51**, 601–617.

Kozlowski, T.T. & Pallardy, S.G. (2002) Acclimation and adaptive responses of woody plants to environmental stresses. *Botanical Review* **68**, 270–334.

Lütz, C. (2010) Cell physiology of plants growing in cold environments. *Protoplasma* **244**, 53–73.

Molina, J., Sikora, M., Garud, N., Flowers, J.M., Rubinstein, S., Reynolds, A., Huang, P., Jackson, S., Schaal, B.A., Bustamante, C.D., Boyko, A.R. & Purugganan, M.D. (2011) Molecular evidence for a single evolutionary origin of domesticated rice. *Proceedings of the National Academy of Science, USA* **108**, 8351–8356.

Munns, R. & Tester, M. (2008) Mechanisms of salinity tolerance. *Annual Review of Plant Biology* **59**, 651–681.

Provart, N.J., Gil, P., Chen, W., Han, B., Chang, H-S., Wang, X. & Zhu, T. (2003) Gene expression phenotypes of *Arabidopsis* associated with sensitivity to low temperature. *Plant Physiology* **132**, 893–906.

Ruddiman, W.F. (2005) How did humans first alter global climate? *Scientific American* **292**, 46–53.

Ruellanda, E. & Zachowski, A. (2010) How plants sense temperature. *Environmental and Experimental Botany* **69**, 225–232.

Ryan, P.R., Tyerman, S.D., Sasaki, T., Furuichi, T., Yamamoto, Y., Zhang, W.H. & Delhaize, E. (2011) The identification of aluminium-resistance genes provides opportunities for enhancing crop production on acid soils. *Journal of Experimental Botany* **62**, 9–20.

Sevillano, L., Sanchez-Ballesta, M.T., Romojaro, F. & Flores, F.B. (2009) Physiological, hormonal and molecular mechanisms regulating chilling injury in horticultural species. Postharvest technologies applied to reduce its impact. *Journal of the Science of Food and Agriculture* **89**, 555–573.

Stewart, G.R. & Lee, J.A. (1974) The role of proline accumulation in halophytes. *Planta* **120**, 279–289.

Stewart, G.R., Larher, F., Ahmad, I. & Lee, J.A. (1979) Nitrogen metabolism and salt-tolerance in higher plant halophytes. In: Jefferies, R.L. & Davy, A.J. (eds.) *Ecological Processes in Coastal Environments*. Blackwell, Oxford.

Sweeney, M. & McCouch, S. (2007) The complex history of the domestication of rice. *Annals of Botany* **100**, 951–957.

Toothill, E. (ed.) (1984) *The Penguin Dictionary of Botany*. Penguin Books, London.

Venketesh, S. & Dayananda, C. (2008) Properties, potentials, and prospects of antifreeze proteins. *Critical Reviews in Biotechnology* **28**, 57–82.

Wang, W., Vinocur, B. & Altman, A. (2003) Plant responses to drought, salinity and extreme temperatures: towards genetic engineering for stress tolerance. *Planta* **218**, 1–14.

Wyn Jones, R.G. (1981) Salt tolerance. In: Johnson, C.B. (ed.) *Physiological processes limiting plant productivity*. Butterworths, London. pp. 271–292.

Yamaguchi, T. & Blumwald, E. (2005) Developing salt-tolerant crop plants: challenges and opportunities. *Trends in Plant Science* **10**, 615–620.

Ziemienowicz, A., Tinland, B., Bryant, J., Gloeckler, V. & Hohn, B. (2000) Plant enzymes but not *Agrobacterium* VirD2 mediate T-DNA ligation *in vitro*. *Molecular and Cellular Biology* **20**, 6317–6322.

CHAPTER 11
Biotic Stresses

In chapters 9 and 10, we looked at how plants are affected by, and adapt to, abiotic stresses imposed by the physical environment. In the current chapter, we will move on to consider biotic stresses – how plants interact with other organisms. We will begin with plant/plant competition, and then go on to plant interactions with animals. Finally, we will give brief coverage to the large topic of plant pathology, to fungal, bacterial and viral infections and plant resistance mechanisms. We will also give some prominence to those biotic stresses that have economic importance and the ways that agriculturalists have sought to limit their negative impacts.

11.1 Plant/plant competition

Thus far in this book we have largely treated plants as isolated entities, only interacting with their physical surroundings. Obviously, in the natural environment this is not the case, and plants frequently interact with each other. In this section, we will first consider how plants compete for resources and the related topic of **allelopathy**. There will follow sections on parasitic higher plants, on weeds – whose interactions with crop plants are of great significance in agriculture – and on the major means of controlling weeds, herbicides.

11.1.1 Competition between plants for resources

Competition happens when two or more plants of the same or different species are in close proximity and the supply of a factor or factors needed for plant growth is less than the combined needs of the plants. If a plant is a successful competitor, it will acquire a disproportionate amount of the resources available for growth, which will result in the reduction of its competitor's growth.

Plants cannot choose their neighbours, but they show many competitive strategies to improve their chances of prospering in a given environment. Some plants avoid their neighbours, thereby minimizing competition. Others aggressively confront their neighbours and maximize competition. A third strategy involves tolerating competition from neighbours. These different strategies will be adopted at different times and the three options are not mutually exclusive.

Plants essentially compete for three major resources: water, nutrients and light. Root competition involves a decrease in the availability of water and nutrients to roots due to the presence of other roots. This is known as scramble or exploitation competition, and it is caused by resource depletion in the rhizosphere. Plants also compete for light by shading those below them in the canopy and decreasing light availability (see Chapter 9, section 9.7.1). In addition, roots engage in contest or interference competition by inhibiting other roots from having access to resources.

A key way of inhibiting the growth of the roots of other plants is known as allelopathy, which has been defined as '*the release of a chemical by a plant that inhibits the growth of nearby plants and thus reduces competition*' (*The Penguin Dictionary of Botany*, 1984).

Phytotoxins can be released from decomposing leaf and root tissue. In some cases, leachates or volatile substances are released by the living leaves. For example, shrubs growing in the arid Californian chaparral (e.g. *Artemisia* and *Salvia*) release oils that inhibit the germination and growth of herbaceous plants. A large range of low molecular weight compounds are also exuded by plant roots into the rhizosphere, and many of these have allelopathic effects.

The phytotoxins produced by roots have many different chemical structures and include flavonoids, quinones,

Functional Biology of Plants, First Edition. Martin J. Hodson and John A. Bryant.
© 2012 John Wiley & Sons, Ltd. Published 2012 by John Wiley & Sons, Ltd.

quinolines and hydroxamic acids. Their modes of action also vary, but they can have inhibitory effects on processes including photosynthesis, respiration, membrane transport, germination and plant growth. The effects will depend on the concentration of the substances exuded. Often, toxicity is confined to other species but, in autotoxicity, inhibitory effects are noted within one species.

It has also become clear that allelopathic interactions in the rhizosphere are not only of the root-to-root type but include those between roots and microbes, fungi and insects.

11.1.2 Parasitic higher plants

In many respects, the parasitic higher plants are an extreme example of plant/plant competition. Sixteen angiosperm families contain 3,000 parasitic species – about one per cent of all angiosperms. These plants have a considerable number of lifeforms, including trees, shrubs, vines and herbs, and they can be found around the world in many different plant communities. Some species, such as mistletoe (*Viscum album*), colonize the shoot of the parasitized plant (Figure 11.1).

Mistletoe is an example of a hemiparasite that obtains water and mineral elements from the host plant, but is photosynthetic and obtains its own carbon. In contrast, more than 400 species of parasitic higher plants in 87 different genera lack chlorophyll, are not photosynthetic and also obtain carbon from their host plant

Figure 11.2 Tall broomrape (*Orobanche elatior*), an obligate parasite, here seen off a path between White Horse Hill and Wayland's Smithy, Berkshire, England.
Photo: MJH.

(e.g. *Striga* and *Orobanche*). Figure 11.2 shows tall broomrape (*Orobanche elatior*), an obligate parasite. When seeds of such plants germinate in the soil, they form a structure called a **haustorium**, which attaches itself to the roots of host plants. This extracts nutrients from the affected plants, leading to decreased growth or death of the host.

Some parasitic higher plants have severe economic impacts on crop yields and are regarded as important weeds. Among the worst of these are *Striga* spp., which affect African cereal and legume crops particularly badly. *Orobanche* spp. parasitize many crops, including crucifers, legumes, sunflower and tomato, and are important weeds in parts of Europe, Africa, the Middle East and the Indian subcontinent. We will consider weeds next.

11.1.3 Introduction to weeds

Weeds are a good example of plant/plant interactions. They are very important in agricultural contexts but are not confined to those contexts. A weed can be simply

Figure 11.1 European mistletoe (*Viscum album*), a hemi-parasitic plant, here growing on an apple tree near Woolhope, Herefordshire, England.
Photo: MJH.

defined as, 'any plant growing where it is not wanted' (*The Penguin Dictionary of Botany*, 1984). Thus it is humans who decide what a weed is, and this can often lead to some confusion. For example, poppies can be grown as pretty flowers in a garden, but they can also be a serious weed in crops.

Weeds have certainly been with humans since the very beginnings of agriculture and, indeed, have been widely written about. In the famous parable of the sower in the New Testament of the Bible, a farmer spread seed about a field and noted the various fates that awaited it: '*Other seed fell among thorns which grew up and choked the plants*' (Matthew 13:7).

In all probability, most weeds were originally colonizers of open habitats. These might be disturbed places or areas where open habitat was maintained by the physical environment, such as seashores and mountaintops. Today they are colonizers of disturbed habitats that have been created by humans. Weeds often have to tolerate more extreme physical environments than other plants. For example, exposed ground is likely to be hotter and drier than ground under vegetation.

Although weeds have a variety of effects, their most important is decrease in crop yield. It is very difficult to work out precisely the losses due to weeds and to disentangle them from the effects of pests and diseases. A weed infestation may weaken a crop and allow pests and diseases to flourish, but the reverse may also be true. Exact figures on global losses of crop production due to weeds are not available, but it is certain that this is far less in highly developed agricultural systems, where weeds are controlled by herbicides, than in peasant agriculture in the developing world, where herbicides and other modern methods are little used.

Competition with crop plants is the most important weed characteristic. The major factors being competed for are, again, water, nutrients and light. However, competition for these factors does not remain the same throughout the season. In the case of an annual crop at the beginning of the season, its seedlings are small but so are the weed seedlings, and competition is likely to be weak. In a similar way, a crop towards the end of its life cycle, when it is senescent and setting seed, is less likely to be harmed by weeds. It is between these two times that the 'critical period' for weed competition occurs.

Under good conditions, most weeds have a high output of seeds. For example, an individual poppy (*Papaver rhoeas*) can produce up to 20,000 seeds. This usually has a great selective advantage, as only a few of these seeds need to produce plants for the species to multiply. The disadvantage of this strategy is that production of large numbers of seeds means producing small seeds, which often contain only small reserves and are very susceptible to drought. However, large numbers of small seeds are not a precondition to success as a weed, and sunflower (*Helianthus annuus*), which can become an important weed in some situations, is one exception.

Seed dormancy (see Chapter 4, section 4.10) and longevity are crucial to the success of many weeds. The numbers of weed seeds in topsoil can be very large. Moreover, weed seeds are often capable of germinating after very long periods in the soil, some for 100 years or more. Seeds brought to the surface after many years germinate and produce plants. However, not all seeds will germinate straight away, and weed seeds show great variation in the length of time they take to germinate. It is obviously of some advantage not to have all seeds germinating at once, even in ideal conditions. Crops have been selected for seed that will germinate immediately and evenly, which is an advantage to the farmer, but a weed with this behaviour would be very easy to control.

Weed seeds will often not germinate for a long time, even in conditions that seem suitable, due to seed dormancy. Perennial weeds may use vegetative reproduction – examples including bracken, couch grass and bindweed. All of these are serious weeds that are difficult to remove, as they have persistent underground structures that are resistant to cultivation practices.

11.1.4 Weed spread and control

There are many ways in which humans spread weeds. Contamination of crop seed by weed seeds is almost inevitable. One example of a weed problem that originated in this way is tumbleweed (*Salsola tragus*), which was probably brought to the United States in flax seed by Ukrainian farmers in the 1870s (Figure 11.3).

Weed seed contamination can also occur in animal feeds and straw. Farm animals and machinery, when moving from one area to another, can carry seed with them. The spread of many weeds has resulted from the deliberate introduction of plants into a new country as garden plants, herbs and curiosities. The Oxford ragwort (*Senecio squalidus*) came to the United Kingdom from southern Italy and escaped from Oxford Botanic Garden.

Box 11.1 Purple loosestrife

Purple loosestrife (*Lythrum salicaria*) is a major weed of waterways and wetlands in Northern America, having been introduced from Northern Europe (Figure 11.4). This species was brought over as a garden plant, and also came in as a contaminant of mud used as ballast in sailing ships 200 years ago. It is a native wetland plant in northern Europe but, in North America, where it has no natural enemies, it took over many wetlands. There it provides food for very few species, out-competes the native species and lowers the water table of any wetland that it invades.

Attempts have been made to control this pest using biological control. In April 1992 the United States Department of Agriculture released the long-nosed beetle (*Hylobius transversovittatus*), a European resident, whose larvae are known to eat loosestrife roots. This was soon followed by the release of two leaf beetles in the genus *Galerucella*, as these feed on the leaves. Finally, the flower-feeding weevil (*Nanophyes marmoratus*) was introduced in New York and Minnesota in 1994. All four introduced species have established populations in the United States, and in some cases purple loosestrife infestation has been reduced by as much as 95 per cent. It is still uncertain how effective these biological control programmes will be in the longer term.

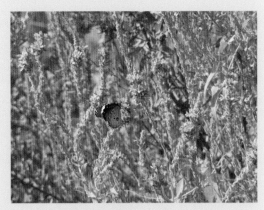

Figure 11.4 Purple loosestrife (*Lythrum salicaria*) is native to Europe, Asia and northwest Africa, and was introduced into North America, where it has become a serious weed of wetlands. Here it is growing in the Hula Valley Nature Reserve in northern Israel. The visiting butterfly is a plain tiger (*Danaus chrysippus*). Photo: Margot Hodson.

The late John Harper once said, '*it seems to me worthwhile to point out how potentially dangerous botanic gardens are*'.

We will now briefly consider methods of controlling weeds without using herbicides. Cultivation involves controlling weeds by burying them, cutting them off near soil level, desiccation of roots and rhizomes brought near the surface and continually cutting back deep-rooted perennials. In mulching, straw, peat, bark chips or sawdust are

Figure 11.3 Tumbleweed (*Salsola tragus*) is an important introduced weed in the United States. It is here photographed near Peoria, Arizona, USA.
Photo: Peter Hodson.

spread on the soil surface to control annual weeds. Soil sterilization using methyl bromide, dazomet (a fumigant that releases amines and mercaptans in the soil) and steam controls weeds, pests and pathogens, but it is very expensive and generally only used on high value crops, particularly under glass.

When herbicides first became available, they were thought to provide the answer to weeds, but it soon became apparent that this was not the case. Thus new concepts evolved, including integrated pest management, organic farming, and biological control (see Box 11.1). Since the Second World War, the method of choice for weed control in the developed countries has undoubtedly been the use of herbicides, and we will turn to this topic next.

11.1.5 Herbicides

The first known account of a herbicide was from Marco Terentius Varro (116–27 BCE), who recommended the application of 'Amurca' to weeds. This was the residue left after extraction of olive oil from olives and it was quite toxic, probably because of its high salt content. Salt itself became an accepted herbicide in the 17th century.

Until the Second World War, farmers protected their crops by applying various chemicals, including petroleum

Table 11.1 Some commonly used groups of herbicides.

Herbicide group	Common examples	Date introduced	Mode of action
Hormones	2,4-D; 2,4,5-T and MCPA	1940s	Interfere with hormonal balance
Carbamates and thiocarbamates	Propham	1956	Cell division and growth; inhibit lipid synthesis
Quaternary ammonium compounds	Paraquat and diquat	1960	Inhibit photosystem I
Ureas	Diuron, linuron, fluometuron,	1965	Inhibit photosystem II
Diphenyl ethers	Nitrofen	1970	Inhibits photosystem I; reduces plant respiration
Triazines	Atrazine, simazine	1971	Inhibit photosystem II
Organophosphorus compounds	Glyphosate	1976	Inhibit aromatic amino acid synthesis

oils, sulphuric acid, copper sulphate and sodium arsenate, to control broad-leaved weeds in cereals. A nitrophenol compound, DNOC (4,6-dinitro-o-cresol), was introduced in 1932 as the first selective organic compound, and this was also used for killing weeds in cereals. During the 1930s, the auxin plant hormones were discovered (see Chapter 3, section 3.6.2), and this led directly to the development of hormone herbicides in the 1940s.

From the 1950s to the 1970s, several other important groups of herbicides came onto the market (see Table 11.1), with varying modes of action. Since that time, herbicide research has continued, but no major additional groups have been added and the chemicals that are most used today are from the groups discovered before 1980.

Herbicides can be categorized into those that are selective and those that are non-selective. Selective herbicides are intended to kill only weeds, leaving the crops intact, and include the hormone herbicides and triazines. Non-selective herbicides kill all vegetation, and include paraquat and the ureas. Another important distinction is between contact (non-systemic) herbicides and translocated (systemic) herbicides. Non-systemic herbicides (e.g. paraquat) are effective against annual weeds but are often of little use against perennials that have extensive underground systems that cannot be touched. To remove perennial weeds, systemic herbicides such as the hormones and glyphosate should be used.

The timing of application is also important. Pre-sowing or pre-planting applications for the control of annual weeds are made to an area before the crop is planted. Pre-emergence applications are completed prior to the emergence of the crop, after planting. Post-emergence applications are made after the crop has emerged from the soil, when it will be necessary to use selective herbicides.

The development of herbicides has undoubtedly been one of the most important advances in plant science in the last fifty years. In the developed world, they have markedly reduced crop losses due to weeds. Writing in 2007, Gianessi and Reigner concluded an article assessing the importance of herbicides to crop production in the United States with these words:

'The value of herbicides can also be seen in the results of aggregate studies that simulate the impacts of their non-use. Crop production in the United States would decline by 20 per cent even with the substitution of tillage and hand weeding labor for herbicides. By controlling weeds effectively, herbicides do the work of 70 million laborers.'

Gianessi & Reigner (2007).

Although herbicides have been very successful, they have not been without their problems. Pollution of waterways and aquifers with runoff from herbicide operations has been one difficulty, while the occurrence of herbicide-resistant weeds has been another. The latter has been the topic of a considerable amount of research by plant scientists, which we will now briefly outline.

The first instance of herbicide resistance, to triazines, was reported in 1968 in the state of Washington, USA. By 2011, resistance to at least 16 different chemical classes of herbicides had been reported in 197 weed species (115 dicots and 82 monocots).[i] In the UK, the major

[i]For up-to-date information on herbicide-resistant weeds, see: http://www.weedscience.org/

problems in arable crops are with resistant grass weeds such as black-grass (*Alopecurus myosuroides*), wild oats (*Avena* spp.) and Italian rye-grass (*Lolium multiflorum*). Even those herbicides where resistance had been thought unlikely, such as the hormones and glyphosate, have now produced resistant weeds. The main feature of all these cases is that the herbicide in question has been used at a site over a number of years. Weeds typically become resistant to herbicides when the same herbicide is used repeatedly for several (4–10) years in a row. It seems that if the selection pressure becomes too great in any one place, then the evolution of tolerant weeds is inevitable.

Genetically modified (GM) herbicide-resistant crops have rapidly become one of the most widespread of all GM crops. In 1996 Roundup-Ready soybeans were introduced by Monsanto, and these were tolerant to Roundup and other glyphosate-based herbicides. Soon, other crops were transformed to produce Roundup-Ready maize, cotton and oilseed rape. Glyphosate has become the most used herbicide in the world, and herbicide-resistant crops are one of the two largest categories of GM crops that are currently in production (the other main trait is insect resistance – see Box 11.2).

As we have seen in Chapter 10, the genetic modification of plants to increase characteristics such as drought and salt tolerance has proven to be much more difficult than was expected, as these are very complex phenomena controlled by many genes. Herbicides, however, have specific target sites in plant biochemistry, and resistance is usually controlled by a single dominant gene.

Therefore, technically, it is much easier to produce herbicide-resistant crops. Many herbicides act on a single enzyme, blocking its activity by binding to it. For example, glyphosate's mode of action is to inhibit the enzyme 5-enolpyruvylshikimate-3-phosphate synthase (EPSPS), which is involved in the synthesis of the aromatic amino acids. A small change in the structure of an enzyme that prevents the binding of the herbicide increases the herbicide tolerance of the crop. This is relatively easy to attain, but it may have a cost in that the enzyme may not be as efficient after this change.

Alternatively, some microorganisms have a version of EPSPS that is resistant to glyphosate inhibition. The type used in GM Roundup-Ready soybean was isolated from *Agrobacterium* strain CP4 (CP4 EPSPS), which was resistant to glyphosate. There have been persistent reports of some yield drag in herbicide resistant crops when

Figure 11.5 Oilseed rape (*Brassica napus*), an important source of vegetable oil, a source of biodiesel, and a GM crop that is extensively grown in North America. Photographed in early May near Fawley, Berkshire, England.
Photo: MJH.

compared to their progenitors, but this is usually much less than the decrease in yield caused by a weed infestation. Herbicide-resistant crops allow the wider use of effective herbicides, as the crop is unharmed but the weeds are killed. It is also possible to increase application rates of selective herbicides, leading to better weed control.

One further difficulty with herbicide-resistant crops is the possibility that resistance genes from crops could transfer to related weedy taxa. This possibility appeared to come closer to a reality in 2011, when Schafer *et al.* reported that genes for glyphosate and glufosinate resistance had 'escaped' from transgenic oilseed rape (*Brassica napus*, known as 'canola' in North America – see Figure 11.5) into feral populations in North Dakota, USA. Devos *et al.* (2011) reviewed the literature on GM oilseed rape and indicated that GM feral plants were quite common in areas where the crop was grown. There is now also good evidence that *B. napus* and the weed *B. rapa* have hybridized in the wild, but at low frequencies. We will consider the advantages and potential problems of GM crops in more detail in Chapter 12, section 12.4.

11.2 Plant/animal interactions

The majority of animals differ from most plants in two key respects: they are not photosynthetic; and they have greater mobility. Of course there are exceptions, and we cover some of these in this book. Plants are mostly (though not entirely) autotrophs that produce complex organic compounds from simple inorganic molecules using photosynthesis (Chapter 7, section 7.4). All animals are heterotrophs, and some eat autotrophs as food to obtain energy and raw materials. Herbivorous animals gain energy by breaking down organic molecules obtained in plant food, while carnivorous animals also rely on autotrophs, as the nutrients obtained from the heterotrophs they eat come from autotrophs lower down the food chain.

The main emphasis of this section, therefore, will inevitably be on the negative impacts of herbivory on plants and on the defence mechanisms that plants use to mitigate or prevent these impacts. We will also consider some of the implications for agriculture of animal pests and briefly look at some more positive plant/animal interactions.

11.2.1 Herbivory

When animals eat plants, this is known as herbivory. In some cases, for example with an infestation of locusts, the herbivory is very severe and all the soft tissues of the plant will be eaten. In other instances, the effects of herbivory may be relatively minor.

The two main types of herbivory are due to biting and sucking. Mammalian herbivores, such as cattle and sheep, simply cut off leaves or other plant parts with their teeth and ingest them. Many insects and their larvae (e.g. caterpillars) also bite off plant tissue with their mandibles. Leafhoppers, bugs and mites pierce cell walls with needle-like stylets, then inject digestive enzymes and suck the cell contents out. Sap feeders, such as aphids, insert a stylet into the phloem and suck the sugary contents out (Figure 11.6). Often these insects then excrete honeydew globules onto the leaf surface, and sooty moulds may develop on these. Some animals, like locusts, feed on many different plants, and are thus polyphagous. The majority of herbivores are oligophagous, demonstrating some selectivity, and attack one family of plants. Only a few herbivores are monophagous, only eating one plant species.

Damage to plants caused by biting can be indiscriminate or may be targeted at a particular organ, such as the leaves, the buds or the flowers. Some insects remove tissue from one side of a leaf, leading to 'windowing', where the opposite epidermis remains intact. A number

Figure 11.6 Black bean aphid (*Aphis fabae*) on a parsnip (*Pastinaca sativa*) stem. This aphid has a wide host range of over 200 species of cultivated and wild plants.
Photo: MJH.

of weevils only attack the edges of leaves, making distinctive notches. Other insects inject toxic material into leaves, causing deformation or death of the tissues. In many cases, herbivory itself is a less of a problem for the plant than is the herbivore being the vector for an important disease. This is particularly so with aphids and other vectors of plant viruses (see section 11.3.3).

The resistance mechanisms of plants against herbivory are usually put into two categories: avoidance and tolerance. Plants have evolved a number of types of avoidance defences against herbivory: phenological, behavioural, structural and chemical. We will now consider each of these in turn.

Phenological defences often involve avoidance of herbivory through the timing of the life cycle. Other plants show fast turnover of plant organs that are vulnerable to herbivory. Many grasses have growth points near ground level that quickly replace leaf material lost above through grazing. Plants such as *Mimosa pudica* show rapid leaf movements in what may be considered a behavioural response to touch (see Chapter 6, section 6.14.2 and Figure 6.17), and it is believed that these deter predation.

The structural nature of plant tissues can deter predation. Structural defence can be defined as '*any morphological or anatomical trait that confers a fitness advantage to the plant by directly deterring herbivores from feeding on it*' (Hanley *et al.*, 2007).

There are several types of structural defence, which can be broadly divided into *spinescence, pubescence, sclerophylly* and *mineral deposition*.

Figure 11.7 The spines of cacti often seem to have a dual role of increasing leaf reflectance and defence against herbivory. Here is a close-up view of the spines of the Saguaro cactus (*Carnegiea gigantea*), photographed in Peoria, Arizona, USA.
Photo: Ruth McFadden.

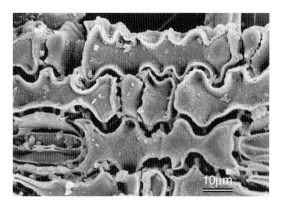

Figure 11.8 Scanning electron micrograph of a leaf surface from Mitchell grass (*Astrebla lappacea*) after the organic matter has been removed by a microwave digestion method. Silica phytoliths are visible *in situ* with a variety of morphologies showing remarkable articulation.
Photo: Prof. Leigh Sullivan, Southern Cross University, Lismore, NSW, Australia.

• Spinescence involves the production of thorns, spines and prickles. These are generally large structures, and spinescence is usually considered as a better defence against large vertebrate grazers than against small invertebrates. Thus the African savannah, which has many large vertebrate grazers, also has many thorny plants. Experiments involving the removal of thorns from plants have nearly always confirmed that thorns and spines are an effective deterrent to vertebrate grazers.

• In Chapter 10 (section 10.4.2) we saw that surface hairs (pubescence) increase leaf reflectance and protect the boundary layer over the leaf surface in many xerophytic plants, but they also appear to have a role in defence against herbivory (Figure 11.7). Stinging nettles (*Urtica dioica*) have siliceous hairs on their leaves that act like hypodermic needles in injecting a toxin into animals coming into close contact. With few exceptions it appears that pubescence is an effective structural defence.

• Sclerophylly is a term derived from the Greek meaning 'hard-leaved', and it seems to have evolved partly to support leaves, particularly against wilting (see Chapter 10, section 10.4.2). However, sclerophylly also appears to be a defence against herbivory. Leaves and shoots that are scleromorphic show reduced palatability and digestibility. Combinations of cellulose, hemicellulose, lignin and silica together make plant material very tough and give it mechanical properties that decrease palatability.

• Many plant species deposit solid minerals, mostly silica (see Chapter 2, section 2.2.4) and calcium, in their tissues. For example grass leaves accumulate silica, absorbing monosilicic acid from the soil, and then transporting it to the cell lumina and walls of shoot cells where it is deposited in the form of silica bodies known as phytoliths (Figure 11.8). The grass phytoliths increase plant toughness and, thus, resistance to stem-boring insects and leaf-eating invertebrates, including molluscs. Vertebrates such as the North American prairie vole (*Microtus ochrogaster*) and the European field vole (*Microtus agrestis*) are also known to be deterred by grasses containing phytoliths. Silica increases tooth wear in vertebrates and led to the evolution of hardened, high-crowned teeth, known as hypsodont dentition, by ungulates (mammals with hooves) and macropods (kangaroos and their relatives), and of continuously growing teeth in rodents. It seems that in the late Miocene (approximately 11.6–5.3 million years ago), hypsodont dentition rapidly evolved in the large ungulates in response to the spread of silica-containing grasses. Calcium minerals, including calcium oxalate crystals, also accumulate in plant tissues and are known as a deterrent against both invertebrate and vertebrate herbivores.

Many plants have chemical defences against herbivory – for instance, the manufacture of phenolics that discourage herbivores from sustained feeding after the first bite. Other compounds used as chemical defences include alkaloids, glucosinolates, lectins and brassinosteroids (which are also plant hormones – see Chapter 3, section 3.6.7). Members of the Solanaceae, including the potato and tomato, accumulate chaconine and solanine, which are steroidal glycoalkaloids. These compounds are highly toxic to most insect herbivores, but the Colorado beetle (*Leptinotarsa decemlineata*), an important pest of potatoes, is not deterred and ingests the compounds before excreting them.

Some defences against herbivory are inducible and are produced in response to damage within hours, reducing again after the attack has ceased. One example is a wound-signalling mechanism which is induced by herbivore attack in tomato leaves. Herbivory triggers the release of systemin, a plant peptide hormone, which activates the jasmonate pathway (see Chapter 3, section 3.6.9 and section 11.3.9 below). This then increases the activity of over fifteen genes, one of which is responsible for the production of a proteinase inhibitor that decreases herbivory. Mutants that manufacture less proteinase inhibitor allow greater herbivore feeding and growth.

Some plants use volatile chemical signals to attract predators that help control herbivores. For example, Kessler & Baldwin (2001) worked on volatile emissions from *Nicotiana attenuata* during attack on the leaves by three insect herbivores. Three compounds (*cis*-3-hexene-1-ol, linalool, and *cis*-a-bergamotene) were released by the plants when under herbivore attack. When these compounds were applied to leaves, a generalist predator (*Geocoris pallens*) showed increased egg predation rates. Thus, the predator had been attracted to the leaves by the volatile compounds and it ate the eggs of the herbivores. Application of linalool or a mixture of the compounds decreased oviposition of adult *Manduca* moths. The number of herbivores was reduced by more than 90 per cent when volatiles were released.

11.2.2 Galls caused by animals

A gall is defined as '*an abnormal localized swelling or outgrowth produced by a plant as a result of attack by a parasite*' (*The Penguin Dictionary of Botany*, 1984). Galls can be produced by interactions with bacteria, fungi, animals or their combinations. Animals that are known to produce galls include the true bugs, gall midges, gall wasps, sawflies and gall mites. At least 13,000 species of insects are known to produce galls, and here we will confine ourselves to a few examples.

Mites, such as the pear leaf blister mite (*Eriophyes. pyri*), overwinter as adults and invade the buds or unfurling leaves of pear trees in the spring. *E. pyri* forms blister-like

Figure 11.9 Oak apple caused by oak apple gall wasps (*Biorhiza pallida*). Photographed at North Gower, Wales. Photo: Prof Paul F. Brain © University of Swansea, Wales. Image from Centre for Bioscience (Higher Education Academy) ImageBank.

galls on the upper surface of the leaves, changing colour from yellow to black as they mature. Infestation of oak (*Quercus* sp.) leaves by oak apple gall wasps (*Biorhiza pallida*) leads to the formation of galls called oak apples (Figure 11.9). Female wasps lay an egg in the midrib of a growing leaf in the spring. The wasp larva causes the leaf to grow abnormally and a gall is formed. Inside the gall, the larva grows and pupates, eventually emerging as an adult wasp.

Stems can also be targets for gall formation, and the woolly aphid (*Eriosoma lanigerum*) forms large cankerous galls on apple trees. This species is of North American origin, but has been introduced elsewhere, and is a major pest in Europe. Galls are not confined to the shoots of plants, and an example of a root gall is cereal root-knot nematode (*Meloidogyne naasi*), which affects the roots of a variety of plants, including sugar beet. Similarly, larvae of the turnip gall weevil (*Ceutorhynchus pleurostigma*) produce marble-like galls on the roots of turnips and swedes.

In many cases, galls do relatively little harm to the plant, although they can be unsightly. The one major exception is grape phylloxera (*Viteus vitifoliae*), a small insect (less than 3 mm) that was the cause of the great wine blight in France after its introduction from North America in 1863. This led to the destruction of over two million hectares of vines. Grafting of vines onto American rootstocks, rather than those from Europe, has achieved considerable control over grape phylloxera, but has not eradicated it.

The molecular basis of gall induction is only under-stood in bacterial systems, including crown gall, which is caused by *Agrobacterium tumefaciens* (see Chapter 3, Box 3.1, and section 11.3.2 below), and root nod-ules brought about by nitrogen-fixing *Rhizobium* spp. (Chapter 5, section 5.9.1). *Agrobacterium* exports to its host a section of its plasmid DNA, and galls form as the host expresses bacterial genes that encode enzymes involved in auxin and cytokinin synthesis. *Rhizobium* exports signal molecules, termed 'nod factors', which are responsible for nodule formation.

It is clear that gall-forming insects also export stimuli, including those contained in saliva injected by aphids feeding, the secretions injected by sawflies during ovipo-sition and in secretions from gall wasp larvae. As yet, the active compounds in these secretions have not been iden-tified, and how they interact with plant developmental pathways is highly uncertain. Not surprisingly, plant hor-mones, including the auxins and cytokinins, have been proposed to be involved, as have amino acids, proteins and mutualistic viruses.

Finally, we should consider the various hypotheses on the adaptive significance of galls: the *Nutrition hypoth-esis*, the *Microenvironment hypothesis* and the *Enemy hypothesis*.
• In the Nutrition hypothesis, galls are seen as providers of increased nutrition when compared to other feeding modes.
• Gall tissues may also protect the animal from unfavourable abiotic conditions, such as desiccation, and this is the basis of the Microenvironment hypothesis.
• Protection of animals from attack by predators is the idea behind the Enemy hypothesis. It should be noted, however, that galls are not totally secure environments for their hosts, and a variety of specialist enemies, including fungi, and the larvae of parasitoid wasps, beetles, moths and flies, do penetrate galls.

11.2.3 Pests and pesticides

Pests have been defined as *'an animal or plant that has become too abundant for man's comfort or convenience, or has even become a danger to his interests'* (Imms, 1947). Clearly, this definition is related to those for weeds (Section 11.1.3) and plant pathogens (Section 11.3.5), but there are also animal pests that are herbivores. Much herbivory occurs in natural environments or in situations where it is not detrimental to human interests, but herbivores can also be important pests.

Animal pests have been key determinants of crop productivity since humans first engaged in agriculture, and we have already covered some of their features in the material above. Plagues of locusts were evidently well known in Biblical times. There are nine Hebrew words for locust, one in Greek, and 56 references to these insects in the Bible. Probably the most famous example of a locust infestation was the eighth Plague of Egypt (Exodus 10), where the swarms 'darkened the sun'.

In many cases, the worst pest species are those that have been introduced from elsewhere and are non-indigenous. As we saw in the case of the purple loosestrife (Box 11.1), this is often because introduced species arrive without their natural predators. One example of a rapidly spread-ing pest is the horse chestnut leaf miner (*Cameraria ohridella*), which attacks the leaves (Figure 11.10) of the horse chestnut tree (*Aesculus hippocastanum*). *C. ohridella* was first seen in Macedonia in the 1970s, was named in 1985, and it spread rapidly across Europe, arriving in the UK at Wimbledon in 2002. By the end of 2010, it had continued to spread into most parts of England and into Wales. In severe cases by mid-August, most of the leaves on a horse chestnut tree turn brown. This is unsightly, but most photosynthesis occurs in the tree leaves before the majority of the mines are formed. However, smaller

Figure 11.10 Horse chestnut leaf miner (*Cameraria ohridella*) on the leaves of the horse chestnut tree (*Aesculus hippocastanum*). This shows a fairly badly affected leaf after the first generation of the moth in mid-July.
Photo: Dr Michael Pocock © University of Bristol.

Box 11.2 Bt plants

As we saw in section 11.1.4, herbicide resistant GM crops are one of the main types of modified crops that have been used by farmers so far. The others are so-called 'Bt crops'. In fact, these two types (herbicide-tolerant and Bt crops) presently make up 99 per cent of the GM crops in agricultural usage.

The Gram-positive bacterium *Bacillus thuringiensis* (Bt) is found in the soil. Its spores, and the Cry toxin extracted from them, have been used for many years as a natural insecticide. Because it is 'natural', Bt is one of the few insecticides allowed in organic agriculture. Many Bt strains produce δ-endotoxins in the form of crystal proteins, and these have insecticidal action.

More recently, GM insect-resistant plants have been engineered to produce the Cry toxin from *B. thuringiensis* – first in potato, but then in maize and cotton; the latter two have become the two dominant Bt crops. In general, these have improved yields when compared to non-transformed plants growing in the same conditions, ranging up to about a 30 per cent increase. Economic performance was also improved in 74 per cent of trials of insect-resistant crops because of lower expenditure on insecticides. Quite frequently, herbicide resistance and insect resistance are now 'stacked' in the same plant, and the resultant plants have both resistances. By 2009, over 28 per cent of the global total acreage of GM crops had plants with stacked traits.

Although there have been some concerns that Bt crops might affect non-target insects attempting to eat above-ground plant parts, and organisms living in the soil, generally the evidence suggests that Bt is relatively benign in its effects. A more serious potential problem is that there have been a number of reports of pests gaining resistance to Bt.

conkers have been observed as a result of infestation. About 30 species of parasitoid wasps are known to attack *C. ohridella* and these are currently the subject of active research.[ii]

It is difficult to calculate the global economic cost of pests, but Pimentel and colleagues estimated that in the year 2000, a total of 50,000 non-indigenous species (including plants, animals and microbes) cost the economy of the United States $137 billion. Arthropod crop pests alone were responsible for $14.4 billion. These figures give some indication of the scale of the pest problem but, as they only include non-indigenous pests and not those native to the country, they must be an underestimate of the total pest problem in the United States.

It is beyond the scope of this book to cover the considerable range of pesticides that are used to protect plants from pests. However, it is certainly the case that insects are the most significant pests and insecticides are the most important pesticides used. In recent years, the transformation of crops to express Bt toxin (an insecticidal protein from *Bacillus thuringiensis*), and thus the creation of insect-resistant crops, has reduced the need for insecticides in some parts of the world. This development is covered in Box 11.2.

[ii] In 2011 in the UK, scientists were enlisting the help of the general public and school children in this work. See: http://www.ourweboflife.org.uk/

11.2.4 Plant/animal mutualisms

Not all interactions between plants and animals are negative for the plants concerned; some are mutualisms, where both benefit. In Chapter 8 (section 8.6) we covered pollination, in which the animal gains nutrition (e.g. pollen, nectar) and the plant gains from being pollinated. In a similar way, some seed dispersal is carried out by animals (Chapter 4, section 4.8), which gain nutrition in return (e.g. from eating fruits). Here we will briefly highlight a few other examples of plant/animal mutualisms, where plants appear to benefit from close contact with animals.

Perhaps one of the most well known of mutualisms are 'ant plants', technically known as **myrmecophytes**. In many of these, the plants provide nectar as a reward for some protection of the plant by ants against herbivory. However, over 100 genera of tropical angiosperms have species with specialized pre-formed nesting structures – 'domatia' – for housing ants. These include sites in hollow stems, thorns (e.g. *Acacia*), petioles or leaf pouches, and the relationships are closer to symbiotic associations.

Two other examples of mutualisms concern pitcher plants. These are well known for trapping insects (mostly ants) that fall into the digestive fluid at the bottom of the pitcher (see Figure 11.11). The largest families of pitcher plants are the Sarraceniaceae (New World) and the Nepenthaceae (Old World). Pitchers are, like other insectivorous plants (e.g. the Venus flytrap; see Chapter 6, section 6.14.2), an adaptation to nitrogen-poor environments.

Figure 11.11 The cobra lily (*Darlingtonia californica*) is a pitcher plant that is native to Northern California and Oregon.
Photo: MJH. Used with the permission of Oxford Botanic Gardens, Oxford, UK.

Nepenthes lowii (Nepenthaceae) is a pitcher species from the mountains of Borneo, where ant densities are typically low, so this adaptation would seem to be less beneficial. However, in *N. lowii*, only the pitchers produced by immature plants are used to catch insects, and those from the mature plants are visited by tree shrews, which feed on exudates accumulated on the pitcher lid and then defecate into them afterwards. Tree shrew faeces accounts for 57–100 per cent of leaf nitrogen in mature *N. lowii* plants. This interaction between animal and plant appears to be a mutualism involving the exchange of food resources that are limited in the Borneo Mountains.

Another Borneo pitcher species, *N. rafflesiana* var. *elongata* is also poor at trapping insects, but Hardwicke's woolly bats (*Kerivoula hardwickii hardwickii*) roost within the pitchers of this plant. About a third of leaf nitrogen is derived from the faeces of the bats, and this is another example of an unusual mutualism.

11.3 Plant pathology

For the remainder of this chapter, we will focus on plant pathology, considering the effects of fungi, bacteria, viruses and various other microorganisms on plant health, growth and reproduction.

Microorganisms have three basic modes of nutrition:
• Saprophytes feed on decaying organic matter.
• Parasites feed on living organisms, and do not benefit the host.
• Symbionts have a mutually beneficial relationship with the plant.

Here we are mainly interested in parasites, although some parasites also live as saprophytes at times. We covered two major plant symbioses – mycorrhizae and nitrogen-fixing root nodules – in Chapter 5.

Parasites can be divided into two categories:

1 Obligate parasites (also known as biotrophs) can live and reproduce only by feeding on living plant material and are thus very difficult to culture. The rust fungi (*Puccinia* spp.) and the viruses are good examples.

2 Facultative parasites, which can survive either by feeding on decaying material as saprophytes or by feeding on living plants. An example of such an organism is the oomycete *Pythium*, which causes damping off of seedlings. Some facultative parasites kill their host and then feed on the contents; these are known as necrotrophs.

We will now cover each of the major types of pathogenic microorganisms, before considering their mechanisms of pathogenicity. Then we will turn to fungicides, an example of a method for disease control, and to the important topic of breeding resistant crops. The parasitic higher plants, which are sometimes included in plant pathology texts, were dealt with in section 11.1.2.

11.3.1 Fungi

The fungi are without doubt the most important plant pathogens. About 100,000 species have so far been described by taxonomists, but it is estimated that there may be as many as 1.5 million species in total. Only about 50 fungi cause human diseases, while more than 8,000 are known to cause disease in plants.

Mycology is the branch of biology dedicated to the study of fungi. It is often taught within botany courses, but actually fungi are more allied with animals than with plants (see Chapter 1, section 1.4). One major feature is that the cell walls of fungal cells contain chitin, while those of plants have cellulose (see Chapter 2, section 2.2).

The typical fungus consists of a vegetative body, the mycelium, which is made up of individual branches called hyphae. Reproduction in fungi is complex, involving both sexual and asexual processes – both lead to the production of spores. Asexual spores tend to be more common and are responsible for most epidemics, but sexual spores are useful in fungal classification systems.

Molecular techniques have allowed great advances in fungal taxonomy, enabling Hibbett and colleagues to construct the currently accepted classification (Hibbett *et al.*, 2007). This allows for one kingdom (Fungi), one sub-kingdom (Dikarya), seven phyla (Chytridiomycota, Neo-callimastigomycota, Blastocladiomycota, Microsporidia, Glomeromycota, Basidiomycota and Ascomycota – see Figure 11.12), ten subphyla, 35 classes, 12 subclasses and 129 orders.

11.3.2 Bacteria

Bacteria are prokaryotes with a very simple cellular organization and no nuclei. They are nearly all unicellular and have one of three shapes: round, rods or spiral. Bacterial reproduction is by binary fission, where the cell divides when a wall is formed in the middle of a cell – an asexual process. All of the plant pathogenic bacteria are rod-shaped. There are about 1,600 known bacterial species, about 200 of which are recognized plant

Figure 11.12 Tar spot (*Rhytisma acerinum*) is a plant pathogen in the Ascomycota. It does not normally have a serious effect on plant health. It is here photographed on sycamore leaves (*Acer pseudoplatanus*) in Brussels, Belgium.
Photo: MJH.

pathogens. Many of the rest are saprophytic, but some are important pathogens of humans and other animals. The bacteria causing plant diseases are all facultative parasites that can be grown in culture.

The taxonomy of bacteria is complex, but it now recognized that six genera contain species that cause plant diseases. These are the Gram-negative genera *Agrobacterium*, *Erwinia*, *Pseudomonas* and *Xanthomonas* and the Gram-positive Actinobacteria, which include *Corynebacterium* and *Streptomyces*. The taxonomic problems are made worse by the existence of pathovars of a number of important species.

A pathovar is defined as '*a subdivision of a species distinguished by common characters of pathogenicity, particularly in relation to host range*' (Talboys *et al.*, 1973). One example of this is *Pseudomonas syringae*, which has approximately 120 pathovars, causing diseases in a wide range of crops including beans and fruit trees. *Xanthomonas* species cause at least 350 different plant diseases, and the bacterial blight caused by *Xanthomonas oryzae* pv. *oryzae* is a major disease of basmati rice in tropical Asia. *Agrobacterium tumefaciens* is a soil bacterium, causing galls in over 200 dicot species, and it is now used to genetically modify plants (see Chapter 3, Box 3.1). *Streptomyces scabies*, the Common Scab of potato, is an important pathogen, not so much because it causes decreased growth, but because badly affected potatoes are difficult to sell. Streptomycin, the antibiotic, takes its name from *Streptomyces*, and many other important antibiotics (e.g. chloramphenicol) have been produced from related species.

11.3.3 Viruses

Hollings (1983) defined a virus in two ways: '*An obligate parasite of sub-microscopic size, with one dimension smaller than 200 nm*'; or '*A set of instructions to a suitable host organism to synthesize more virus.*'

Viruses are very small particles consisting of DNA or RNA, surrounded by a protein coat. They take over the genetic machinery of plant cells and make it produce more viruses. Organisms known as *vectors* are responsible for the spread of virus from plant to plant. Insects are often vectors, but some fungi and nematodes also act in this way. Often the most economic way to control plant viruses is to kill the vectors.

Virus taxonomy is very complex, but there are over 700 known plant viruses, classified into three families and

32 groups. Plant viruses are still most commonly known according to the host they were first isolated from and the most striking symptoms that they produce. Here we will confine ourselves to a few examples.

Tobacco mosaic virus (TMV) was the first virus to be discovered. It is an RNA virus that infects tobacco and other members of the Solanaceae, causing mottling and discoloration of the leaves. TMV has been much used as a model virus, because it does not infect humans or animals and it is easy to obtain large quantities. The barley yellow dwarf viruses all belong to the Luteovirus group of RNA viruses and infect over 150 grass species, including many important cereal species. They are transmitted by aphids and are very host-specific. Geminiviruses are DNA viruses which consist of distinctive paired virus particles. They are responsible for such diseases as beet curly top and maize streak (Figure 11.13), both of which are transmitted by various leafhopper species.

11.3.4 Other plant pathogens

In addition to the fungi, bacteria and viruses, there are a number of other groups of microorganisms that contain important plant pathogens. We will briefly consider some of these below.

The water moulds (Oomycetes) are now seen as taxonomically distinct from the Fungi. They lack chitin in their cells walls, are mainly diploid in karyotype and have biflagellate zoospores. The Oomycetes include some important plant pathogens, such as the *Phytophthora* group, the *Pythium* group, the downy mildews and the white blister rusts. *Phytophthora infestans* was the cause of the Irish Potato Famine, which began after the introduction of the organism to Europe in 1845. It is estimated that a million Irish people died in the famine and a further 1.5 million emigrated from Ireland to the United States. This event is often credited as leading to the establishment of plant pathology as a separate scientific discipline.

The Phytomyxea, better known as the plasmodiophorids, are another group that used to be classified within the fungi or slime moulds. Genetic and ultrastructural studies have now shown that they are protists (see Chapter 1, section 1.5) within the Cercozoa. They grow within plant cells, causing gall or scab formation. Important diseases caused by phytomyxeans include club root in cabbage (caused by *Plasmodiophora brassicae*), and powdery scab in potatoes (caused by *Spongospora subterranea*).

Figure 11.13 Maize streak geminivirus on host mature maize (*Zea mays*) plants. Photograph taken in Uganda.
Photo: Prof. Michael Shaw © University of Reading. Image from Centre for Bioscience (Higher Education Academy) ImageBank.

Phytoplasmas are mycoplasma-like-organisms (MLOs) that are plant pathogens. They are prokaryotes, but unlike bacteria they have no cell walls. MLOs include the smallest known cells (between 300 nm and 1 μm) that can multiply independently of other cells. They were discovered in 1967 in the phloem of plants suffering from 'yellows' type diseases.

Several hundred diseases are caused by phytoplasmas, including some important in agriculture. Coconut lethal yellowing disease had a devastating impact on coconut growers in Ghana. The diseases caused by phytoplasmas are again often named after their most striking symptoms, including aster yellows, clover phyllody, citrus stubborn and potato witches' broom. Phytoplasmas require vectors for transmission; for example, clover phyllody is transmitted by the strawberry leafhopper (*Aphrodes bicinctus*), which is well known to infest clover plants.

11.3.5 Key concepts in plant pathology

Agrios (1978) defined plant disease as '*Any disturbance brought about by a pathogen (organism which causes disease), or an environmental factor which interferes with the manufacture, translocation, or utilization of food, mineral nutrients, and water in such a way that the affected plant changes in appearance and/or yields less than a normal, healthy plant of the same variety.*' This is quite a complex definition and it suggests that disease could be caused by one or a combination of the following:

1 Nutrient deficiencies or excesses – see Chapter 5 (section 5.5).

2 Toxic materials in the soil or atmosphere – see Chapter 9 (section 9.6).

3 Infestation by animals, including pests – see Section 11.2.

4 Colonization by parasitic flowering plants – this was covered in section 11.1.2.

5 Infection by microorganisms or viruses – this can be found in the following sections of this chapter.

Diseases generally lead to symptoms, including discolorations, abnormal growth, rots and physiological wilts. Some diseases, however, do not give obvious symptoms at all stages of their life cycle. Moreover, many symptoms are associated with factors other than invasion by a pathogen; examples include poor weather conditions, spray damage from agrochemicals, pest damage and nutrient deficiencies. This leads us on to the requirement for proof that an organism is pathogenic and that it is involved in causing the observed symptoms. Proof of pathogenicity was first recognized as a fundamental of plant pathology by Koch in 1884. He set forward the famous Koch's postulates:

1 The organism must be consistently associated with the symptoms of the disease.

2 The organism must be isolated and grown in pure culture.

3 The organism must be inoculated onto healthy hosts of the same species from which it was originally isolated, and must reproduce the same symptoms as originally observed.

4 The organism must be re-isolated and have the same characteristics as the original isolate.

These postulates are still useful today, but they do have a number of problems. As we have already seen, many pathogenic organisms simply cannot be grown in pure culture. Moreover, some symptoms are caused by simultaneous infection by several organisms.

11.3.6 Factors affecting disease development

Plant disease is a complex interaction of factors associated with the host, the pathogen and the environment. Disease will not build up unless there is an active pathogen, a susceptible host and suitable environmental conditions.

The ultimate success of a plant pathogen depends on several factors:

1 The quantity of the initial inoculum. Potential sources of inoculum include infected seed, pollen, vegetative material and introduced plants. Most fungal pathogens produce many spores on the surface of their hosts, and these are spread to new hosts on the wind. However, viruses and phytoplasmas require vectors for transport and to gain entrance into a new host plant. Some pathogens infect plants but do not produce further inocula until the crop matures and the affected plant dies. An example of this is *Sclerotinia sclerotiorum*, the cause of stem rot in oilseed rape. Others pathogens, for example the powdery mildews and rusts, produce more inocula very quickly, and thus disease rapidly spreads throughout the plant population.

2 Ability of a pathogen to compete with other nearby microorganisms. This is very important in soil-borne pathogens (e.g. *Pythium*) that are often facultative parasites and must be able to compete with saprophytes for dead and decaying material. Competition from soil microorganisms has been used successfully as a technique for biological control.

3 The ability of a pathogen to infect and colonize plant tissue. We will cover this in the following sections.

Environmental factors are also important in determining the spread of plant diseases. Most diseases occur only in the warm parts of the year, and quite often after rain or during a humid period. The major environmental factors we should consider are temperature, moisture and those associated with the soil:

1 *Temperature.* Low temperatures will inhibit the growth of both the host and the pathogen, and very low temperatures will kill both (see Chapter 9, section 9.2). Most fungi grow better and cause more infections at

Table 11.2 The effect of temperature on the latent period of *Puccinia graminis* of wheat.

Temperature (°C)	0	4.5	10.5	12.5	19	21	24
Latent period (days)	85	22	15	12	9	7	5

Data from Stakman & Harrar (1957).

higher temperatures. An example is shown in Table 11.2, where the rust fungus *Puccinia graminis* is growing on wheat. This demonstrates the latent period, which is the time taken for the fungus to penetrate wheat tissue, colonize and produce fresh pustules. The latent period decreases as temperature increases, so summer infections and epidemics are much more likely. Usually, increasing temperature also increases bacterial infections, but viruses show more complex patterns.

2 *Moisture.* Water is an important factor in disease development. Potato blight, as an example, is caused by *Phytophthora infestans*, and it will not develop at all when relative humidity is below 90 per cent. This disease is therefore more prevalent in wet summers and in areas of high rainfall. The most significant influence of moisture on fungi seems to be on fungal spore germination. Nearly all fungal and bacterial infections are increased by damp conditions. In diseases affecting plant roots, severity of disease is often related to soil moisture content. For example, *Pythium*, the cause of damping off, is spread mostly by motile zoospores that swim in soil moisture, so it prefers waterlogged conditions. Waterlogged plants are also under stress (Chapter 9, section 9.3.2), which contributes to their susceptibility to disease.

3 *Soil.* Soil pH can affect the pathogenicity of some microorganisms. For example, *Plasmodiophora brassicae*, the casual agent of club root disease of brassicas, is a major problem at pH 5.7 or below, but is totally inhibited at pH 7.8. Thus, using lime (the primary active component of agricultural lime is calcium carbonate) to increase soil pH has long been used as a method of controlling club root. In contrast, common scab of potatoes, caused by the bacterium *Streptomyces scabies*, can be severe above pH 5.2, but in more acid soils it is inhibited. Likewise, *Phytophthora cinnamomi* is inhibited below pH 4.0. Soil nutrient status can also have an effect, and cereals over-fertilized with nitrogen fertilizers are more susceptible to mildew (*Erysiphe graminis*).

Rice blast fungus (*Magnaporthe grisea*) is major disease of the most important crop in the world. Its infection mechanism and cycle has been worked out and its genome sequenced. Figure 11.14a shows the disease infecting rice in an experimental paddy in Japan. In an adjacent paddy, the rice was supplied with two tons per hectare of silicon fertilizer, and the disease was much less serious (Figure 11.14b). Silicon is an important element in rice nutrition and is known to improve resistance of plants to pathogens (Chapter 5, section 5.5), often by forming a physical barrier to infection (section 11.3.7 below).

Weeds can be a potential source of pathogens (section 11.1). Any wild plant growing in the neighbourhood of a crop is likely to act as a host to pathogens that could later affect the crop. Thus, eliminating weeds is likely to cause a decrease in disease. The most famous instance of this is stem rust of wheat (*Puccinia graminis* f. sp. *tritici*). In 1926, Craigie discovered that this rust has two hosts and a complicated life cycle involving wheat and the barberry (*Berberis vulgaris*) bush. A campaign to remove barberry from wheat growing areas helped partially control the rust in the USA. The rust genus *Puccinia* has a number of economically important pathogenic species, and here we illustrate wheat brown rust (*Puccinia triticina*) pustules (uredia) on a leaf surface (Figure 11.15) and a uredium containing urediospores (Figure 11.16).

11.3.7 Physical and chemical defences against pathogens

The main resistance mechanisms of the potential host plant can be classified as constitutive or active. *Constitutive* resistance is present in the normal healthy plant and is not a response to an invading pathogen. Surface leaf waxes and the cuticle can form a physical barrier to pathogen penetration, as can cell walls. Stomata and lenticels are the main entry point of many pathogenic organisms, but some resistant plants have structures or exhibit behaviour (e.g. closure at the time spores are germinating) that restricts entry. Bark provides physical protection against pathogens and its chemical properties are also likely to be important. In the root, the endodermis, which is often thickened and may be silicified in grasses, can be a very effective barrier to invading pathogens from the soil.

There are a number of constitutive chemical factors that deter plant pathogens. The nutrients in the plant cell sap (e.g. amino acids, sugars), may not be supplied in

(a)   (b)

Figure 11.14 (a) Rice blast fungus (*Magnaporthe grisea*) on rice plants. Note discoloured panicles and leaves. (b) Rice in adjacent field treated with two tons per hectare of silicon fertilizer shows far less infection.
Both photographs taken at Cherry Hills experimental station, Shonai plains, Yamagata prefecture, Japan. Photos: MJH.

Figure 11.15 Wheat (*Triticum aestivum*) leaf covered in pustules (uredia) of wheat brown rust (*Puccinia triticina*).
Photographed in England. Photo: Prof. Michael Shaw © University of Reading. Image from Centre for Bioscience (Higher Education Academy) ImageBank.

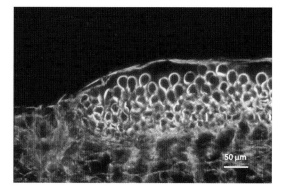

Figure 11.16 A light micrograph of a section through a wheat brown rust (*Puccinia triticina*) pustule (uredium) containing urediospores.
Photo: John Sibley.

sufficient quantity to the pathogen by the host. Moreover, the pH of the sap may not be in the optimum range for pathogen growth. Plants may also contain toxic compounds (e.g. phenolics) which can act as deterrents to pathogens. These constitutive chemical compounds are known as **phytoanticipins**. Finally enzyme inactivators are known to occur which inhibit pathogen enzymes that have been secreted to break down host tissue.

Active mechanisms operate specifically in response to pathogen invasion, and are thus an inducible form of resistance. Induced structural resistance is an attempt by the host to deter infection and colonization by the production of physical barriers. Thus, callose deposits may form around the penetrating pathogen. Callose is a carbohydrate, a β-glucan, also found around the plasmodesmata in phloem sieve elements (Chapter 6, section 6.5.2), and it is a general response to wounding. In other cases, lignin, suberin or deposited silica will form a wound barrier, which is often sufficient to stop the pathogen from penetrating further into the plant tissue. Tyloses are specialized structures produced mainly in response to fungal vascular wilt diseases (e.g. *Verticillium albo-atrum*). Here, living cells produce balloon-like intrusions into the lumen of a xylem vessel through a pit in the xylem wall. This restricts the growth of the fungus within the xylem.

Phytoalexins are antimicrobial, low molecular weight secondary metabolites produced in plants as a response to infection. The phytoalexin hypothesis was first proposed in 1940 by Müller and Borger, making it the longest studied response of plants to infection. Phytoalexins are generally present in low concentrations in the plant, but they may accumulate around the infection site. Chemically, phytoalexins are very diverse, including phenylpropanoids, flavonoids, isoflavonoids, sesquiterpenes and polyketides. Some well-known examples are given in Table 11.3.

Finally, pathogenesis-related (PR) proteins have also been found to accumulate in response to infection with fungi, bacteria, viruses or oomycetes. These proteins have been classified into 17 families, which appear to be common across all species. Other PRs are more specific and only occur in some plant species. PR proteins have a wide range of antimicrobial activities and include chitinases, proteinase inhibitors, peroxidases and oxalate oxidases. Most PRs are induced by salicylic acid, jasmonate or ethylene (see section 11.3.9 below). It is now recognized that some PR proteins are multi-functional and are induced by wounding or low temperature stress, and some have antifreeze activity (Chapter 10, section 10.2.2).

11.3.8 Hypersensitivity and programmed cell death

Hypersensitivity involves the very rapid death of a cell penetrated by the pathogen, and sometimes those around it. The hypersensitive response is triggered within minutes, although cell collapse may take several hours. The response is another form of programmed cell death (PCD) (see Chapter 3, section 3.5.2). After cell collapse, lignin is laid down, phytoalexins accumulate in the area and other defence mechanisms are activated. The net effect of the hypersensitive response is to deprive the pathogen of any food supply.

Reactive oxygen intermediates (ROIs) are very important in plant defences against attack. In a similar manner to abiotic stresses (see Chapter 9, section 9.1.1), increased activity of plasma membrane NADPH oxidases, cell wall peroxidases and apoplastic amine oxidases leads to the production of ROIs in infected plant cells. Hydrogen peroxide then diffuses into cells and, in conjunction with the plant hormone salicylic acid (SA) and the signalling molecule NO (Chapter 3, section 3.6.12), it stimulates the plant defences, including PCD.

Ascorbate peroxidase and catalase activities are decreased by SA and NO during this response, so more ROIs are produced and the cell also decreases its capability to scavenge hydrogen peroxide, leading

Table 11.3 Some well-known phytoalexins and some plants that accumulate them.

Phytoalexin	Class of compound	Plant Species
Rishitin	Sesquiterpene	Potatoes
Phytuberin	Sesquiterpene	Potatoes
Resveratrol	Stilbenoid (Flavonoid)	Grapevine
Kievitone	Isoflavonoid	French beans
Phaseolin	Prenylated pterocarpan (Phenolic)	French beans
Pisatin	Pterocarpan (Phenolic)	Peas
Chlorogenic acid	Hydroxycinnamic acid (Phenolic)	Potato, carrot and sweet potato

to ROI accumulation and induction of PCD. This is in contrast to abiotic stress (Chapter 10, section 10.1) where scavenging mechanisms decrease the concentrations of ROIs in the plant.

Plants that are under abiotic stress are more susceptible to infection, and this may partly be related to differences in ROI metabolism between abiotic and biotic stresses. Therefore, stressed tobacco plants that have higher levels of antioxidative enzymes show less PCD than control plants. Moreover, plants that overproduce catalase have less resistance to fungal invasion.

11.3.9 The molecular basis of defence against pathogens

At the molecular level, a four-part model for resistance to plant disease has now been developed (Figure 11.17).

In addition to the constitutive physical and chemical barriers outlined in Section 11.3.7 above, the parts are as follows:

1 Detection of pathogen molecular components by plant immune systems rapidly triggers basal immunity. This is stimulated by fairly non-specific signals of pathogen presence, which are termed *elicitors*. An example is the recognition of bacterial flagellin by extracellular receptor-like kinases (RLKs). Other examples of what are termed PAMPs (pathogen associated molecular patterns) include lipopolysaccharides, fungal chitin, oomycete Pep-13 (a peptide fragment within the cell wall glycoprotein, GP42) and heptaglucosides. The basal immunity that is then triggered comprises signalling via MAP kinase cascades and transcriptional reprogramming brought about by plant WRKY proteins. WRKY transcription factors are restricted to the plant kingdom. They are named from the 60 amino acid region of the WRKY domain that has a conserved amino acid sequence WRKYGQK at the N-terminal (see the Glossary for a list of the single-letter abbreviations for amino acids). After the pathogen has successfully infected the plant, and the disease is established, its further spread is inhibited by these basal defences. Plants also use basal resistance mechanisms after a hypersensitive response (Section 11.3.8), or during a successful infection, to prevent secondary infections from other pathogens. One type of basal defence is known as systemic acquired resistance (SAR). Salicylic acid (SA), a hormone (see Chapter 3, section 3.6.8), is required for SAR induction, and it builds up in plants before resistance is invoked. Transgenic plants expressing salicylate

hydroxylase, which metabolizes SA, are unable to establish SAR. It is now known that SAR induction involves a complex signalling network that includes numerous factors affecting basal disease resistance. In addition, plants use a number of other pathways to transduce pathogenic signals to bring about the hypersensitive response, SAR, and other resistance mechanisms. So SAR induced by SA is not the only pathway that can lead to basal disease resistance. It is now clear that jasmonate (JA) and ethylene are alternative signals that can induce resistance to pathogens. Ethylene has many roles as a plant hormone (Chapter 3, section 3.6.6) and JA (Chapter 3, section 3.6.9) is well known to be important in the wound responses of plants, including those to herbivory (Section 11.2.1).

2 Some microorganisms evolve virulence factors, turning into adapted pathogens of a particular species by repressing components of the basal defence systems in these plants. For example, Gram-negative bacteria use the type III secretion system (a needle-like protein appendage) to transport multiple effector proteins that target host proteins. These lower the basal immune responses, allowing bacteria to accumulate in the plant apoplast.

3 When the host species evolves specific resistance (R) genes, then adapted biotrophic pathogens will be prevented from invading. Thus, in addition to the basal defence pathways outlined above, more pathogen-specific innate immunity is developed. Interactions between Avr (avirulence) genes of the pathogen and the corresponding R genes govern the battle between plants and pathogens. Resistance to disease will occur if matching R and Avr genes are there in host and pathogen; disease results when either is inactive or not present. The chain of signal transduction events results in plant defence mechanisms being activated, decreasing pathogen growth, calls for the products of R genes identifying Avr-dependent signals. R gene products, plant resistance proteins, identify pathogen effector activity and resistance is reinstated through the resulting immune responses. There is some genetic overlap between basal and R-mediated resistance, so it is possible that R-mediated signalling is responsible for more rapidly activating the defence systems that are common to both pathways. In many cases, a rapid hypersensitive response (see section 11.3.8) occurs to prevent infection as a result of R-mediated resistance.

4 If the pathogen undergoes further evolution, it may escape recognition by R gene products by modifying or

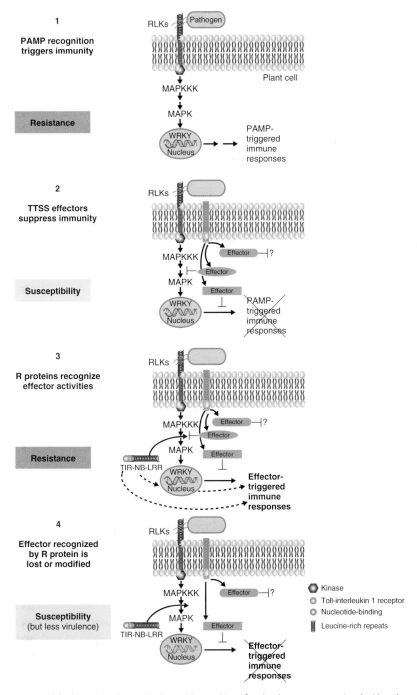

Figure 11.17 Four-part model for bacterial resistance in plants: **1** Recognition of molecular patterns associated with pathogens (PAMPs) by receptor-like kinases (RLKs) triggers basal immunity, including signalling through MAP kinase cascades (MAPKKK) and transcriptional reprogramming mediated by plant WRKY transcription factors. **2** The type III secretion system (TTSS) is used by bacteria to deliver multiple effector proteins targeting host proteins leading to suppression of basal immune responses. This allows bacteria to accumulate in the plant apoplast. **3** R gene products, including a TIR-NB-LRR protein, restore resistance through effector-triggered immune responses. **4** The pathogen avoids R gene-mediated defences by modifying or eliminating the effector(s) that triggers those defences.
From Bent & Mackey, 2007. (Reproduced with permission of Annual Reviews, Inc.)

eliminating virulence factors or by decreasing the induced defences. This part of the model is similar to item **2** above, apart from the pathogen needing to modify or lose an effector protein or produce another effector protein.

Work by Wang *et al.* (2011) illustrates how complex the whole area of molecular plant pathology is, and that it can have some surprising connections. They worked on *Arabidopsis thaliana* and its defence against *Hyaloperonospora arabidopsidis*, an oomycete pathogen that causes downy mildew disease. Programmed cell death (PCD) is a major part of the resistance mechanism. Mutants lacking 22 genes showed decreased R-gene-mediated PCD and also lowered basal resistance, indicating a connection between the two defence mechanisms. These defence genes were shown to be under circadian control by the regulator, CIRCADIAN CLOCK-ASSOCIATED 1 (CCA1) (see Chapter 6, section 6.10.2, for more on circadian rhythms). The defence genes are all expressed in the evening, allowing plants to 'anticipate' infection at dawn, when dispersal of the pathogen spores occurs.

11.3.10 Fungicides

Plant pathogens cause crop yield reductions of nearly 20 per cent worldwide, hence the importance of attempts to reduce this figure. The only aspects of the control of plant diseases that we will cover are the use of fungicides and breeding for resistant plants (section 11.3.11).

Fungicides have a long history. In 1802 the royal gardener, Forsyth, used lime-sulphur to control the mildew in his fruit trees. Prevost was the first to use copper sulphate as a seed treatment to control bunt in wheat in 1807. 'Bordeaux mixture', which contained copper sulphate and lime, was shown by Millardet in 1885 to be an effective agent to control downy mildew when sprayed on vines. In 1913, Riehm introduced organomercurial seed treatments for the control of bunt in wheat, and these later had other applications. They act by non-specific inhibition of enzymes, but are toxic to animals, and mercury also accumulates in the environment. After five decades of use, they have largely been phased out.

The Du Pont Company discovered the first truly 'organic' fungicides in 1934 – the dithiocarbamates. In a similar way to the earlier substances used as fungicides, the dithiocarbamates were protectant or surface fungicides. These do not penetrate the plant cuticle and are not translocated within the plant. They therefore cannot cure diseases and are subject to weathering.

Protectant fungicides are best applied before the fungal spores reach the host surface. A good protectant will kill the spores quickly and not harm the plant. The protectants must stick to the plant surface, and not be markedly affected by weathering. The best known of these are still the dithiocarbamates, which include substances such as mancozeb, maneb and zineb. They are used on a wide range of crops to control diseases such as late blight and downy mildew. These compounds have multiple sites of action on plant metabolism, and there is generally thought to be low risk of fungi developing resistance to them.

In the late 1960s, the first fungicides which could enter plants and be translocated within them were developed. These 'systemic' fungicides have considerable advantages over protectants. They have low phytotoxicity and can be applied to one part of the plant (e.g. roots) and be effective elsewhere, and they can also protect new plant growth. They are not subject to weathering and they can be effective after infection has occurred, even having a curative action. Their major long distance transport system is the xylem; very few seem to be phloem-mobile.

The benzimidazoles were introduced in 1968 and are systemic fungicides against a broad range of diseases. Benomyl has the widest range of fungitoxic activity of all, being used against *Botrytis, Sclerotinia, Rhizoctonia*, powdery mildews and apple scab. One other interesting compound in this group is carbendazim, which was introduced in 1973. This is injected into the trunks of trees infected with Dutch elm disease (Figure 11.18), and has a slow curative action. Toxicity of the benzimidazoles is due to interference with ß-tubulin assembly in mitosis (Chapter 2, section 2.12.2). Unfortunately, the development of resistance to the benzimidazoles by a number of fungi has been a serious problem.

The final group of fungicides we will consider are the strobilurins which were developed in the 1990s. Strobilurin A is a natural fungicidal product found in the Basidiomycete fungus *Strobilurus tenacellus*. Quite a number of natural products have been found to have fungicidal activity, and these discoveries often lead chemists to produce similar compounds in the same groups which are also fungicides. The strobilurins are inhibitors of mitochondrial respiration in fungi. Oxidation of ubiquinol is blocked at the Q_0-site of the cytochrome bc1 complex (Chapter 2, section 2.14.2), which is located in the inner mitochondrial membrane. Azoxystrobin is used to control pathogens from a wide range of fungal groups,

Figure 11.18 English Elm tree (*Ulmus procera*) infected with Dutch Elm disease which is caused by a fungus, *Ceratocystis ulmi*. The fungus is spread by bark beetles.
Photo: Mr Ken Redshaw © University of Leeds. Image from Centre for Bioscience (Higher Education Academy) ImageBank.

on many crops worldwide. Spore germination is strongly inhibited, it shows strong preventative activity, and it is also capable of curative actions.

Various fungal species have developed resistance to the strobilurins; site mutations in the fungal cyt b gene are often the cause. Cross-resistance has been shown between all members of this group of fungicides. This means that the development of resistance to one fungicide will give resistance to other fungicides that share similar modes of action.

11.3.11 Breeding for plants that are resistant to pathogens

Humans have been selecting crops for their desirable qualities for as long as they have grown them. It was,

however, only with the development of genetics in the 20th century that breeding specifically for resistance to pathogens began. In 1904, Rowland Biffen, working in Cambridge, UK showed that the inheritance of susceptibility and resistance of wheat varieties to yellow rust (*Puccinia striiformis*) followed Mendel's laws of heredity. By 1910, he had produced a new rust-resistant wheat variety, 'Little Joss', which was widely grown in the UK for the next 40 years. Biffen's work opened up the prospect of uniting plant genetics with plant pathology.

For most of the last century, breeding for pathogen resistant varieties was conventional, involving crosses between carefully chosen parent plants, and then the selection of the best plants from the offspring to be used for further selection. Conventional breeding was a slow process and only allowed for crosses between related species, but it was highly successful. In the second half of the 20th century, a whole variety of additional techniques became available: protoplast fusion, genomics, marker-assisted breeding and proteomics. These have speeded up the process, increased its accuracy and allowed some inter-specific crosses.

The use of disease-resistant varieties is now very important in modern agriculture and can often reduce the need for pesticide application (particularly fungicides). However, there is a continual race between the plant breeder breeding new resistant varieties and the pathogens, which often produce new virulent strains very rapidly indeed.

The development of genetic modification techniques since the 1980s has added another potential tool for increasing resistance to pathogens. What are the main targets for this work? Section 11.3.9 outlined the current understanding of molecular plant pathology, and many of the ideas arise from this recent work.

It might be possible to improve plant recognition of surface structures of pathogens (PAMPs). Genetic inactivation of pattern recognition receptors in *Arabidopsis* resulted in increased pathogen susceptibility. This suggests that these receptors could be of use for the engineering of plant immunity. Another approach might be to enhance plant defences. We have seen that plants use a variety of responses to pathogen infection, including the manufacture of phytoalexins and ROIs and rapid PCD. An obvious idea is to engineer over-expression of some components of this response and, not surprisingly, synthesis of phytoalexins has been a target for genetic modification.

Tobacco, tomato and rice plants have been transformed to over-produce the phytoalexin resveratrol after the introduction of a grapevine stilbene synthase that is responsible for its synthesis. The transformed tomato plants were more resistant to *Phytophthora infestans* infection, and modified rice was more resistant to the rice blast fungus (*Magnaporthe grisea*). In general where this has been tried, however, the increased resistance has often been fairly weak or effective only against some pathogens. This suggests that increasing parts of the defence response may not be the best way to engineer basal resistance. The transfer of R genes into unrelated plant species has generally not been very successful, but it may work in some cases.

A more promising approach may be to activate immunity-specific signal transduction, thus increasing the whole immune response rather than just individual parts. This has been tried for systemic acquired resistance (SAR), by over-expressing NPR1, a key regulator of SAR. This procedure has been shown to lead to increased immunity in *Arabidopsis*, rice, tobacco, apple and wheat.

Interfering with pathogen effector function is a possible scheme to control infection, but it is pathogen-specific. White mould (*Sclerotinia sclerotiorum*), a fungal pathogen with a large host range of over 400 species, secretes oxalic acid as a virulence factor. One successful approach has been to engineer resistance through enzymatic degradation of the produced oxalate.

Finally, a biotechnological approach that does not involve genetic modification is to increase immunity by chemical or biological stimulation. A range of PAMPs have been shown to trigger defences against pathogens. Exposing plants to these substances could potentially stimulate plant immunity, and this has been shown to be effective. For example, N-deacetylated chitin or laminarin glucans that are similar to oomycete surface glucans are being used in agriculture to increase resistance to oomycete pathogens.

Despite the considerable progress that has been made in understanding the molecular basis of plant pathology, genetically modified crops that are resistant to pathogens have so far hardly been used. Writing in 2010, Gust and colleagues concluded their review of the potential uses of biotechnology in increasing plant resistance to pathogens thus: '*The impressive array of available strategies for pest management in current agriculture is in obvious contrast to the limited number of successful field applications. This may partly be due to the fact that many industrial enterprises focus on establishing broad-spectrum disease resistance in important crops without negative impact on crop yield or other important plant traits.*' (Gust *et al.*, 2010).

They went on to suggest that this may be difficult goal to attain. We will return to this topic in section 12.4 of our final chapter.

Selected references and suggestions for further reading

Agrios, G.N. (1978) *Plant Pathology*. 2nd ed. Academic Press, New York.

Alford, D.V. (2011) *Plant Pests*. Collins, London.

Anderson, W.P. (1996) *Weed Science. Principles and Applications*. 3rd ed. West Publishing, St. Paul, MN.

Bais, H.P., Weir, T.L., Perry, L.G., Gilroy, S. & Vivanco, J.M. (2006) The role of root exudates in rhizosphere interactions with plants and other organisms. *Annual Review of Plant Biology* **57**, 233–266.

Bent, A.F. & Mackey, D. (2007) Elicitors, effectors, and R genes: The new paradigm and a lifetime supply of questions. *Annual Review of Phytopathology* **45**, 399–436.

Carpenter, J.E. (2010) Peer-reviewed surveys indicate positive impact of commercialized GM crops. *Nature Biotechnology* **28**, 319–321.

Clarke, C.M., Bauer, U., Lee, C.C., Tuen, A.A., Rembold, K. & Moran, J.A. (2009) Tree shrew lavatories: a novel nitrogen sequestration strategy in a tropical pitcher plant. *Biology Letters* **5**, 632–635.

Dangl, J.L. & Jones, J. D. G. (2001) Plant pathogens and integrated defence responses to infection. *Nature* **411**, 826–833.

Devos, Y., Hails, R.S., Messéan, A., Perry, J.N. & Squire, G.R. (2011) Feral genetically modified herbicide tolerant oilseed rape from seed import spills: are concerns scientifically justified? *Transgenic Research* (in press) DOI 10.1007/s11248-011-9515-9.

France, P. (1986) *An Encyclopedia of Bible Animals*. Steimatzky, Tel Aviv, Israel.

Gianessi, L.P. & Reigner, N.P. (2007) The value of herbicides in U.S. crop production. *Weed Technology* **21**, 559–566.

Grafe, T.U., Schöner, C.R., Kerth, G., Junaidi, A. & Schöner, M.G. (2011) A novel resource-service mutualism between bats and pitcher plants. *Biology Letters* **7**, 436–439.

Gust, A.A., Brunner, F. & Nürnberger, T. (2010) Biotechnological concepts for improving plant innate immunity. *Current Opinion in Biotechnology* **21**, 204–210.

Hammerschmidt, R. (1999) Phytoalexins: What have we learned after 60 years? *Annual Review of Phytopathology* **37**, 285–306.

Hanley, M.E., Lamont, B.B., Fairbanks, M.M. & Rafferty C.M. (2007) Plant structural traits and their role in anti-herbivore defence. *Perspectives in Plant Ecology, Evolution and Systematics* **8**, 157–178.

Heil, M. & McKey, D. (2003) Protective ant-plant interactions as model systems in ecological and evolutionary research. *Annual Review of Ecology Evolution and Systematics* **34**, 425–453.

Hibbett, D.S., Binder, M., Bischoff, J.F., Blackwell, M., Cannon, P.F., Eriksson, O.E., Huhndorf, S., James, T., Kirk, P.M., Lücking, R., Lumbsch, H.T., Lutzoni, F., Matheny, P.B., McLaughlin, D.J., Powell, M.J., Redhead, S., Schoch, C.L., Spatafora, J.W., Stalpers, J.A., Vilgalys, R., Aime, M.C., Aptroot, A., Bauer, R., Begerow, D., Benny, G.L., Castlebury, L.A., Crous, P.W., Dai, Y.C., Gams, W., Geiser, D.M., Griffith, G.W., Gueidan, C., Hawksworth, D.L., Hestmark, G., Hosaka, K., Humber, R.A., Hyde, K.D., Ironside, J.E., Kõljalg, U., Jyrtznabm C.P., Larsson, K.H., Lichtwardt, R., Longcore, J., Miadlikowska, J., Miller, A., Moncalvo, J.M., Mozley-Standridge, S., Oberwinkler, F., Parmasto, E., Reeb, V., Rogers, J.D., Roux, C., Ryvarden, L., Sampaio, J.P., Schüssler, A., Sugiyama, J., Thorn, R.G., Tibell, L., Untereiner, W.A., Walker, C., Wang, Z., Weir, A., Weiss, M., White, M.M., Winka, K., Yao, Y.J. & Zhang, N. (2007) A higher-level phylogenetic classification of the Fungi. *Mycological Research* **111**, 509–547.

Hollings, M. (1983) Virus diseases. In Johnston, A. & Booth, C. (eds.) *Plant Pathologist's Pocketbook*. Commonwealth Agricultural Bureaux, Slough, UK.

Imms, A.D. (1947) *Insect Natural History*. Collins, London.

Kessler, A. & Baldwin, I.T. (2001) Defensive function of herbivore-induced plant volatile emissions in nature. *Science* **291**, 2141–2144.

Knight, S.C., Anthony, V.M., Brady, A.M., Greenland, A.J., Heaney, S.P., Murray, D.C., Powell, K.A., Schulz, M.A., Spinks, C.A., Worthington, P.A. & Youle, D. (1998) Rationale and perspectives on the development of fungicides. *Annual Review of Phytopathology* **35**, 349–372.

Menges, M. & Murray, J.A.H. (2007) Plant D-type cyclins: structure, roles and functions. In: Bryant, J. & Francis, D. (eds.) *Eukaryotic Cell Cycle*, Society for Experimental Biology Seminar Series 59. Taylor and Francis, Abingdon, UK, pp 1–28.

Mittler, R. (2002) Oxidative stress, antioxidants and stress tolerance. *Trends in Plant Science* **7**, 405–410.

Novoplansky, A. (2009) Picking battles wisely: plant behaviour under competition. *Plant, Cell and Environment* **32**, 726–741.

Parry, D.W. (1990) *Plant Pathology in Agriculture*. Cambridge University Press, Cambridge, UK.

Pennings, S.C. & Callaway, R.M. (2002) Parasitic plants: parallels and contrasts with herbivores. *Oecologia* **131**, 479–489.

Pimentel, D., Lach, L., Zuniga, R. & Morrison, D. (2000) Environmental and economic costs of nonindigenous species in the United States. *BioScience* **50**, 53–65.

Pocock, M., Evans, D., Straw, N. & Polaszek, A. (2011) The Horse-chestnut leaf-miner and its parasitoids. *British Wildlife* **22**, 305–313.

Schafer, M.G., Ross, A.A., Londo, J.P., Burdick, C.A., Lee, E.H., Travers, S.E., Van de Water, P.K. & Sagers, C.L. (2011) The establishment of genetically engineered Canola populations in the U.S.. *PLoS ONE* **6**(10): e25736. doi:10.1371/journal.pone.0025736

Stakman, E.C. & Harrar, J.G. (1957) *Principles of Plant Pathology*. Ronald Press, New York.

Stamp, N. (2003) Out of the quagmire of plant defense hypotheses. *The Quarterly Review of Biology* **78**, 23–55.

Stone, G.N. & Schönrogge, K. (2003) The adaptive significance of insect gall morphology. *Trends in Ecology and Evolution* **18**, 512–522.

Strange, R.N. (2003) *Introduction to Plant Pathology*. John Wiley, Chichester, UK.

Talboys, P.W., Garrett, M.E., Ainsworth, G., Pegg, G.F. & Wallace, E.R. (1973) A guide to the use of terms in plant pathology. *Phytopathological Paper* no. 17. Commonwealth Agricultural Bureaux, Slough, UK.

Toothill, E. (ed.) (1984) *The Penguin Dictionary of Botany*. Penguin Books, London.

van Loon, L.C., Rep, M. & Pieterse, C. M. J. (2006) Significance of inducible related proteins in infected plants. *Annual Review of Phytopathology* **44**, 135–162.

Wang, W., Barnaby, J.Y., Tada, Y., Li, H., Tör, M., Caldelari, D., Lee, D., Fu, X-D. & Dong, X. (2011) Timing of plant immune responses by a central circadian regulator. *Nature* **470**, 110–115.

CHAPTER 12

Plants and the Future

We have considered plant function at the molecular and cellular levels and looked at how those functions are integrated and coordinated in the growth, development and life of plants and in their adaptation to their environments (including stresses of various types). We now look to the future to see how our increasing knowledge of plants may be applied to some of the problems facing humankind in the 21st century.

Many people say that the biggest global problems that we face are climate change, biodiversity loss, human population increase and resource depletion (this would include shortages of both water and fossil fuels). All of these factors interact and some would add other problems to the list, but they do give us a basis to work from. Plants will have major roles to play in the future, just as they have in the past, and they are potentially important in providing solutions to some of our environmental problems. Here we will choose a few major topics to illustrate this, well aware that we cannot cover everything.

First, we will consider two environmental issues that are keys to the future, climate change and biodiversity loss, particularly concentrating on the plant aspects. Then we have selected two potential 'solutions' to some of the difficulties we face – the use of plants to produce biomass and biofuels, and genetically modified crops. Finally, we will briefly consider the feeding of the world's human population this century.

12.1 Climate change

Climate change is potentially the most serious environmental problem that the world is facing, and it is certainly in the news almost every day. Here we focus on aspects concerned with plants, but first we will put the topic into context.

12.1.1 The basics of climate change

Since the Industrial Revolution began, some 200 years ago, the burning of fossil fuels has steadily increased the carbon dioxide concentration in the atmosphere from around 284 ppm in 1832 to 390 ppm in 2010. At present, the concentration is rising by about 2 ppm every year. At the same time, methane concentration in the atmosphere has also increased. The reasons for this are more complex, but the main anthropogenic sources are rice paddies (see Chapter 10, Textbox 10.1), the methanogenic bacteria in the gut of ruminants and emissions from landfill.

Methane, CO_2, water vapour and a number of other gases are known as 'greenhouse gases' because they trap heat in the atmosphere. If they were not there, the global temperature of the atmosphere would be around $-15°C$ and life on Earth would not be possible. As the concentrations of CO_2 and methane have risen, the global temperature has also risen (by about $0.8°C$ in the last century). The main contributor to anthropogenic global warming is CO_2 (63 per cent) followed by methane (24 per cent), nitrous oxide (10 per cent) and other minor components (3 per cent). It is very likely that we will see a greater temperature rise in the current century.

In 2007, the Intergovernmental Panel on Climate Change (IPCC) released its fourth assessment report. By the end of the 21st century they predicted that global temperature would probably increase by $1.8–4.0°C$ – and possibly by as much as $6.4°C$. Temperatures will not rise equally everywhere, however. Figure 12.1 shows some global annual mean surface warming predictions from the IPCC (2007) under three different emissions scenarios (B1, low; A1B, medium; and A2, high), and at three different times. The Arctic has already been relatively badly affected by temperature increases, leading to the melting of much of the ice cap and permafrost, and this will continue.

Functional Biology of Plants, First Edition. Martin J. Hodson and John A. Bryant.
© 2012 John Wiley & Sons, Ltd. Published 2012 by John Wiley & Sons, Ltd.

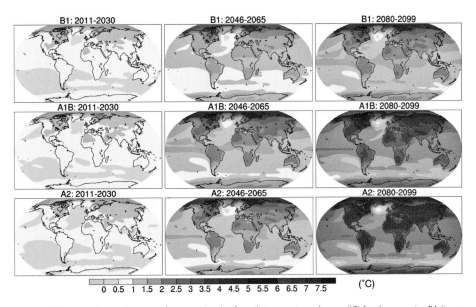

Figure 12.1 Multi-model mean of annual mean surface warming (surface air temperature change, °C) for the scenarios B1 (top row- low emissions), A1B (middle row- medium emissions) and A2 (bottom row- high emissions), and three time periods, 2011 to 2030 (left column), 2046 to 2065 (middle column) and 2080 to 2099 (right column). Anomalies are relative to the average of the period 1980 to 1999. From: Climate Change 2007: The Physical Science Basis. Working Group I Contribution to the Fourth Assessment Report of the Intergovernmental Panel on Climate Change, Figure 10.8. Cambridge University Press.

The IPCC predicted that sea level will rise by 28–43 cm, but this excluded ice melt, and it now seems likely that we could see a rise of a metre or more this century. Tropical storm intensity will probably increase, and droughts and floods will be more frequent. Some areas will become drier (e.g. the Mediterranean) and some wetter (e.g. Antarctica, where the icecap is getting thicker because of increased precipitation). Precipitation increases are very likely in high latitudes, while decreases are likely in most subtropical land regions. It has been suggested that sea level rise, together with all of the climatic changes noted above, will lead to 200 million people becoming environmental refugees by 2050.

12.1.2 The effects of climate change on plants and agriculture

We would expect that, as the climate changes in this century, there will be considerable effects on plants and also on agricultural production. In Chapter 9, we covered the stresses that are likely to increase in intensity as the climate warms. High-temperature stress (see section 9.2.3) is the most obvious stress to increase, and there will almost inevitably be more fires (section 9.2.4). With more energy in the atmospheric system, we can expect

not only more flooding and waterlogging (section 9.3) but also more drought (Section 9.4). Salinity problems (Section 9.5) will also probably increase, due to greater evaporation rates, increased use of irrigation and sea level rise in coastal areas. Finally, with more intense and extreme weather events such as hurricanes, which involve high winds, we can also expect more plants to be blown over or damaged by wind. Thus, a combination of all of these effects might be expected to lead to decreased plant growth and agricultural yield.

One topic that is much raised by those sceptical of human impacts on global climate is the CO_2 fertilization effect, as increasing concentrations of carbon dioxide in the atmosphere might be expected to have positive effects on plant growth. As we saw in Chapter 7, increasing the CO_2 concentration should lead to both an increase in photosynthetic rate and a decreased rate of transpiration from the leaves. The increase in CO_2 fixation is due to repression of photorespiration (Chapter 7, section 7.5), particularly in C_3 plants, and also because substrate supply is increased. Water loss is decreased because the stomata partially close (Chapter 7, section 7.10). However, the longer-term effects on physiology and growth of plants are less certain. The stimulation of photosynthesis can fade away, as negative feedback effects frequently occur.

There are some circumstances where positive effects of CO_2 fertilization have undoubtedly been observed. For example, in glasshouses, CO_2 is routinely released to increase yields of crops such as tomato. However, while one might assume that a rise from a pre-Industrial Revolution CO_2 concentration of 284 ppm to 390 ppm in 2010 would have led to an increase in arable crop production in a similar way to glasshouse crops when fertilized with CO_2, this is not the case. So far, the positive effects from CO_2 fertilization appear to be more than outweighed by the negative effects of drought, flooding and extreme weather events.

Predicting future trends in agricultural production as the climate warms is proving to be difficult. Jaggard *et al.* (2010) have attempted predictions for 2050, by which time they estimate that atmospheric CO_2 concentration will be 550 ppm, leading to a global increase in temperature of about 2.0°C. Free air carbon dioxide enrichment (FACE) experiments have suggested that 550 ppm CO_2 will increase yields of C_3 crops by about 13 per cent but will have no effect on C_4 species. Increased CO_2 will also decrease water consumption due to partial stomatal closure, so rain-fed crops should be less affected by water stress. The temperature increase will speed up crop development, probably increasing the yields of species like sugar beet that do not flower before harvest. However, crops like wheat and rice, which do flower before harvest, will potentially show decreased yields because flowering is particularly sensitive to high temperatures. Moreover, the rate of evapotranspiration will also increase as temperature rises, counteracting the positive effect of CO_2 on water consumption.

Extreme weather events are likely to become more common, and a 1-in-20 year high temperature event in 2010 is likely to be about 3°C hotter by 2050. These extreme events are likely to have serious impacts on crop yield but are difficult to predict, and plant breeders are unlikely to want to breed for rare events. Increased temperatures are also liable to increase both weed growth and problems due to pests and disease.

In addition, climate change is likely to cause changes in the distribution of plants. In general, as the climate warms, plants will be expected to move northwards in the northern hemisphere and southwards in the southern hemisphere. They will also change their altitudinal ranges and move up mountains. The changes for tree species are likely to be complex but, in the northern hemisphere, it seems that net ecosystem production may be increased in the north as conifer forests spread, and decreased in the south by the loss of deciduous forests. We saw in Chapter 6 (Box 6.2) that tree height and, thus, timber production is likely to diminish in a warmer climate. Farmers will also almost certainly have to adapt by growing different crops as the climate changes.

12.1.3 Plants and biogeochemical cycles in a time of climate change

We have seen above that climate change is likely to have major effects on plant growth and crop yield. Plants can, however, have an impact on the climate, and we will now briefly survey some of these.

One particularly good example of how plants are involved in global and regional climate change concerns the Amazonian forests. In 2001, these forests covered 5.4 million km^2 – about 87 per cent of their initial area, due to previous logging and agricultural activity. Since the Cretaceous period (145–65 million years ago), they have had an important role in Earth system functioning and biogeochemical cycling. The temperature has recently been increasing in the Amazonian forests at a rate of around 0.25°C per decade; at the end of the last ice age, the rate was only 0.1°C per century. Further temperature rises are predicted of about 3.3°C in the 21st century, assuming a medium range of carbon emissions. However, this figure might reach 8°C under some scenarios where major forest dieback occurs, affecting regional climatic phenomena.

Amazonian forests are important as evaporation and condensation above them drives global atmospheric circulation. This affects precipitation across the continent and also in the northern hemisphere. The Amazon is particularly affected by changes in precipitation during the dry season. Northern Amazonia has been on a steadily drying trend since the mid-1970s, and this is predicted to continue this century. Figure 12.2 suggests that the southeastern part of Amazonia is particularly vulnerable to further drying. The drought in 2005 was very serious and was calculated to be a 1 in 100 year event but, in many respects, the 2010 drought was even worse.

The Amazonian forests account for about 15 per cent of global terrestrial photosynthesis and the intact forests are

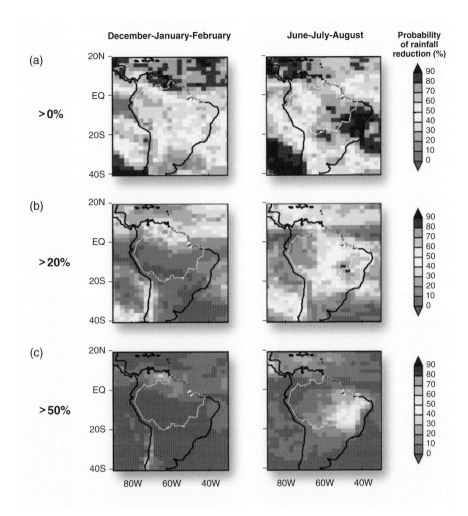

Figure 12.2 The probability of enhanced drought in Amazonia. The percentage of 23 climate models that show a decline in rainfall between 1980 to 1999 and 2080 to 2099 under midrange global greenhouse gas emissions scenarios. (a) Any decline (rainfall decline greater than 0 per cent); (b) substantial decline (rainfall decline greater than 20 per cent); (c) severe decline (rainfall decline greater than 50 per cent). Dry season rainfall is particularly important. Left column: December-January-February (dry season in north); right column: June-July-August (dry season in central and southern Amazonia).
From Malhi *et al.*, 2008 (Reproduced by permission of the American Association for the Advancement of Science).

a major carbon sink, storing about 120 Pg C in biomass carbon, at a rate of 0.6 Pg C per year in most years. However, an average of 0.5 Pg C per year was released in the 1990s due to deforestation. Thus, deforestation is a cause of climate change – it both reduces the amount of CO_2 absorbed, as there is less forest, and increases the amount released due to burning and other phenomena. Major drought years, such as those in 2005 and 2010, cause widespread plant death, and at these times the forests cease to be sinks for carbon and become sources. Thus

the worry is that, with increased temperatures and more droughts, this century we may see even greater releases of carbon and a positive feedback loop might become established, contributing to runaway climate change.

The importance of plants in affecting climate change at a global level was emphasized in a modelling study by Cao *et al.* (2010). They assumed a doubling of atmospheric CO_2 from 400 to 800 ppm, and showed that the radiative effect of CO_2 (i.e. trapping of heat) would cause an increase of $2.86°C$ in land surface air temperature. However, this increase in CO_2 would cause greater stomatal closure (Chapter 7, section 7.10), lower transpiration, and less cooling effect, and thus temperature would increase by an additional $0.42°C$.

Global runoff is predicted to increase by 5.2 per cent as the radiative effect increases continental precipitation, but decreased evapotranspiration would lead to a further increase of 8.4 per cent. Relative humidity is predicted to decrease as increased CO_2 causes stomatal closure and reduces transpiration. It thus appears likely that the effects of increasing atmospheric CO_2 on plant physiology will lead to additional global warming beyond the radiative effects of CO_2.

The final topic we will briefly cover in this section is carbon offsetting. In simple terms, this usually involves a reduction in emissions of CO_2 to compensate for emissions made elsewhere. There are many different types of offsets, but they all fit under two major categories:

1 In the compliance market, companies and governments are capped on the total amount of CO_2 they are allowed to emit, and they buy carbon offsets in order to reach these targets. This carbon trading market exists under the Kyoto Protocol and the European Union Emissions Trading Scheme.

2 In the much smaller voluntary market, individuals, companies and governments procure offsets to mitigate the effects of their CO_2-producing activities. The most well known voluntary offset is when a person pays an additional sum to compensate for the environmental damage caused when travelling by air.

Both types of carbon offsets are highly controversial but, interestingly, both frequently involve plants. Deforestation currently accounts for about 20 per cent of carbon emissions. It can be reduced by paying either to preserve forests or to provide alternatives to forest-based products such as wood. At the international climate change meetings in Copenhagen (2009) and Cancun (2010), there was much discussion of REDD schemes (reducing emissions from deforestation and forest degradation). The idea is that developed nations pay developing nations to protect their native forests, as a component of a post-Kyoto Protocol agreement. Many of the voluntary carbon offsetting schemes involve the planting of trees, often in developing countries. Some schemes are better than others, and tree planting is generally positive for the environment and the trees will absorb some CO_2, but often these benefits are not enough to compensate for the environmental impacts of air travel.

12.2 Loss of plant biodiversity

Biological diversity means the variability among living organisms from all sources including, inter alia, *terrestrial, marine and other aquatic ecosystems and the ecological complexes of which they are part; this includes diversity within species, between species and of ecosystems.*[i]

Biodiversity is a relatively new word, first suggested by W.G. Rosen in 1985 as a short version of 'biological diversity'. It includes all species, but here we will confine the discussion almost entirely to plants. First, we will consider the total number of plants and where they are distributed. Then we will look at the reasons why we are rapidly losing biodiversity and why we should be concerned about this. Our final topic in this section will concern methods to prevent loss of plant biodiversity and to conserve it for the future.

12.2.1 Plant numbers and distribution

We do not know how many plant species there are in the world, but the current best estimate is 370,000. Even known plants are, not infrequently, named differently in different countries, and the use of Latin binomials is ignored. This makes it difficult to estimate numbers of species. At present there is much work going on to produce definitive lists of plant species, where local names can all be matched against the agreed scientific names. By 2007, about 53 per cent of known species were listed, and it was hoped that this figure would reach 85 per cent by 2010 and 100 per cent soon after.

However, there are still many species that we have not recognized. Most of these are relatively small, but quite large plants that were previously unknown to science are still being discovered. One example is the Wollemi Pine (*Wollemia nobilis*). Prior to 1994, this tree was only known from the fossil record and was thought to be extinct until, in September of that year, David Noble discovered a population of them in a remote canyon in Wollemi National Park, New South Wales, Australia (Figure 12.3). There are less than a hundred mature trees in the wild.

---

[i]Convention on Biological Diversity (1992), signed at the Earth Summit in Rio de Janeiro.

Figure 12.3 Wollemi Pine (*Wollemia nobilis*), from Wollemi National Park, New South Wales, Australia.
Photo: MJH. Used by the permission of the National Botanic Gardens, Glasnevin, Dublin, Ireland.

The world's biodiversity is not evenly spread; tropical rainforests (Figure 12.4) are more diverse than temperate habitats, which in turn have higher biodiversity than Arctic or Antarctic habitats. Biodiversity hotspots have very high numbers of organisms present. Myers *et al.* (2000) found that 44 per cent of land plants were concentrated in 25 hotspots, which only accounted for 1.4 per cent of the Earth's surface.

12.2.2 Biodiversity loss

Extinction of plant species is a natural process, but the rate at which organisms have become extinct has undoubtedly increased, possibly to 100 or 1,000 times faster than the natural rate. Since 1600, about 654 known plant species have become extinct – but this is almost certainly an underestimate, as many species could have become extinct without our knowledge. Approximately 60 per cent of ecosystems in the world are degraded or not used sustainably. In the European Union, only 17 per cent of

habitats and species and 11 per cent of key ecosystems are in a favourable state.

There are several major causes of biodiversity loss, including habitat loss, pollution, invasive species and overexploitation. Undoubtedly, the loss of natural habitats was the biggest cause of biodiversity loss in the 20th century, with increased use of land for agriculture, logging for forestry and urban development mostly responsible. About 25 per cent of the Earth's land surface area is now under agriculture. A further one per cent is urban, but that is a rapidly growing figure.

In section 12.1.3, we noted that tropical rainforests – and particularly the Amazon – are important carbon sinks, but that recent droughts could herald a time when they will become carbon sources. These rainforests, which are also habitats for other plants and animals, are being cut down rapidly; about 13 million hectares of tropical forests are removed every year. Forests are often logged for their hardwoods and then turned over to agriculture.

We covered many of the effects of pollution on plants in Chapter 9, section 9.6, and there is little doubt that these have led to major loss of biodiversity. The problems of invasive species in causing biodiversity loss are more often linked to animals rather than plants. There have been instances where introduced animals have had major impacts on plant biodiversity (e.g. the rabbit in Australia). Moreover, there have also been cases where introduced plants have out-competed native species. In Chapter 11 (Box 11.1), we mentioned the example of purple loosestrife (*Lythrum salicaria*), which was introduced into North America from Europe and became a major pest species on wetlands.

Again, overexploitation is often linked to animals (e.g. whales, fish), but many tree species, particularly in the tropics, have been logged in an unsustainable manner. Climate change is also likely to cause major impacts on biodiversity in the next century, '*with species distributions and relative abundances shifting as their preferred climates move towards the poles and higher altitudes.*'[ii] The problem is that the speed of the change will bring the likelihood of many extinctions. Maclean & Wilson (2011) used a meta-analysis in an attempt to quantify likely losses of biodiversity this century. They predicted about a ten per cent loss across all taxa, with vertebrates being worse affected than plants and invertebrates.

[ii]GEO-4 (2007).

Figure 12.4 Aerial photograph of rainforest. Province Orientale in the Democratic Republic of Congo boasts the second largest rainforest in the world, which spreads across the centre of the country. This is virgin rainforest, virtually uninhabited by humans for thousands of square miles. Photo: Geoff Andrews.

12.2.3 The importance of biodiversity and how to conserve it

If there are 370,000 plant species on Earth, will the loss of a few or even the 10 per cent currently predicted this century have any effect? Of course, some biodiversity could be useful to humans. Plants have often been the source of new drugs, and there is every reason to believe that there are still more discoveries to be made. It is also possible that there are new crops or species related to our present crops that could be used in plant breeding.

The other major concern is maintenance of ecosystem services. The Millennium Ecosystem Assessment (2005) suggests there are four types of ecosystem services: provisioning (e.g. food), regulating (e.g. climate), supporting (e.g. pollination) and cultural (e.g. aesthetic enjoyment). If too much biodiversity is lost, then it is possible that ecosystems may become seriously degraded or may even collapse altogether. The worry is that if this were to occur, then human survival might also be threatened.

Of course, the reasons stated so far for conserving biodiversity are anthropocentric ones, with the good of humanity at their core. It is interesting that the Millennium Ecosystem Assessment recognized, '*that biodiversity and ecosystems also have intrinsic value- value of something in and for itself, irrespective of its utility for someone else- and that people make decisions concerning ecosystems based on consideration of their own well-being and that of others as well as on intrinsic value.*' They also noted that more biodiversity would be preserved if intrinsic value was taken into account.[iii]

What is being done and what *should* be done in the future to conserve and preserve plant biodiversity? There is little doubt that some of the positive actions already taken by governments, non-governmental organizations, communities, businesses, industry and individuals have conserved biodiversity, and that as a consequence there is more biodiversity left than would otherwise have been the case. It is quite evident, however, that these actions have not been enough, and that further measures will be needed to decrease the current rate of biodiversity loss.

[iii] See Hodson & Hodson (2008) pp 30–35 and 70–75 for a more detailed account.

As of 2011, there are 224 National Nature Reserves in England, covering 94,400 hectares, or about 0.6 per cent of the land surface of the country. Similarly, in the United States, about 39 million hectares have been designated as National Wildlife Refuges. Protected areas have undoubtedly been important in conserving biodiversity, but it has become increasingly recognized that these islands of high biodiversity need to be connected if they are to function properly. The concept of wildlife corridors to provide these connections is now appreciated.

Often this can be achieved by the agriculture and forestry sectors changing their management practices to incorporate biodiversity conservation. In the UK, a variety of government-sponsored schemes have been introduced to give incentives for farmers to protect biodiversity. For example, John Neal, the farmer at Manor Farm in Warmington, near Banbury, entered a 20-year environmental scheme in 1991 and has since farmed in a very environmentally sensitive way. He writes: '*The result has been good; for five years little change was noticed but then, as the effect of fertilizers wore off, natural grasses, herbs and wild flowers slowly spread from the hedgerows.*'[iv]

One piece of land has been free from fertilizer, with only spot spraying of weeds for 22 years (since 1989). In about 2002, a single common spotted orchid (*Dactylorhiza fuchsii*) began to grow there (Figure 12.5). For seven years, this was the only orchid present, but from 2009 the numbers started to increase to the 19 counted in 2011. Also in 2011, the same orchid species was noted in three other places on the farm. This well illustrates the value of such schemes in increasing biodiversity in agricultural land.

For species threatened with extinction, special protection measures may also be needed. For example, in the case of the Wollemi pine mentioned in section 12.2.1 above, the area in which it was discovered has been given special protection status and a breeding programme was initiated to ensure that the plant did not become extinct. In some cases, biodiversity has also been enhanced by ecosystem restoration schemes, such as those involving the clean-up of contaminated land.

Finally, considerable worldwide investment has been made into creating gene banks. These are repositories where plant material can be safely stored, often at low

iv See Hodson & Hodson (2008) pp 139–141 for more on the Manor Farm environmental schemes.

Figure 12.5 Common spotted orchid (*Dactylorhiza fuchsii*) growing at Manor Farm, Warmington, near Banbury, UK. Photo: MJH.

temperatures ($-80°C$) to conserve genetic diversity for the future. One example is the National Plant Germplasm System (NPGS) in the USA, which was established by Congress after World War II. This now consists of 26 repositories with about 500,000 individual collections. The NPGS preserves agricultural biodiversity and genetic resources that are required for future environmental considerations and food security.

12.3 Biomass and biofuels derived from plants

We will now turn to the use of plants and their products as energy sources. At the moment, 13.4 per cent of global primary energy supply is obtained from plants. That represents about 46 EJ per year (1 EJ = 1 Exajoule = 10^{18} joules) of energy. The total production achievable by plants globally is uncertain, but 2–400 EJ per year is the suggested figure. This implies that we could produce far more energy from plants than we do at the moment if we had suitable technology.

Crops can be grown as energy sources for three different purposes: solid fuels for power generation (electricity, heat), liquid fuels for transport and production of 'biogas'. In the literature, 'biomass' and 'biofuel' have often been used in a confusing way, where some have used the terms interchangeably and others have implied they are different. Here we will use the term 'biomass' for solid products and 'biofuel' for liquids.

12.3.1 Traditional biomass

Humans have been burning plant material for thousands of years. Until the Industrial Revolution, when the use of coal became widespread, wood was the major source of energy for cooking and heating. In many parts of the developing world, it is still the major source of energy. In 2004, this use of 'traditional biomass' represented 8.5 per cent of the total world energy consumption. This, of course, contributes to deforestation, and inefficient cooking stoves also produce much smoke that is the cause of human respiratory problems. There has therefore been considerable interest in the production of more efficient cooking stoves for the developing world that burn more efficiently, use less wood and produce less smoke.

12.3.2 The modern use of biomass

With increasing energy demand, biomass is again being considered as part of the energy mix by developed nations. At present, however, it only tends to make up a small fraction of the energy supply for developed nations. For example, in 2005, the United Kingdom obtained only 3.5 per cent of its electricity and 0.6 per cent of its heat from biomass. There is clearly much potential for improving figures like these.

Two main types of plants are grown as biomass fuels. These are perennial rhizomatous C_4 grasses, including switchgrass (*Panicum virgatum*) and various *Miscanthus* species, and fast-growing trees from the family Salicaceae, including poplars (*Populus*) and willows (*Salix*).

The grasses shoot from the base each spring and produce long stems, which can be harvested later in the year. Figure 12.6 shows stems of *Miscanthus giganteus* (often known as elephant grass) in December after the season's growth. Dedicated harvesting equipment is required to harvest the stems (in this case in the following April), which are then formed into bails using a bailer (Figure 12.7).

Figure 12.6 *Miscanthus giganteus* (elephant grass) stems in December after the seasons growth. Photograph taken at Manor Farm, Haddenham, Buckinghamshire, UK. Photo: MJH.

Figure 12.7 After harvesting the *Miscanthus giganteus* stems in April they are then formed into bails using a bailer. Photograph taken at Manor Farm, Haddenham, Buckinghamshire, UK. Photo: MJH.

Often the material is co-fired with coal at power stations (see below), but it may also be turned into briquettes for domestic consumption (Figure 12.8).

The poplars and willows are generally harvested after three to four years. Traditionally, these species were coppiced, cutting stems to the ground, and this approach is therefore continuing, albeit with more modern harvesting equipment (Figure 12.9a & b).

At present, switchgrass, *Miscanthus*, willow and poplar are the materials of choice for thermal technologies as they are low in simple carbohydrates, which makes them unsuitable for manufacture of liquid biofuels. Their inorganic fraction is low in nitrogen or sulphur-containing

Figure 12.8 *Miscanthus* can be turned into briquettes for domestic consumption.
Photo: JAB.

(a)

(b)

Figure 12.9 Willow (*Salix*) trees being harvested in March at Manor Farm, Haddenham, Buckinghamshire, UK. (a) One specially adapted tractor cuts the willow at near ground level, then chips it before firing it into a trailer pulled by a second tractor. (b) A close-up of the specialized cutting equipment mounted on the front of a tractor adapted for harvesting willow.
Photos: MJH.

compounds, which means that, when burned, they produce low levels of NO_x and SO_2 in the exhaust gasses.

Biomass is often burned in the form of chips or pellets. Co-firing with coal is very common at power stations. Alternative solid fuel sources are wood pellets made from compacted sawdust, which are a by-product of sawmilling and other wood transformation activities. The pellets are very dense, have low water content and can be burned with high combustion efficiency. They are mostly used to supply energy for central heating systems.

12.3.3 Concerns over energy security

Although solid fuels are very useful for certain purposes, liquid fossil fuels (petrol and diesel) are easier to transport and have been the fuel of choice for most vehicles for many years. However, they are not renewable resources, and there are increasing concerns that supplies may not meet demand in the future.

For many, years environmentalists have been concerned by the phenomenon known as 'Peak Oil'. This revolves around the time when global oil production peaks. All agree that this *will* happen, but environmentalists tend to put the date in the near future, while the oil companies consider that it is still some decades away. In 2009, the UK Energy Research Council carried out a major survey of all the available data and concluded: '*On the basis of current evidence we suggest that a peak of conventional oil production before 2030 appears likely and there is a significant risk of a peak before 2020. Given the lead times required to both develop substitute fuels and improve energy efficiency, this risk needs to be given serious consideration.*' UKERC (2009).

Although solid fuels such as coal can be converted to liquids, the processes involved are expensive and not very efficient, leading to increased CO_2 emissions. If oil production peaks and demand is still increasing (a very likely scenario, with China and India rapidly developing), then the price of oil would certainly increase. Our developed economies have been very much based on plentiful supplies of oil, which we use to run our vehicles, heat our buildings, manufacture plastics and medicines, and a lot more.

Of particular concern to plant scientists and agriculturalists is the fact that our nitrogen fertilizers are produced using the Haber process, which requires fossil fuels to run. Therefore, scientists and governments around the world have steadily come to the conclusion that we need to move away from burning fossil fuels, to avoid carbon emissions and the associated climate change and to increase energy security.

12.3.4 Biogas

Biogas is produced by the biological breakdown (fermentation) of organic matter in the absence of oxygen. Many different biodegradable materials can be used, including plant material and crop residues. The latter can be sourced from many crops, including maize, winter rye, wheat, barley, oilseed rape, sugar beet and sunflower. Water hyacinth (*Eichhornia crassipes*) has also been used for biogas production (see Chapter 10, Figure 10.5).

Biogas mostly consists of methane and carbon dioxide. This can be combusted with oxygen, releasing energy and allowing biogas to be used as a fuel. Biogas can be used for heating and cooking, or it can be used to produce electricity.

12.3.5 First generation biofuels

The idea for biofuels is older than might be thought. Rudolf Diesel had intended vegetable oil to be the fuel source for the diesel engine, and his early work concentrated on biofuel. At the World Exhibition in Paris in 1900, Diesel ran his engine on peanut oil. Likewise, Henry Ford anticipated that his cars would be run on ethanol from maize. However, as fossil fuel oil became more widely available, it was obvious that petrol would be a cheaper and easier source of liquid fuel.

It was not until the 1970s that the idea of using biofuels gained much further attention. The Arab oil embargo of 1973–1974 and the Iranian Revolution (1978–1979), together with decreased domestic oil production in the USA, drove oil prices upwards. Steadily, in the late 20th century, problems with carbon emissions, climate change and fuel security forced developed nations to reconsider liquid biofuel options. Initially, biofuels were regarded as an environmentally friendly option and a solution to energy problems that was 'carbon neutral'. Both governments and environmental groups were very positive about their introduction.

The two most common biofuels are bioethanol and biodiesel. Bioethanol is produced by fermentation of sugars or starches, using microorganisms and enzymes. Globally, bioethanol is the most widespread biofuel and Brazil is the largest user. It is produced most commonly from wheat, maize, sugar beet and sugar cane.

In Europe, biodiesel is the most common biofuel. This is similar in chemical composition to diesel obtained from fossil oil and is produced from oils or fats by transesterification (this involves exchanging an organic group of an ester with the organic group of an alcohol). Biodiesel sources include soybean, oilseed rape (see Chapter 11, Figure 11.5), jatropha, sunflower and palm oil.

Palm oil has become the major source of biodiesel in the tropics, and its production has been somewhat controversial. Oil palm trees (*Elaeis guineensis*) are shown in Figure 12.10, and in a dense plantation in Figure 12.11.

Oil palm is native to the humid tropics of West Africa, but has been introduced elsewhere, including

Figure 12.10 Oil palm (*Elaeis guineensis*) trees in Bas-Congo (Democratic Republic of Congo).
Photo: Paul Latham.

Figure 12.11 Oil palm (*Elaeis guineensis*) plantation in Borneo where the trees are grown specifically for biodiesel. Photo: Dr David Stafford.

South America, Indonesia and Malaysia. After flowering, the fruits mature between five to nine months later. The female inflorescences (Figure 12.12a) can bear up to 200 fruits, and technically these are drupes (Chapter 4, section 4.8.4). Trees may produce 1,000 fruits in a season. Within the fruit, the fleshy orange mesocarp surrounds the kernels, which consist of a stony black endocarp with the seeds inside (Figure 12.12b).

From the fruits, two distinct products are extracted: palm oil and palm kernel oil:
• Palm oil is pressed from the fleshy mesocarp and is bright orange in colour, due to its high ß-carotene content (Figure 12.12c). It is one of the few highly saturated vegetable fats and it is semi-solid at room temperature. Palm oil is much used in cooking – particularly for frying.
• Palm kernel oil is white or pale yellow in colour and has been used in the production of soap and margarine.

Both palm oil and palm kernel oil can be converted to biodiesel.

Many governments have introduced targets for increased biofuel production. For example, the USA has mandated a national target of 36 billion US gallons (136 billion litres) by 2022, while the European Union has instituted a target of a minimum of 10 per cent of biofuels in transport uses by 2020. Global production of biofuel in 2005 was estimated at 45 billion litres of ethanol equivalents, but this is predicted to rise to 314 billion litres by 2030. However, it has been found that 'first generation' biofuels are not without their problems, which we will outline below.

One of the key problems with first generation biofuels has been the impact on biodiversity, particularly in the case of oil palm plantations in the tropics. In Thailand, 108 bird species were found in the natural tropical forests, but only 41 species in adjacent plantations. Fewer than 10 per cent of birds and mammals found in Sumatran primary forests survive in oil palm plantations, and bat species decreased in numbers by over 75 per cent.

Another difficulty with first generation biofuels has been competition with food supplies. Production of grain ethanol consumed 25 per cent of the maize (corn) grown in the USA in 2007, and greater than 30 per cent in 2008. It is likely that this will continue for at least the next decade, due to the current infrastructure investment in grain ethanol refineries. Using grain for biofuel production tends to push up grain prices worldwide, and this can be a major problem for poor, developing nations.

There have also been concerns that the presently available biofuels are not energy-efficient. This is a complex area, and results of analyses can vary considerably. Sometimes it has been claimed that more energy is used to make some biofuels than is obtained when it is eventually burned in a vehicle. It is now recognized that the whole life cycle of the biofuel product has to be taken into account. Their manufacture requires energy from many sources:
• farming activities
• the production and application of fertilizers
• pesticides, herbicides and fungicides
• transport to the site where conversion to biofuel takes place
• conversion to the final product
• transport of the liquid biofuel to the place of use.

It has been increasingly recognized that the first generation biofuels described so far have too many problems to be a long-term option.

12.3.6 Second generation biofuels

We have seen that the first generation biofuels have some serious problems, which almost certainly indicate that they must eventually be replaced by fuels that have fewer impacts on biodiversity, do not compete with food requirements and are more energy efficient. One solution is to use the parts of the crop plant that are not eaten (e.g. the stems) rather than the grain, as is presently the case. There would thus be no competition with food.

Cell walls make up most of the dry weight of a plant, and about 75 per cent of cell walls consist of polysaccharides (see Chapter 2, section 2.2.1). If these polysaccharides could be broken down to monosaccharides, they could be fermented to ethanol. On this basis, it is estimated that there is a similar amount of sugar in the stems of wheat (*Triticum aestivum*) as there is of starch of the grains. So far, crop residues such as straw have not been used for biofuel production but, with demand increasing, these agricultural by-products will only be able to supply a fraction of the total requirements. Therefore, using dedicated biomass crops for biofuel production is also being considered. These include all those mentioned as biomass crops in Section 12.3.2 above. Several of these crops (e.g. switchgrass) can be grown on degraded land, and thus they have the advantage that they will not use

land that is rich in biodiversity, nor will they take land that could be used for growing food crops. It is even possible to grow some of these crops as poly-cultures with several species, enhancing biodiversity. Thus there is a real possibility that the 'second generation' biofuels might show major improvements on the first generation. However, the major – and as yet unsolved – problem is that ligno-cellulosic material is tough and needs breaking down. This is the reason why the easily digested starch in maize and wheat grain has been the feedstock of choice thus far.

Around the world, many laboratories are working on the problem of how to break down tough ligno-cellulosic

(a)

(b)

(c)

Figure 12.12 Oil palm (*Elaeis guineensis*) inflorescences and fruit from Malaysia. (a) Intact female inflorescence. Each inflorescence can bear up to 200 fruits, each of which is about 4 cm long and 2 cm broad. (b) Inflorescence split open, in which the fruits (drupes) can be seen as fleshy orange mesocarp surrounding the kernels which consist of black (endocarp) with white centres (seeds). (c) Left to right: the kernels, still surrounded by mesocarp after they have been separated from the inflorescence; the seeds surrounded by black endocarp; and the bright orange oil extracted from the fleshy mesocarp in a beaker.
Photos: Dr Greg Hodson.

material. There are two potential ways to tackle this problem: thermochemical processing and biochemical processing.

Thermochemical processing involves heating biomass with differing concentrations of oxygen. In air, the material is ignited to produce heat energy (this was covered in section 12.3.2). When biomass is heated in the absence of oxygen, however, *pyrolysis* occurs. This generates organic liquids that can be refined to produce liquid biofuel. Between these two is a process called *gasification*, in which heating happens at low oxygen concentrations, leading to the production of hydrogen and organic gases, which can then be changed into liquid biofuels by the Fisher-Tropsch process. Thermochemical processing has the advantage that it converts all biomass organic components, while biochemical processing mostly concerns the polysaccharides. However, the disadvantage of the thermochemical route is that high temperature processing involves considerable start-up and plant maintenance costs.

Biochemical processing entails converting the lignocellulosic biomass into sugars before fermentation to produce alcohols. Its major advantage is that start-up and maintenance costs are much lower than for thermochemical plants. However, saccharification, the conversion of biomass to sugars, is a major difficulty in obtaining liquid fuels by biochemical processing. As we saw in Chapter 11, section 11.3.7 cell walls are a major defensive barrier against pathogens, and they have evolved to be very resistant to enzyme attack. This is obviously to the advantage of plants, but it is very much to the disadvantage of biochemists seeking to release the cell wall sugars.

There are two ways around this problem. The first is to improve the plant biomass so that it is easier to convert to sugars. There has been progress on identifying the genetic loci which will make saccharification easier. Once this is done, it is hoped that conventional breeding or genetic engineering can be used to produce plants with cell walls that are easier to break down (although care will need to be taken that such an approach does not then produce plants that are more susceptible to pathogens). The second idea is to find enzymes and conditions that are optimal for the breakdown of cell walls.

Finally in this section, we should briefly consider what many call 'third generation' biofuels. Some would place the genetic modification of plants to make cell wall sugars more easily available in this category. However,

the use of microalgae to produce biofuels is certainly 'third generation'. Although higher plants can make a valuable contribution to biofuel production, it is generally recognized that they cannot provide all of the liquid fuels that are presently required without seriously competing with food production; the land area required would almost certainly be far too great. Microalgae, however, produce a wide range of potential chemicals that could be converted into biofuels. Many of these species could be grown in very harsh conditions (e.g. *Dunaliella parva* grows in the Dead Sea), so it might be possible to grow such algae in hot desert areas which cannot be used for agriculture and have low biodiversity. Pilot projects have shown that such schemes are possible, but much further work is needed before they can make a contribution to our biofuel requirements.

12.4 Genetically modified crops

'The possible application of novel genetic techniques to enhance the physiological and biochemical capability of plants exposed to stress was addressed at the symposium with a general consensus that in the area of osmoregulation in higher plants considerable progress has been made in describing the phenomenon and as a process it is reasonably well understood. It was suggested that a more basic understanding of the metabolic and genetic control of the process will be required before genetic manipulation of this characteristic is realized.'

Rains & Valentine (1980).

Throughout this book, we have frequently mentioned GM (genetic modification) as a research tool and in the production of new strains of crops. Great hopes have been pinned on these crops for increasing global food production, but they have also been the subject of major controversy. In Chapter 3 (Box 3.1), we considered the methodology for producing GM plants, and we have also had cause to discuss them regularly elsewhere. The question here is, 'where are we now with GM crops?'

Table 12.1 summarizes progress in this area up to 2011. It is not a complete inventory of all work on GM crops, although it does cover most areas. As we write, two types of GM crops are dominant; herbicide-resistant and insect-resistant (Bt) plants, and these represent around 99 per cent of the total of the GM crops that are grown

Table 12.1 Progress on GM crops.

Type of transformation	Present status	Section in this book
Introducing nitrogen fixation genes into plant species that do not normally fix nitrogen	Experimental stage	5.9.4
Decrease photorespiration in C_3 plants	Experimental stage	7.5.2
Introducing the machinery for C_4 photosynthesis into C_3 plants	Experimental stage	7.7.6
Increased resistance to chilling temperatures	Experimental stage	10.2.1
Increased resistance to freezing temperatures	Experimental stage	10.2.2
Increased drought tolerance	Experimental stage	10.4.3
Golden Rice (production of beta-carotene in the grain)	Release for agricultural use in 2012	Box 10.1
Increased salt tolerance	Experimental stage	10.5.3
Increased metal tolerance	Experimental stage	10.6.2 and 10.6.3
Herbicide resistance	Widespread agricultural usage	11.1.5
Insect resistance (Bt plants)	Widespread agricultural usage	Box 11.2
Increased resistance to pathogens	Limited agricultural usage	11.3.11
Modification of plant cell walls to make saccharification easier for the production of biofuels	Experimental stage	12.3.4

Table 12.2 The top ten countries growing GM Crops in 2010.[v]

Country[*]	Area of GM crops (million hectares)	Main GM crops
USA	66.8	Maize, Soybean, Cotton, Oilseed Rape, Sugar Beet, Alfalfa, Papaya, Squash.
Brazil	25.4	Soybean, Maize, Cotton.
Argentina	22.9	Soybean, Maize, Cotton.
India	9.4	Cotton.
Canada	8.9	Oilseed Rape, Maize, Soybean, Sugar Beet.
China	3.5	Cotton, Tomato, Poplar, Papaya, Sweet Pepper.
Paraguay	2.6	Soybean.
Pakistan	2.4	Cotton.
South Africa	2.2	Soybean, Maize, Cotton.
Uruguay	1.1	Soybean, Maize.

*The largest European grower of GM Crops is Spain at 0.1 million hectares.

commercially. The area planted with these two types is increasing every year and this looks set to continue. In 2010, the global area planted with GM crops exceeded one billion hectares for the first time (Table 12.2).

After a long saga, it appears that Golden Rice (Chapter 10, Box 10.1) will shortly be released for agricultural usage. Almost all of the other types listed in Table 12.1 are still largely at the experimental stage, with plant scientists conducting laboratory investigations and field trials to determine if they are viable.

What appear to be the biggest technical blocks to advances in this area?

• It has proved to be much easier to work on simple transformations that involve one or a few genes (e.g. herbicide resistance, Bt crops) than to transform plants where many genes are being changed. The idea that genes can be introduced to change whole plant structures (e.g. the formation of root nodules for nitrogen fixation) has largely been shelved.

• Even if transformation is successful and a useful trait is introduced, there may be other consequences which

[v] Data extracted from James (2010).

mean that the project is not feasible, and expression of the transgene may have 'side-effects' on plant growth. For example, many attempts to increase stress tolerance have led to slower plant growth and decreased yield. The 1980 quote from Rains and Valentine at the head of this section is now over thirty years old and is just as relevant today. We still need more basic research before engineering for complex traits such as stress tolerance is likely to succeed.

• Following on from above, there may be energetic limitations as to what is possible. Many stress-tolerant plants can survive in very stressful environments, but they do so by growing very slowly. It appears that much energy in such plants is directed into survival mechanisms. Can we really expect to produce plants that are very tolerant to stressful environments *and* grow quickly and have high yields? Or do the energetics just not add up?

The technical limitations to GM crops are real, but some major agricultural crops are now GM crops. What other issues surround the use of GM crops, and why are they so controversial?[vi] The reasons are outlined briefly below:

1 Some see GM techniques in general, and GM crops specifically, as in some way 'morally wrong' and dislike scientists 'playing God'. Of course, humanity has been changing plants by selective breeding for as long as agriculture has existed. Even so, some worry about transformations involving genes from totally unrelated species.

2 Possibly the biggest concern about GM crops among the general public (particularly in Europe) is food safety. There are occasional reports in the scientific literature questioning the safety of GM crops, but the huge scientific consensus suggests that the presently available crops are safe to eat. Certainly, they have been widely eaten in North America for many years with no obvious problems. Many Europeans, rightly or wrongly, are concerned about the food safety issues surrounding GM crops, and the food industry is well aware of this. No supermarket will stock GM produce if they think it will get adverse reactions from their customers.

3 GM crops, along with any newly bred crop, may pose environmental problems. Each new crop type needs to be investigated carefully, as they will all differ in their environmental impacts. Thus far we only have major experience with herbicide-resistant and Bt crops.

With herbicide-resistant crops, the problems seem to be escape of genes into related weed species and overuse of herbicides. With Bt crops, the biggest worry has been effects on non-target insects interacting with plant shoots, and more recently that the Cry toxin protein may affect soil organisms or even get into freshwater systems. All of these are real concerns, but agriculture always has an impact on the local environment. In each individual case, we need to assess whether the impact of GM crops is significantly greater than conventional crops and, if it is, then do the benefits in increased yield outweigh the environmental problems?

4 Possibly the biggest, and certainly the most difficult, problem to get a handle on concerns what may be termed 'globalization issues'. The wide-scale production of GM crops of necessity involves large companies and organizations, as the sums of money needed to carry out the research, bring a product to market, produce it and sell it are vast. Since the potential of GM crops was first recognized in the early 1980s, the large agrichemical companies gradually acquired small seed companies and 'cutting-edge' biotechnology companies. A few large multinational agrichemical companies have now taken over most activities concerning GM crops in the developed West (China and India are producing their own GM varieties, as are some other countries of 'intermediate' status in terms of development). For example, herbicides and the linked seeds for a herbicide-resistant crop are usually sold as a package by the same company. The worry expressed by both environmental and world development groups is that these large multinationals are more concerned with making profits than with feeding the world or environmental problems. Scientists often defend GM crops on their scientific merits but fail to address the concerns in this area and, indeed, may be ill-equipped to do so. Obviously this is a huge topic, and to go into it in detail would be well beyond the scope of the present book.

In conclusion, GM crops have great potential but are also somewhat controversial. They have largely been accepted unopposed in North America, and vast areas have been planted with GM crops. However, Europe remains mostly opposed, and very few GM crops are grown. It is uncertain what the future holds with regards to GM crops. Some would argue that their considerable advantages will eventually win through, even in largely sceptical Europe. Others might suggest that soon what

---

[vi]See Hodson & Hodson (2008), pp 182–188 and Bryant *et al.* (2005) pp 90–101 for more detailed discussions.

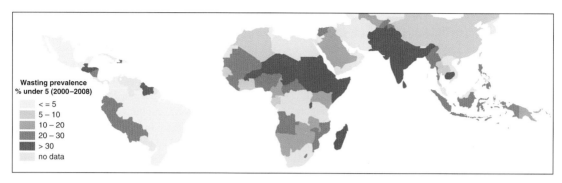

Figure 12.13 Wasting prevalence. Wasting is an indicator of a severe level of malnutrition. Prevalence of child malnutrition (wasting) is the percentage of children under age 5 whose weight for age is more than two standard deviations below the median for the international reference population (ages 0–59 months). Low weight for age (underweight) indicates chronic food insecurity.
From: Ericksen *et al.*, 2011 (Reproduced by permission of the Consultative Group on International Agricultural Research- CGIAR).

they see as the major disadvantages of the technology will become apparent and GM crops will become a thing of the past. Whichever of these turns out to be true, it should be remembered that at the very least, genetic modification is a valuable research tool.

12.5 Conclusion

In the present chapter, we have outlined some of the major problems facing humanity this century: climate change, biodiversity loss and resource depletion. Many would say that behind all these problems lies the phenomenon of rapid human population growth. Our population is approaching seven billion in 2011 and is projected to reach nine billion by 2050. It has grown very rapidly, more than doubling in the lifespans of the present authors. There is some debate as to whether human population will level off after 2050 or keep rising. However, it is clear that there must be an upper limit, the carrying capacity, to the number of humans that the planet can hold; what is uncertain is what that limit is.

Already there are many people in the world who do not have enough to eat (Figure 12.13), so essentially we need to feed a growing population without causing further biodiversity loss, at a time when some level of climate change is inevitable and resource limitations are beginning to grip.

What can be done? This is a massive topic, and one that could well take up another text book; it ventures into social, political and economic issues as well as science. Rather than attempt a summary of this material, we would

refer the interested reader to two major international reports which both appeared in 2009: IAASD and CGIAR.

Our book has mostly concerned what one might call 'academic' plant science, and we have considered the structure, physiology, biochemistry and molecular biology of plant processes in some detail. Some might think that, although this body of knowledge is interesting, it has little practical use. However, if you have done more than just skimmed the surface of this book, you will have seen just how important plant science has been in increasing agricultural production. Areas in which it has done so include our knowledge of plant mineral nutrition aiding fertilizer development, the development of reliable herbicides and crop breeding in all its forms. The list could go on.

Without basic plant science research, none of this would have been possible, and the great increase in agricultural yield we have seen would not have happened. Now, however, behind all the science lurks a major problem – how we feed the world in this century. Writing in June 2011 in the *Financial Times Magazine*, Clive Cookson listed 'plants to feed to world' as one of 'science's ten hottest fields.' (Cookson, 2011). Undoubtedly, plants must have a major role in the future and plant science, despite its rather unfashionable image nowadays in some quarters, will once again be a key discipline.

Selected references and suggestions for further reading

Bryant J.A., La Velle L.B. & Searle, J. (2005). *Introduction to Bioethics.* Wiley, Chichester.

Cao, L., Balab, G., Caldeira, K., Nemanid, R. & Ban-Weissa, G. (2010) Importance of carbon dioxide physiological forcing to future climate change. *Proceedings of the National Academy of Sciences, USA* **107**, 9513–9518.

CGIAR (2009) *Climate, agriculture and food security: A strategy for change*. The Consultative Group on International Agricultural Research (CGIAR).

Chisti, Y. (2007) Biodiesel from microalgae. *Biotechnology Advances* **25**, 294–306.

Cookson, C. (2011) Science's 10 hottest fields. *Financial Times Magazine* June 24, 2011.

DEFRA (2007) *UK Biomass Strategy*. Published by the Department for Environment, Food and Rural Affairs.

Ericksen, P., Thornton, P., Notenbaert, A., Cramer, L., Jones, P. & Herrero, M. (2011) Mapping hotspots of climate change and food insecurity in the global tropics. *CCAFS Report no. 5. CGIAR Research Program on Climate Change, Agriculture and Food Security (CCAFS)*. Copenhagen, Denmark. Available online at: www.ccafs.cgiar.org.

European Commission (2011) Our life insurance, our natural capital: an EU biodiversity strategy to 2020. See: http://ec. europa.eu/environment/nature/biodiversity/comm2006/ pdf/2020/1_EN_ACT_part1_v7%5b1%5d.pdf

GEO-4 (2007) *The global environmental outlook: environment for development*. United Nations Environment Programme (UNEP).

Gomez, L.D., Steele-King, C.G. & McQueen-Mason, S.J. (2008) Sustainable liquid biofuels from biomass: the writing's on the walls. *New Phytologist* **178**, 473–485.

Hadley Centre for Climate Prediction and Research (2005) *Climate change and the greenhouse effect*. The Meteorological Office, UK.

Hodson, M.J. & Hodson, M.R. (2008) *Cherishing the Earth: How to Care for God's Creation*. Monarch, Oxford.

Houghton, J. (2009) *Global Warming: the Complete Briefing*. 4th ed. Cambridge University Press, Cambridge, UK.

IAASTD (2009) *International assessment of agricultural knowledge, science and technology for development (IAASTD). Synthesis report with executive summary: a synthesis of the global and sub-global IAASTD reports* Ed. BD McIntyre *et al.*

IPCC (2007) *Reports of the Intergovernmental Panel on Climate Change: Climate Change 2007*. Also available at www.ipcc.ch

Jaggard, K.W., Qi, A. & Ober, E.S. (2011) Possible changes to arable crop yields by 2050. *Philosophical Transactions of the Royal Society of London, Series B* **365**, 2835–2851.

Johnson, R.C. (2008) Gene banks pay big dividends to agriculture, the environment, and human welfare. *PLoS Biology* **6**, e148, 1152–1155. doi:10.1371/journal.pbio.0060148

James, C. (2010) *ISAAA Report on Global Status of Biotech/GM Crops*. International Service for the Acquisition of Agri-biotech Applications (ISAAA) http://www.isaaa.org

Karp, A. & Shield, I. (2008) Bioenergy from plants and the sustainable yield challenge. *New Phytologist* **179**, 15–32.

Kole, C. (ed.) (2011) *Wild Crop Relatives: Genomic and Breeding Resources. Legume Crops and Forages*. Springer Verlag, Berlin.

Lewis, S.L., Brando, P.M., Phillips, O.L., Van der Heijden, G.M.F. & Nepstad, D. (2011) The 2010 Amazon Drought. *Science* **331**, 554.

Maclean, I.M.D. & Wilson, R.J. (2011) Recent ecological responses to climate change support predictions of high extinction risk. *Proceedings of the National Academy of Sciences of the United States of America* (in press) doi:10.1073/pnas.1017352108.

Malhi, Y., Roberts, J.T., Betts, R.A., Killeen, T.J., Li, W. & Nobre, C.A. (2008) Climate change, deforestation, and the fate of the Amazon. *Science* **319**, 169–172.

Millennium Ecosystem Assessment (2005) *Ecosystems and human well-being: Biodiversity synthesis*. Washington, DC: World Resources Institute. www.millenniumassessment.org

Myers, N. (2002) Environmental refugees: a growing phenomenon of the 21st century. *Philosophical Transactions of the Royal Society of London, Series B* **357**, 609–613.

Myers, N., Mittermeier, R.A., Mittermeier, C.G., Da Fonseca, G.A.B. & Kent, J. (2000) Biodiversity hotspots for conservation priorities. *Nature* **403**, 853–858.

Nelson, G.C., Rosegrant, M.W., Koo, J., Robertson, R., Sulser, T., Zhu, T., Ringler, C., Msangi, S., Palazzo, A., Batka, M., Magalhaes, M., Valmonte-Santos, R., Ewing, M. & Lee, D. (2009) *Climate Change–Impact on Agriculture and Costs of Adaptation*. International Food Policy Research Institute, Washington, DC.

Poorter, H, Navas, M-L (2003) Plant growth and competition at elevated CO_2: on winners, losers and functional groups. *New Phytologist* **157**, 175–198.

Rains, D.W. & Valentine, R.C. (1980) Biological strategies for osmoregulation. In: Rains, D.W., Valentine, R.C. & Hollaender, A. (eds.) *Genetic Engineering of Osmoregulation. Impact on Plant Productivity for Food, Chemicals, and Energy*. pp 1–5. Plenum Press, New York.

Saxe, H., Cannell, M.G.R., Johnsen, O., Ryan, M.G. & Vourlitis, G. (2001) Tree and forest functioning in response to global warming. *New Phytologist* **149**, 369–400.

Secretariat of the Convention on Biological Diversity (2009) *The Convention on Biological Diversity Plant Conservation Report: A Review of Progress in Implementing the Global Strategy of Plant Conservation (GSPC)*, 48 pages.

UKERC (2009) *Global Oil Depletion. An assessment of the evidence for a near-term peak in global oil production*. UK Energy Research Centre's Technology and Policy Assessment. London.

US Department of Energy (2008) *World Biofuels Production Potential. Understanding the Challenges to Meeting the U.S. Renewable Fuel Standard*. Office of Policy Analysis, US Department of Energy, Washington, DC.

Glossary

actinomorphic flowers flowers with radial symmetry.

action spectrum measures the performance of a process (e.g. photosynthesis) over different wavelengths of light.

aerenchyma a tissue with many large intercellular spaces, common in the roots and stems of hydrophytes; may also form as a response to anaerobiosis in the roots of non-hydrophytes.

albumins water-soluble storage proteins.

allelopathy the release by plants of chemicals that inhibit the growth of their neighbours.

aminoacids		single-letter codes	
A	alanine	M	methionine
C	cysteine	N	asparagine
D	aspartic acid	P	proline
E	glutamic acid	Q	glutamine
F	phenylalanine	R	arginine
G	glycine	S	serine
H	histidine	T	threonine
I	isoleucine	V	valine
K	lysine	W	tryptophan
L	leucine	Y	tyrosine

anaplerotic reactions reactions that 'top up' metabolic pathways that have become depleted.

androecium collective term for the plant's male reproductive organs.

ANITA grade a grouping of the most primitive living angiosperms.

antipodal cells three cells of the female gametophyte that are situated at the opposite end of the ovule from the egg cell.

antiporter refers to a carrier that simultaneously transports two different solutes across a cell membrane, one into the cell and one out of the cell.

apomixis the production of seeds asexually.

apoplast refers to all space external to the plasma membrane, including the intercellular space, and space within the cell wall.

aquaporins gated protein channels that allow passage of water through the plasma membrane and tonoplast.

bacteroid a *Rhizobium* bacterium with much increased size in a legume root nodule.

biodiversity the variety of life at all levels.

biofuels here used for the liquid fuels extracted from plants and burnt as energy sources.

biogas gas produced by anaerobic breakdown of organic matter.

biomass here used for the solid plant material burnt as an energy source.

calmodulin major calcium-binding protein that acts as a calcium receptor/sensor.

calyptrogen the meristematic tissue from which the root cap is formed in many species.

calyx the outer whorl of floral organs consisting of sepals.

Functional Biology of Plants, First Edition. Martin J. Hodson and John A. Bryant.
© 2012 John Wiley & Sons, Ltd. Published 2012 by John Wiley & Sons, Ltd.

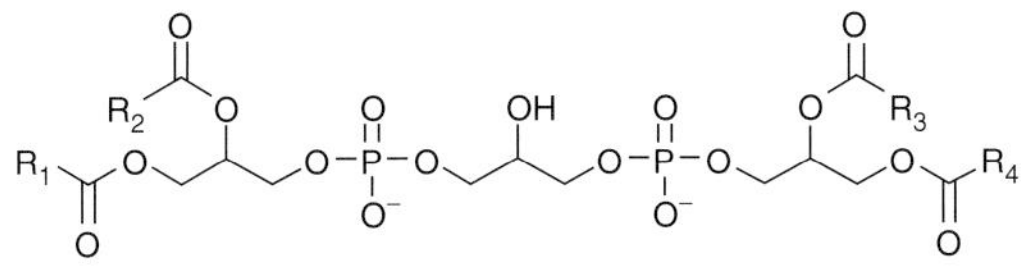

Figure G1 Structure of cardiolipin.

cambium   a lateral meristem in plants that show secondary growth.

cap (in mRNA structure)   modified guanosine added at the 5′ end of mRNA molecules; essential for binding to the ribosome.

cardiolipin   diphosphatidylglycerol. Phosphatidylglycerol is glycerol phosphate combined with two phosphatidic acid molecules and the dimer is formed *via* the phosphate groups at position 3 in the glycerol molecule: (Figure G.1)

Casparian strip   a suberized band in the radial and transverse cell walls of the endodermis and exodermis.

cavitation   air bubbles or partial vacuums that block xylem elements and prevent water flow.

chalaza   the other end of the ovule from the micropyle (see also *antipodal cells*).

chaperone   a protein that accompanies another protein; may assist with folding, with protein-protein interactions and with protein targeting.

chromophore   a light-sensitive molecule that is attached to a photoreceptor protein.

circadian   refers to daily cycles.

climacteric   phenomenon occurring in fleshy fruit, involving a large increase and then a decline in respiration rate.

cohesin   a protein complex that holds sister chromatids together and facilitates the attachment of the spindle microtubules to the chromosome.

compatible solutes   low molecular weight organic solutes that are often involved in adaptation to stress.

corolla   the outer-but-one whorl of floral organs, consisting of petals.

corpus   refers to the tunica-corpus theory of organization of the apical meristem. The corpus gives rise to the internal tissues of the plant body.

cortex   relatively undifferentiated tissue between the epidermis and the stele in the root and stem.

cosuppression   the phenomenon of suppression of a gene's activity by the presence of an extra copy of the same gene. It does not, however, occur with all genes – many genes are successfully 'over-expressed' in GM experiments.

dioecious   bearing male and female flowers on separate plants.

double fertilization   the fusion of one sperm cell with the egg cell (forming the embryo) and of the other sperm cell with the two polar nuclei in the central cell of the ovule (thus forming the endosperm).

dynein   a microtubule motor protein capable of carrying 'cargo' along a microtubule (see also kinesin).

ectotherm   an organism whose body temperature depends on the temperature of the environment.

edaphic ecotype   a distinct population of a species that is adapted to particular soil conditions.

embryophyte   a lifestyle that characterizes land plants, including in particular matrotrophic embryos which, for all or part of their period of existence, are closely associated with maternal tissues.

endocytosis   uptake into cells in a sac formed by invagination of the plasma membrane.

endodermis   a cell layer surrounding the vascular tissues of all roots and of some stems and leaves. They are characterized by having Casparian strips.

endoreduplication   repeated replication of DNA in the absence of mitosis.

endosymbiosis/endosymbiotic theory   the now well-established hypothesis that chloroplasts and mitochondria have evolved from cells that were originally engulfed by another cell.

enhancer   a DNA sequence that increases the level of expression of a gene; may be located some distance from the gene itself.

ephemeral   a plant that completes its life cycle in a short period of time.

epidermis   the outermost single layer of cells on an organ.

epigeal germination   germination in which the cotyledons appear above ground.

epigenetics   genetic control phenomena that are not directly related to DNA sequences but which, nevertheless, may be heritable.

epinasty   nastic growth response in which the upper adaxial surfaces of leaf petioles expand, and the leaves droop as a result.

eudicots   'true' dicotyledonous plants (i.e. plants whose seeds have two cotyledons), as opposed to the primitive dicot plants represented by the ANITA grade and the magnolids.

euphotic zone   in water, the upper zone in which there is enough light for photosynthesis.

exodermis   a layer of cells beneath the root epidermis (hypodermis) when these cells have a Casparian strip.

exine   the tough outer wall of a pollen grain.

exocytosis   the discharge out of cells by fusion of a vesicle (e.g. from the Golgi) with the plasma membrane.

expansins   proteins involved in cell wall extension.

extensins   a class of hydroxyproline-rich glycoproteins forming cross-linked networks in the cell wall; despite the name, they are probably not involved in cell wall extension.

Fibonacci series   the number series 1, 1, 2, 3, 5, 8, 13, 21, 34, 55 . . . Many shoot branching patterns can be described by two successive terms of the series.

funiculus   the stalk that attaches the ovule to the ovary wall.

gametophyte   the haploid, gamete-bearing generation, in angiosperms reduced to eight cells.

globulins   storage proteins soluble in dilute salt solutions.

glucocerebroside   this is glucose linked to sphingosine to which is attached a fatty acid molecule; it is an intermediate in glycolipid metabolism.

glutelins   a sub-class of prolamin storage proteins (see below) that are soluble only in dilute acid or alkali.

gluten   a term used to describe the prolamins of the wheat, barley and rye in relation to their baking properties.

glycophyte   plant adapted to living in 'normal' non-salty habitats.

great oxidation event   the increase in the oxygen content of Earth's atmosphere as a result of the evolution of photosynthesis.

guttation   the exudation of water onto a plant surface through hydathodes under conditions that do not favour transpiration (e.g. high humidity).

gynoecium   collective term for a plant's female reproductive organs.

halophyte   plant adapted to living in habitats with a high salt content.

haustorium   structure of a parasite which attaches itself to the host plant and absorbs water and nutrients.

heat-shock proteins (HSPs)   proteins that are synthesized in response to heat shock; they act as chaperones, protecting other proteins from denaturation. Some also act as chaperones in other, non-stress situations, e.g. in guiding protein-protein interactions.

Hechtian strands   strands that attach the plasma membrane to the cell wall.

helicase   an enzyme that separates the two strands of the DNA double helix.

homeotic genes   genes with major regulatory roles in the formation of organs. The proteins (transcription factors) encoded by the 'classical' homeotic genes contain a specific amino acid sequence – the homeobox.

hydathode   gland that secretes water; found at the tips and edges of leaves often at vein endings.

hydrophytes   plants adapted to living in waterlogged conditions.

hyperaccumulators   plants that take up much higher than normal amounts of metals.

hypersensitivity the very rapid death of a cell when penetrated by a pathogen.

hypodermis layer of one or more cells in thickness that is immediately beneath the epidermis.

hypogeal germination germination in which the cotyledons remain below the soil surface.

imperfect flowers flowers that lack one or more whorls of floral organs.

inflorescence a structure that bears flowers.

integuments the cell layers surrounding the ovule; they eventually form the seed coat.

intine the inner part of the pollen call wall that is mostly composed of cellulose.

iteroparous describes plants that have multiple reproductive events (see also polycarpic).

kinesin a microtubule motor protein capable of carrying 'cargo' along a microtubule (they do so in the opposite direction to dyneins).

Kranz anatomy here plants have two distinctive photosynthetic tissues – the mesophyll and the bundle sheath – and this plays a major role in the C_4 syndrome in many plants.

leghaemoglobin the particular form of haemoglobin that is involved in maintaining low oxygen tension in legume-*Rhizobium* root nodules.

light compensation point the light level at which respiration and photosynthesis are equally rapid.

locule a cavity in an ovary or anther.

lysigeny here a space is formed through the death of cells (e.g. in response to anaerobiosis).

matrotrophic describes embryos that are associated with maternal tissues (see also embryophyte).

meristem regions of plants in which dividing cells are located.

mesophyll the main photosynthetic tissue of the leaf.

mesophyte a plant that is not adapted to extreme environments.

metallothioneins cysteine-enriched metal-binding peptides that are involved in tolerance to heavy metals.

micropyle an opening in the integument layer, through which the pollen tube usually enters before fertilization.

monocarpic bearing flowers only once in a lifetime.

monocot a plant whose seed has one cotyledon.

monoecious bearing bisexual flowers or, if the flowers are unisexual, bearing male and female flowers on the same plant.

mycorrhizae close mutualistic associations between roots and fungi.

myrmecophytes 'ant plants' that provide nectar as a reward for some protection of the plant by ants against herbivory.

nastic movements a response to a stimulus that is not dependent on the direction of that stimulus. These movements can be cause by differential growth or by rapid changes in water potential in certain specialized cells.

nucellus enclosed by the integuments, this is the central tissue of the ovule and contains the embryo sac.

nyctinastic refers to nastic movements in response to the alternation of day and night. Flowers may open and close or leaves change their positions.

origin of replication a site on a DNA molecule at which replication is initiated.

osmotic potential also known as the solute potential, and is the component of water potential that is due to dissolved solutes.

paraheliotropism movement of a leaf to avoid or reduce exposure to sunlight.

parastichies lines drawn through spirals that connect adjacent leaf primordia (see *phyllotaxis*).

parietal cells a sterile ground tissue within which sporogenous cells develop in the anther.

pathogenesis-related (PR) proteins proteins that are synthesized in response to attack by pathogens. Some PR proteins are also synthesized on exposure to other stresses.

perfect flowers flowers possessing all the whorls of floral organs.

pericycle   a layer of cells immediately inside the endodermis.

periderm   a protective tissue that forms in roots and stems when they undergo secondary growth. It consists of the phellogen, the phellem and the phelloderm.

phellem   the cork, a protective tissue replacing the epidermis in stems with secondary growth.

phloem   a living vascular tissue that is principally responsible for the long-distance transport of sugars within the plant and also transports other organic and inorganic solutes.

phospholipids   complex lipids containing at least one phosphate group; major components of most plant membranes.

photolyase   an enzyme which chemically reverses pyrimidine dimer formation in DNA.

photoreactivation   a repair mechanism which is dependent on blue and/or UV light.

phragmoplast   involved in the formation of the cell plate using the ER and microtubules that are reassembled there after disassembly of the spindle at the end of cell division.

phyllotaxis   the spiral arrangement of leaves on a stem.

phytoalexins   toxic compounds that are produced in response to pathogens.

phytoanticipins   toxic compounds that are constitutive deterrents to pathogens.

phytochelatins   metal-complexing peptides involved in tolerance to heavy metals.

phytoremediation   the use of plants to clean up contaminated land or water.

plamina   term applied specifically to the nuclear lamina in plants (which lacks the lamin proteins characteristic of the nuclear lamina in animals).

plasmodesmata   cytoplasmic connections between cells.

plastochron   the period of time that elapses between the development of one leaf primordium and the next.

pneumatophore   a specialized aerial root that is produced by some mangrove species.

poikilotherm   an organism that has a temperature that varies with the environment (almost all plants).

polar nuclei   after division of the megaspore, these are the two haploid nuclei that remain at the centre of the embryo sac.

polycarpic   bearing flowers more than once in a lifetime (in many perennials, flowers are borne annually).

pre-prophase band   the main population of microtubules depolymerizes before the start of mitosis, then reassembles as the band. This encircles the nucleus and probably determines the plane and position of cell division.

primordium (*pl:* primordia)   the earliest developmental stage of an organ or tissue.

prolamins   a large class of storage proteins occurring only in cereals and grasses; soluble in aqueous alcohol.

promoter   a region of DNA that aids in the transcription of a gene by virtue of possessing a specific sequence that is recognized by a transcription factor. Promoters are positioned near the genes they regulate and are nearly always directly upstream.

proteasome   the complex that degrades proteins which have been marked for destruction by ubiquitin (see below).

protein kinase   an enzyme that uses ATP to phosphorylate other proteins.

proteoid roots   also known as cluster roots; an adaptation to soils with low mineral status. The roots have clusters of compact, short lateral rootlets.

protofilament   a row of tubulin units.

pulvinus   a swelling at the base of a petiole that is involved in the movement of a leaf or leaflet in response to a stimulus.

pyrimidine dimer   adjacent pyrimidine bases (i.e. T or C) in a DNA chain that are chemically bonded together; this is a serious type of DNA damage.

quiescent centre   region of the root apical meristem where little or no cell division occurs.

reactive oxygen intermediates (ROI)   reduced oxygen species, including singlet oxygen ($^1O_2$), superoxide radicals (O_2^-), hydrogen peroxide (H_2O_2) and hydroxyl radicals (HO^-). These are often produced by plants under stress.

replicon the tract of DNA replicated from one origin of replication.

retroelements DNA sequences that have arisen by 'reverse transcription' of RNA molecules.

retrotransposons retroelements that are potentially mobile within the genome (see also *transposon*).

rhizosphere the zone within the soil that is influenced by roots and their associated microorganisms.

RNA silencing the silencing of genes by the action of RNA molecules.

seismonastic rapid plant movements in response to shock.

semelparous refers to plants that have a single reproductive event (see also *monocarpic*).

separase and securin separin is a protease that separates two sister chromatids before anaphase. Separase is not functional when bound to securin, its chaperone protein.

serotinous late in developing, flowering or opening.

schizogeny separation of cells to form a cavity or air space (e.g. in the formation of aerenchyma).

signal peptide amino acid sequence at the N-terminal end of a protein that enables it to be taken up into the endoplasmic reticulum.

silencer a DNA sequence involved in switching off a gene.

spliceosome the RNA-protein complex that splices out the introns from pre-mRNA.

sporangiophore a spore-bearing structure.

sporophyte plant that bears spores.

statolith starch grains in the root cap that act as gravity sensors.

sterol a steroid alcohol often found in membranes.

stele the cylinder of vascular tissue.

stomium place where rupture occurs in a sporangium or pollen sac to release spores or pollen.

stroma the component of chloroplasts outside the thylakoid membrane system; effectively, the chloroplast's equivalent of the cytosol.

symplast the cell protoplasts that are linked in the plant by plasmodesmata.

symporter refers to a carrier that simultaneously transports two different solutes across a cell membrane, both in the same direction.

syncytium a single cell containing several nuclei.

synergids two cells that accompany the egg cell, forming the 'egg apparatus' at the micropylar end of the ovule.

tapetum these are nutritive cells forming the innermost layer of the wall of the pollen sac.

temperature compensation point the temperature at which respiration and photosynthesis are equally rapid.

tepals fused petals and sepals.

thigmotropism a directional movement in response to touch.

thylakoid the photosynthetic membrane system in which the light reactions occur in the chloroplast.

tracheids non-living water-conducting element in the xylem formed from a single cell. Tracheids are elongated with tapering ends and pitted walls.

transit peptide a short sequence of amino acids at the end of a protein that is involved in the protein's uptake into an organelle.

transpiration ratio the ratio of the mass of water transpired to the mass of dry matter produced.

transposons sections of DNA that can be moved within the genome.

tunica refers to the tunica-corpus theory of organization of the apical meristem. The tunica layer cells mostly have anticlinal divisions, and these give rise to the external tissues of the plant body.

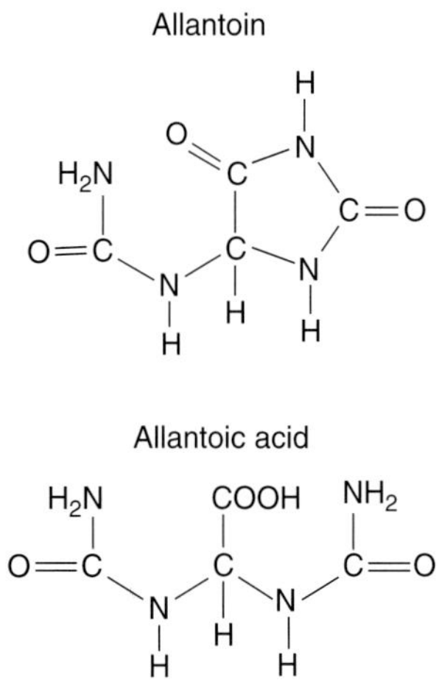

Allantoin

Allantoic acid

Figure G2 Structures of two typical plant ureides, allantoin and allantoic acid.

turgor pressure   refers to the state of cells when water uptake causes the protoplast to swell and put pressure on the surrounding cell wall.

ubiquitin   a small protein which, when attached to a second protein, marks that protein for destruction.

ureide   an acyl derivative of urea (see Figure G.2: allantoin and allantoic acid are typical plant ureides).

vessel elements   the long, dead cells that are joined end-to-end to provide the main routes for water conduction.

water potential   a measure of the energy available in an aqueous solution.

xerophyte   a plant adapted to growth in dry environments.

xylem   the water-conducting tissue in seed plants. It is also responsible for the translocation of mineral elements and some organic compounds.

zygomorphic flowers   flowers with bilateral symmetry.

Index

Functional Biology of Plants, First Edition. Martin J. Hodson and John A. Bryant.
© 2012 John Wiley & Sons, Ltd. Published 2012 by John Wiley & Sons, Ltd.